Thermophysical Properties of Liquid Crystals

Thermophysical Properties of Liquid Crystals

A. L. Tsykalo
Odessa Technological Institute of Refrigeration, USSR

Translated from the Russian by Michael E. Alferieff

GORDON AND BREACH SCIENCE PUBLISHERS
New York Philadelphia London Paris Montreux Tokyo Melbourne

Gordon and Breach Science Publishers

Post Office Box 786
Cooper Station
New York, New York 10276
United States of America

58, rue Lhomond
75005 Paris
France

5301 Tacony Street, Drawer 330
Philadelphia, Pennsylvania 19137
Unites States of America

Post Office Box 161
1820 Montreux 2
Switzerland

Post Office Box 197
London WC2E 9PX
United Kingdom

3-14-9, Okubo
Shinjuku-ku, Tokyo 169
Japan

Private Bag 8
Camberwell, Victoria 3124
Australia

Portions of this volume are revised and expanded from two reviews originally published in Russian in *Obzory po teplofizicheskim svoystvam veshchestv: Teplofizicheskiye svoystva nematicheskik zhidkikh kristalof*, No. 6, © 1978, and *Teplofizicheskiye svoystva smekticheskikh i kholestericheskikh zhidkikh kristalof*, No. 4(24), © 1980, Nauka Publishers, Moscow.

Library of Congress Cataloging-in-Publication Data

Tsykalo, A. L. (Al 'fred Leonidovich)
Thermophysical properties of liquid crystals / Alfred L. Tsykalo : translated from the Russian by Michael Alferieff.
p. cm.—
Includes bibliographical references.
ISBN 2-88124-322-3
1. Liquid crystals—Thermal properties. I. Title.
QD923.T79 1991 90-2983
530.4′29—dc20 CIP

This book is dedicated to my parents

CONTENTS

Preface xi

Introduction xiii

PART I. NEMATIC LIQUID CRYSTALS 1

Chapter 1. General Information on the Structure of Molecules and the Structure and Classification of Liquid Crystals 3

1.1. The Structure and Shape of the Particles Constituting Liquid Crystals 3
1.2. Structure and Classification of Liquid Crystals 6

Chapter 2. Orientational Order and the Molecular Dynamics of Nematic Liquid Crystals 13

2.1. Description of the Orientation Ordering of Molecules 13
2.2. The Characteristic Dynamic Behavior of Molecules of Liquid Crystals 20

Chapter 3. The Computational-Theoretical Study of the Thermophysical Properties and Structure of Nematic Liquid Crystals 25

3.1. Born's Theory and Tsvetkov's Model 25
3.2. The Maier–Saupe Theory and Its Modifications 27
3.3. Onsager's Method, Flory's Model, and Lattice Models 34
3.4. Landau-de Gennes Model, and Bethe's Approximation 36
3.5. Study of Nematic Liquid Crystals by Means of Computer Experiments 41
3.6. Other Methods of Calculation of Structure and Equilibrium Properties 59
3.7. Molecular Theories of Transport Properties 61

Chapter 4. The Thermophysical Properties of Nematic Liquid Crystals and Comparison of Experimental Data and Computational Results 65

4.1. Region of Existence of Nematic Liquid Crystals 65

4.2. Thermal Properties, Surface Tension, and Diamagnetic Susceptibility 70
4.3. Caloric Properties and Sound Velocity 85
4.4. Transport Properties 100

References for Part I **119**

PART II. SMECTIC AND CHOLESTERIC LIQUID CRYSTALS **129**

Chapter 5. Structural Characteristics of Molecules and the Structure and Classification of Smectic and Cholesteric Liquid Crystals **131**
5.1. Structure and Properties of Molecules 131
5.2. Classification and Structural Features of Smectic Liquid Crystals 136
5.3. Structure of Cholesteric Liquid Crystals 140

Chapter 6. Orientational and Translational Order in Smectic and Cholesteric Liquid Crystals and Dynamics of Molecules **145**
6.1. Distribution of the Centers of Mass of Molecules 145
6.2. Orientational and Translational Ordering in Smectic Liquid Crystals 147
6.3. Pitch of the Cholesteric Helix and Ordering of Molecules in Cholesteric Liquid Crystals 152

Chapter 7. Computational-Theoretical Investigations of the Thermophysical Properties, Structure, and Molecular Dynamics of Smectic and Cholesteric Liquid Crystals **159**
7.1. The Kobayashi–McMillan Theory of the Properties of Smectic-A Liquid Crystals 159
7.2. The McMillan, Wulf, and de Gennes Theories of the Properties of Smectic-C Liquid Crystals and the Meyer-McMillan Theory of the Properties of Smectic-C, -B, and -H Liquid Crystals 164
7.3. Other Theories of Smectic Liquid Crystals 168
7.4. Molecular Theories of the Cholesteric Mesophase 169
7.5. Molecular-Dynamics Investigation of Smectic and Cholesteric Liquid Crystals 171
7.6. Theoretical Investigations of Phase Transitions 185
7.7. Theoretical Study of Transport Processes and the Percolation Effect 189

Chapter 8. Thermophysical Properties of Smectic and Cholesteric Liquid Crystals **191**
8.1. Regions of Existence and Thermal Properties of Smectic and Cholesteric Liquid Crystals 191
8.2. Caloric Properties and the Velocity of Sound 199
8.3. Surface Tension 208
8.4. Transport Properties 213

References for Part II **231**

PART III. CHIRAL NEMATICS AND LIQUID CRYSTALS IN SUPERCOOLED STATES **239**

Chapter 9. Properties of Chiral Nematics **241**

Chapter 10. Liquid Crystals in Supercooled (Metastable) States **253**

References for Part III **263**

PART IV. LIQUID CRYSTALS CONSISTING OF DISC-LIKE MOLECULES **265**

Chapter 11. The Molecular Structure and Structural Features of Discotic Liquid Crystals and the Foundations of the Theory of These Substances **267**
11.1. Characteristics of the Molecular Structure and Structure of LC-D 267
11.2. Theoretical Model of a Liquid Crystal Consisting of Disc-Like Molecules 281

Chapter 12. The Results of Computational-Theoretical and Experimental Study of Liquid Crystals Consisting of Disc-Like Molecules **291**
12.1. The Results of Molecular-Dynamics Studies of the Particle Dynamics, Structures and Properties of Liquid Crystals Consisting of Disc-Like Molecules 291
12.2. The Results of Experimental Investigations 317

References for Part IV **325**

PART V. MIXTURES OF LIQUID CRYSTALS **329**

Chapter 13. The Structure of the Solutions of Liquid Crystals **331**

13.1. Experimental-Investigations of Structural Features 331

13.2. Computational-Theoretical Studies of Structure 338

Chapter 14. Thermophysical Properties of Solutions of Liquid Crystals **345**

14.1. Phase Diagrams and Regions of Existence of Liquid Crystal Mesophases 345

14.2. Thermophysical Properties of Solutions of Liquid Crystals and Computational-Theoretical Methods for Determining these Properties 371

References for Part V **397**

Conclusions **403**

List of Abbreviated Chemical Names of Liquid-Crystalline Substances **405**

Subject Index **411**

Author Index **417**

PREFACE

This book is based on two short review monographs written by the author and published in the USSR on the thermophysical properties of nematic liquid crystals (1978) and smectic and cholesteric liquid crystals (1980). In recent years the amount of information on the properties of liquid crystals has increased considerably. Moreover, new areas of research, such as the study of liquid crystals consisting of disc-like molecules, the synthesis and study of chiral nematics, the study of metastable liquid-crystalline states, and comprehensive studies of the properties of solutions, have evolved. A number of fundamental results have been obtained. To name only two examples, new types of structures in systems of disc-like molecules and induced phases in solutions have been discovered. For this reason, in writing the book I deemed it useful to expand the material by adding appropriate sections which would reflect the extensive contributions to the development of this field made by scientists in the USSR, India, France, England, the United States, the GDR, Bulgaria, Hungary, Japan, and a number of other countries.

At all stages of this work I benefitted from the constant support and interest of B.K. Vainshtein, Academician of the USSR Academy of Sciences, A.V. Bogatskii, Academician of the Academy of Sciences of the Ukrainian SSR, and Professor V.P. Belousov, who first introduced me to liquid crystals. Discussions of various problems in physics, physical chemistry, and the thermophysics of liquid crystals with the following individuals were of great help to me: V.N. Tsvetkov, Corresponding Member of the USSR Academy of Sciences and Professors A.P. Kapustin, I.G. Chistyakov, S.A. Pikin, E.I. Ryumtsev, G.W. Gray, S. Chandrasekhar, G.H. Brown, E.I. Kats, V.A. Belyakov and D. Demus. My interest in the statistical theory of condensed systems and in the application of computer simulation to the study of liquid crystals was first stimulated by the work of and personal contact with Professor I.Z. Fisher. Many researchers kindly sent me copies of their papers, which not only made it possible for me to include these results in the book, but also stimulated my own research. To all of these scientists, I express my sincere gratitude.

It is a pleasure for me to thank my Odessa colleagues Drs. A.D. Bagmet, V.I. Kofman, A.I. Eremenko and V.A. Vasserman for assistance in preparing separate materials for this book, and N.E.

Sukhodol'skaya, N.I. Khutornova and S.A. Avrotinskaya for laying out the manuscript and for the graphs, which were prepared with great understanding and patience. Great thanks also go to my mother and my wife, who performed the greater part of verifying and retyping the text.

The publication of this book was made possible by the initiative of Professor Glenn H. Brown, director of the Liquid Crystals Institute (Kent State University, USA), to whom I offer sincere thanks.

A.L. Tsykalo

INTRODUCTION

They form
in cubes, rectangles,
tetrahedrons,
they may have
their own peculiar axes and
molecular arrangements
but they,
like life,
like men,
are twisted by
the places into which
they come.

Loren Eiseley, *Notes of an Alchemist*

Interest in liquid crystals has increased markedly in recent years. This is attributable to the extensive possibilities for practical applications of liquid-crystal systems (information display; computer and television technology; holography; engineering and medical thermography; nondestructive testing; detection of harmful impurities; indication of thermal, acoustic, electric, and magnetic fields.[1-7]) as well as the role of liquid crystals in biological systems.[8-10] These materials have highly unusual structural, physicochemical and thermophysical properties: existence of long-range orientational order in the absence of long-range positional order of the centers of gravity of the molecules; the fine balance of the structure and its dependence on the temperature, pressure, and external fields; the existence of unusual phase transitions; effects occurring in liquid-crystal films; and so on. The study of liquid crystals is thus of great significance for the further development of the theory of the condensed state of matter and for increasing the depth and breadth of understanding of the nature, structure, and properties of possible aggregate states and of the corresponding phase transitions.

The classification, chemistry, and molecular structure and the electro-optical, magneto-optical, acoustical and other properties of liquid crystals and liquid-crystal films are studied in a number of sources (B.K. Vainshtein and I.G. Chistyakov,[11] A.P. Kapustin,[12] de Gennes,[13] L.M. Blinov,[14,15] Stephen and Straley,[16] Chandrasekhar[17]). Unfortunately, however, there are no reviews and monographs on the thermophysical properties of these materials.

Nevertheless, a knowledge of the thermophysical characteristics of liquid crystals is very important. In particular, the study of thermodynamic properties and structural characteristics permits prediction of the regions of existence of liquid-crystal mesophases and a knowledge of the transport characteristics provides the necessary foundation for engineering and medical thermography, creating television systems, and solving other important practical problems.

Specific experimental information on the thermophysical properties of liquid crystals has now been accumulated, but there is a serious omission in the literature with respect to generalizations of computational-theoretical research methods. Clearly, without progress in theoretical research on the thermophysical properties of liquid crystals it is difficult to find new applications of these materials, synthesize new liquid-crystal compounds with predetermined properties, improve the quality of liquid-crystal materials, and interpret and analyze experimental results.

The structure and thermodynamic behavior of liquid crystals cannot, as a rule, be described by the usual computational-theoretical methods used for studying isotropic systems. The general statistical-mechanical approach greatly increases in complexity when applied to liquid crystals, and it is often impossible to obtain in this manner the quantitative results necessary for solving applied problems and determining the structure of the material. A useful theory must permit calculating the thermodynamic and transport properties based on a minimum of experimental data. In addition, it is necessary to determine the structural characteristics that permit distinguishing nematic, smectic, and cholesteric liquid crystals and their various forms and isotropic and crystalline phases and determining the regions of the corresponding phase transitions. An important advantage of the theoretical method is that it elucidates the characteristics of the motion of the constituent molecules of the liquid crystal. It is also very important to have available a theory which permits establishing a relationship between the properties of a liquid crystal mentioned above and the characteristics of the constituent molecules of the crystal.

Part I is concerned with the basic directions for the theoretical study of the structure and most important thermophysical properties of nematic liquid crystals (NLC), which have been studied most completely and often serve as a starting model for constructing theories of other types of liquid crystals. The dynamics of the particles constituting NLC are also discussed based on molecular data. The traditional Maier–Saupe theory as well as its modifications and different variants of the lattice theory are still important, and the perturbation and scaled-particle theories have found application in a number of cases. In recent years, however, computer modeling of liquid crystals (computer "experiments" performed by the Monte Carlo and molecular-dynamics methods) has been playing an increasing role. Considerable attention is devoted in this book to this last direction, which, even now, permits successful study of not only idealized systems of solid rods, solid spherocylinders, etc., but also of quite realistic molecular models. This part of the book also includes experimental results, which are included primarily as a means of elucidating the inherent properties of NLC and in making comparisons between computed and measured quantities.

Part II is concerned with smectic and cholesteric liquid crystals (SLC and CLC, respectively), which are no less interesting than NLC with regard to applications as well as from a theoretical point of view. The role of SLC and CLC in biological systems, their thermophysical and physicochemical properties, and their extensive practical applications in technology and medicine have in recent years stimulated the interest of researchers, engineers, and doctors. Publications in this field are increasing rapidly. To date, nevertheless, there are, as stated previously, no reviews generalizing the existing experimental information on the thermophysical properties of SLC and CLC.

It is very interesting to analyze and generalize the computational-theoretical methods used to determine the thermophysical properties of SLC and CLC. The greater structural complexity of CLC and SLC compared to that of NLC, the presence of different modifications (smectic-A, -B, -C, -G, -H, etc., and levo- and dextrorotary CLC) make them interesting and useful objects of research. Computational-theoretical methods not only provide a foundation for explaining the structural characteristics and the most important properties of SLC and CLC, but they also facilitate the interpretation and generalization of experimental results, whose volume is increasing continuously.

This book covers the basic methods for calculating the structural characteristics and thermophysical properties of SLC and CLC. These include the traditional approaches based on extensions of the Maier–Saupe theory of NLC to SLC (theories of Kobayashi, McMillan, and Meyer–McMillan) as well as Gooseens' theory of CLC, which has been published in recent years. The basic results of the study of phase transitions in liquid crystals by the methods of L.D. Landau and P.G. de Gennes are briefly discussed. Particular attention is focused on the new and promising method of molecular dynamics for studying the nature of the motion of the constituent particles, the structure, and the properties of SLC and CLC.

The basic results obtained in experimental studies of the thermophysical properties of SLC and CLC are examined in this book. Since the experimental data are examined primarily in order to show the inherent qualitative properties of liquid crystals of this type, the more compact graphic (rather than tabular) form of representation is used. For a number of cases, the most reliable experimental values of thermophysical properties are compared with the results obtained using theoretical methods and computer modeling. Based on such comparisons, it is possible to draw important conclusions concerning the role of the different components of the potential energy of molecular interaction in establishing the order characteristic of SLC and CLC and to elucidate the relationship between the molecular properties and thermophysical characteristics.

Part III is concerned with chiral nematics and liquid crystals in supercooled states. The molecular structure of chiral nematics is elucidated and their optical and thermodynamic properties are examined. The properties of liquid crystals that retain their liquid-crystalline properties under rapid cooling (supercooled metastable states) are also discussed. The ability of mesogenic substances to retain their liquid-crystalline properties under deep cooling will probably open up new areas of application of liquid crystals at very low temperatures, which is not only of interest from the theoretical point of view, but is also of great practical interest for refrigeration technology, cryobiology and medicine.

Data on the dynamics of particles and the structure and properties of liquid crystals consisting of disc-like molecules are presented in Part IV. These interesting, recently discovered systems are of great scientific interest and, in the near future, they will undoubtedly find wide practical application. Inasmuch as the experimental data in this area are still sparse, the computational methods examined in this book and the quantitative and qualitative results obtained with their use will play an important role in the study of the properties of these substances.

Part V is concerned with solutions of liquid crystals, whose properties often best meet the requirements of modern technology. The state diagrams of solutions of different types of liquid crystals (including state diagrams of discotic mixtures as well as state diagrams exhibiting regions of induced phases) and information on the density, enthalpy, and other important characteristics are presented.

This book is based primarily on work published in the scientific literature after 1966, a period in which the major role of liquid crystals in science, technology, and medicine was recognized and the frontier of research on these materials was markedly extended.

PART I.

Nematic Liquid Crystals

There are some qualities—some incorporate things,
That have a double life, which thus is made
A type of that twin entity which springs
From matter and light, evinced in solid and shade.
Edgar Allan Poe, Sonnet "To Silence"

Nematic liquid crystals (NLC) have at the present time been studied in greatest detail both experimentally and theoretically. Comparatively complete experimental data on the thermal, caloric, and transport properties of a number of nematic substances exists. Theoretical models which not only have been successfully used to analyze and interpret the observed properties of liquid crystals but have also served as a point of departure for the construction of theories of other types of liquid crystals (smectic and cholesteric) have been proposed originally for NLC. This is not accidental. The intermolecular interaction of the constituent particles of NLC does not have the singularities inherent to the interparticle interaction in smectic and cholesteric liquid crystals, which further complicates the already complex picture of the real anisotropic interaction of particles in the NLC. Moreover, it has been established that for NLC quite simple molecular models (solid rods, ellipsoids, or spherocylinders) can be used as the zeroth-order approximation. Nematics are therefore the most convenient objects which permit explaining many of the general features inherent to the liquid-crystalline state.

With the exception of Chapter 1, which includes information on general types of liquid crystals, Part I is concerned primarily with nematic liquid crystals.

CHAPTER I.

General Information on the Structure of Molecules and the Structure and Classification of Liquid Crystals

What immortal hand or eye
Could frame thy fearful symmetry?
William Blake, "The Tyger"

1.1. THE STRUCTURE AND SHAPE OF THE PARTICLES CONSTITUTING LIQUID CRYSTALS

One of the most important properties of liquid crystals is the combination of long-range orientational ordering of the particles constituting the substance (which is characteristic of crystals) with a total or partial absence of long-range order in the positioning of their centers. The latter is usually characteristic of liquids: short-range order is observed in the positioning of molecules, but long-range order is absent. This feature of a liquid crystal is easily explained if it is assumed that the liquid crystal consists of elongated particles, and the building blocks of well-known liquid crystals are indeed elongated objects. These objects can be small organic molecules which are long helical formations with a characteristic length of 10^2–10^3Å (synthetic polypeptides, viruses, desoxyribonucleic acid), as well as complex structures (amphiphilic substances).[13] The first two types of particles form thermotropic liquid crystals, i.e., systems in which the transition into some liquid-crystalline (mesomorphic) phase occurs as a result of a temperature change. Amphiphilic systems can form both thermotropic and lyotropic liquid crystals. In the latter, the liquid-crystalline state arises with a corresponding change in composition (solutions of some polymers, soaps, dyes, etc.).

Most experimental and theoretical studies are concerned with liquid crystals formed by small organic molecules. Structural formulas are

Table 1.1
Characteristics of Some Nematic, Smectic and Cholesteric Liquid Crystals

Substance	Type of Liquid Crystal	Chemical Formula	Temperatures and Heats of Phase Transitions*
p-azoxyanisole (PAA)	Nematic	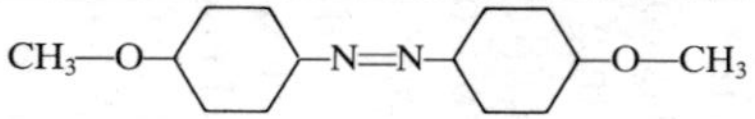	$C \xrightarrow[29,6]{391} NLC \xrightarrow[3]{408} IL$
N-(*p*-methoxy-benzylidene)-*p*-*n*-butylaniline (MBBA)	Nematic	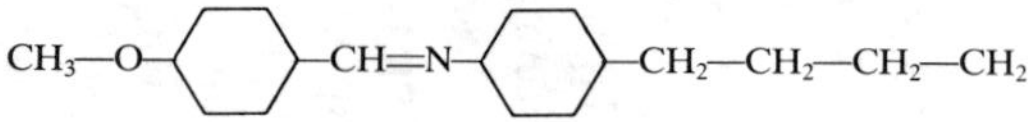	$C \xrightarrow{290} NLC \xrightarrow{319} IL$
4-cyanoben-zylidene-4′-*n*-octyloxyaniline (CBOOA)	Smectic A, nematic	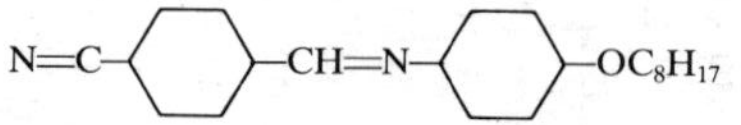	$C \xrightarrow[38,1]{346} SLCA \xrightarrow{356} NLC \xrightarrow[1,1]{382} IL$
Terephthal-bis-(4-*n*-butyl)-aniline (TBBA)	Smectics A, C, B, nematic	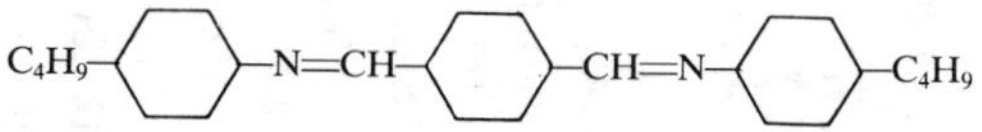	$C \xrightarrow[18,2]{386,0} SLCB \xrightarrow[83,8]{417,3} SLCC \xrightarrow{445,7} SLCA \xrightarrow[0,3]{472,8} NLC \xrightarrow[0,8]{509,7} IL$
Cholesteryl cinnamate	Cholesteric	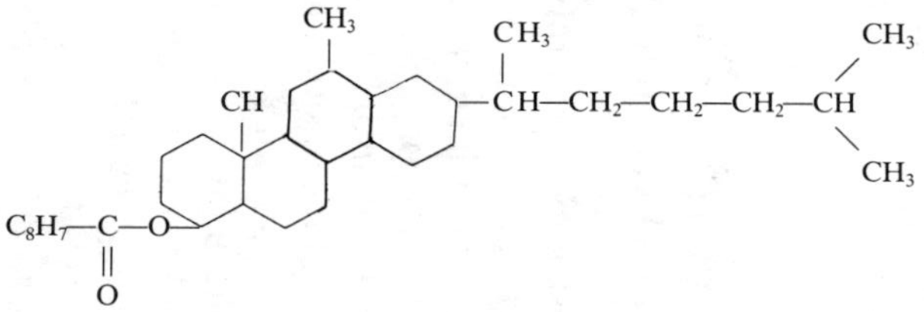	$C \xrightarrow{429} CLC \xrightarrow{470} IL$
N-(*p*-methoxyben-zylidene)-*p*-(2-methyl-butyl)-aniline (MBMBA)	Chiral nematic	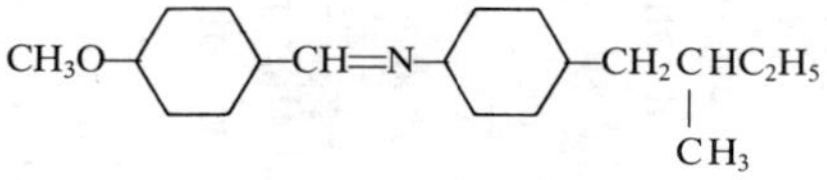	$C \xrightarrow{294} ChN \xrightarrow{297} IL$

*Numbers above lines are the temperatures of phase transitions (K); numbers below lines are the heats of these transitions (kJ/kg) presented in Table 1.1 of some of the molecules constituting the studied liquid crystals (whose properties are often mentioned in this review). Such molecules have an elongated shape (Fig. 1.1) and a characteristic length of 15–30 Å with the length-to-width ratio equal to 2–4. Such a molecule can be represented in the first, roughest approximation as a solid rod (spherocylinder, ellipsoid of revolution). A more exact description must take into account the presence of dispersion, dipole–dipole, dipole–quadrupole, higher-order multipole interactions, the actual shape of the molecule (which, under detailed examination, may turn out to be more complex, for example, more like a twisted rod), the effect of "soft" repulsion, intramolecular motion of the atoms in the molecule, etc. Fortunately, it is usually unnecessary to take into account all details of the molecular structure and the forms of interaction, since even simple models give in many cases a qualitative and even quantitative description of the most important features of the behavior of liquid crystals. For example, the polarity of molecules can usually be neglected, for although molecules in liquid crystals have dipole moments up to 3–4 D, no correlation is observed between the value of the dipole moment and the capability of the substance to make a transition into the liquid-crystalline phase.[14] Amongst the molecular characteristics which determine the possibility of forming a liquid-crystalline phase, the steric repulsion forces, dispersion (van der Waals) interactions, and anisotropic polarizability are the most important.[1,13,14]

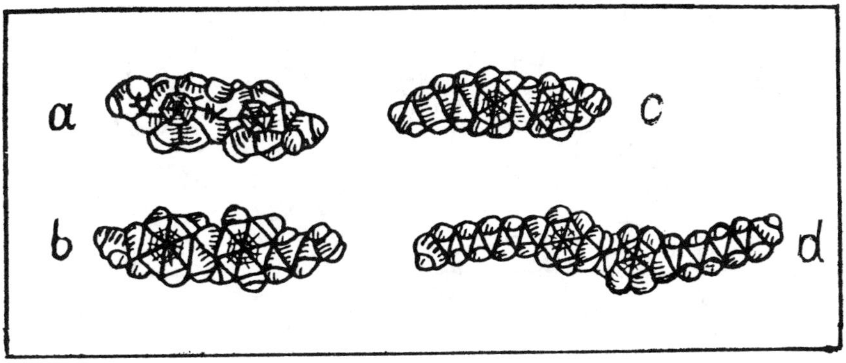

Figure 1.1. The molecule forms of some NLC: a—*p*-azoxyanisole(PAA); b—*N*-(*p*-methoxybenzylidene)-*p*-*n*-butylaniline (MBBA); c—ethyl-*p*-anisalaminocinnamate (EAAC); d—*p*-nonyloxybenzoic acid.

1.2. STRUCTURE AND CLASSIFICATION OF LIQUID CRYSTALS

Liquid crystals simultaneously exhibit the properties of a liquid (inability to preserve shape, fluidity, absence of resistance to shear, tendency for drops to coalesce, etc.) and of a solid crystal (anisotropy of optical, magnetic, electric, and thermophysical properties). Any thermotropic liquid crystal exists in a characteristic temperature interval below which it transforms to a solid state and above which it transforms into an isotropic liquid state. The corresponding phase transitions are usually first-order phase transitions, though changes in the specific volumes and the heat effects of these transitions are small. Transitions from one liquid-crystalline form to another are also

Figure 1.2. The arrangement of molecules in an isotropic liquid. The figure is drawn according to the data in Ref. 22.

first-order phase transitions if they occur below the tricritical point (above this point second-order transitions are observed*).

The general feature of liquid crystals, namely long-range orientational order in the absence of long-range order in the arrangement of the centers of molecules, can occur with different spatial arrangement of the particles (Fig. 1.2). Depending on the arrangement of the molecules, three types of liquid crystals are usually distinguished: nematic (NLC), smectic (SLC), and cholesteric (CLC).** The geometry and symmetry of liquid crystals were studied in detail by B. K. Vainshtein and I. G. Chistyakov,[11] who used and developed A. V. Shubnikov's ideas of symmetry.

In the case of NLC (Fig. 1.3), a definite order in the direction of the long axes of the molecules is observed for the entire sample: there exists some preferential orientation of molecules, characterized by the axis of nematic order and by the unit vector **n** (director).*** Of course, in principle this direction can be arbitrary and is determined by the orienting effect of surfaces bounding the sample or by other weak interactions. In addition, the preferential orientations **n** and **-n** are not distinguishable (all orientations of the ends of the molecules are equally probable). A nematic is an optically uniaxial medium. The absence of long-range translational order in the arrangement of the centers of molecules is indicated by the form of their distribution function (Fig. 1.4), which is the characteristic form for an isotropic liquid.

One of the characteristic molecular properties of nematics is that the right- and left-handed forms are indistinguishable (achirality). In

*The special cases, when second-order phase transitions are possible (for example, smectic-A to nematic), are discussed by de Gennes.[13] S. A. Pikin and A. I. Larkin[18,19] established the existence of first-order phase transitions, close to second-order transitions, in various types of crystals.

**The cholesteric mesophase can be viewed as a form of the nematic phase in connection with the fact that within the cholesteric layer (over distances of the order of several molecular dimensions), the arrangement of the molecules corresponds to the structure of nematics. On the other hand, reports appeared concerning an aggregate pseudolayered structure of some nematic systems.[11] In addition, in a large volume, when the thickness of the layer of the sample is appreciable, the helical structure typical for cholesteric liquid crystals can break down.[20,21] Thus, the separation of liquid crystals into the types mentioned is somewhat arbitrary. Nevertheless, it is widely used in practice and in this review.

***Under the action of an external field, separate regions (domains) with characteristic preferential orientations of molecules, somewhat tilted away from **n** of the sample as a whole, can arise in liquid crystals.

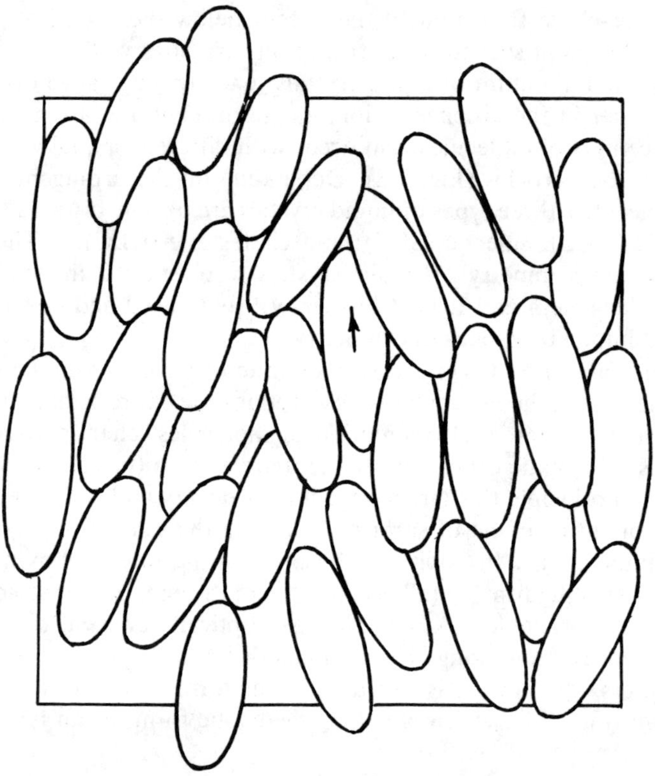

Figure 1.3. The structure of a nematic liquid crystal (a flat layer); the arrow indicates the preferential orientation of the long axes of the molecules (orientation of the director). The figure is drawn according to the data in Ref. 22.

the opposite case, the system must be a racemic mixture of right- and left-handed molecules.

Cholesteric liquid crystals are usually formed by chiral molecules (molecules which differ from their mirror images). The arrangement of the molecules in a CLC layer whose thickness is of the order of the thickness of the molecule is similar to the positioning of the particles in the nematic mesophase. In CLC, as in NLC, there is no long-range translational order. The preferred orientation of the long axes of the molecules changes from layer to layer so that a helix is formed (the director **n** rotates; see Fig. 1.5). Both right- and left-handed helices, whose pitch depends on external factors such as the temperature, pressure, and external field as well as on the molecular properties of the substance, are possible.

It is important to note that the difference between a CLC molecule and its mirror image is usually very small and the angle of rotation of the director from layer to layer is small, while the pitch is appreciable (2000–3000 Å) compared with the dimensions of the molecule.

Helical distortions are also observed in solutions of nematics with chiral molecules and in nonracemic mixtures of right- and left-handed molecules. Cholesteric crystals also include "chiral nematics," systems whose molecules exhibit optical activity but coincide with nematic molecules according to other characteristics.

In smectic liquid crystals, layers situated at definite distances from one another are observed. Thus there exists a long-range translational order in the arrangement of the layers (Fig. 1.6). The orientation of the molecules in the layers can be perpendicular (smectic-A, Fig. 1.6) or inclined (smectic-C) to the plane of the layer. Thus the thickness

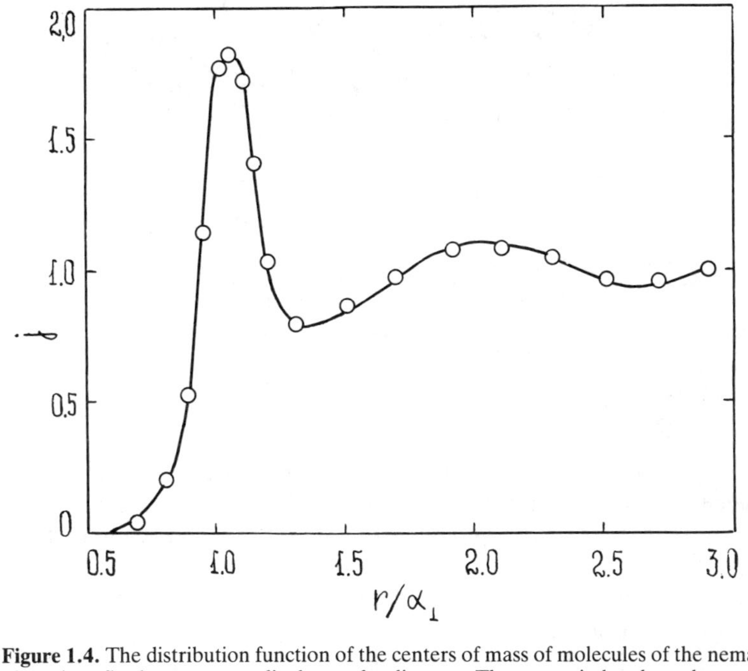

Figure 1.4. The distribution function of the centers of mass of molecules of the nematic phase in a flat layer, perpendicular to the director. The open circles show the results in Ref. 22.

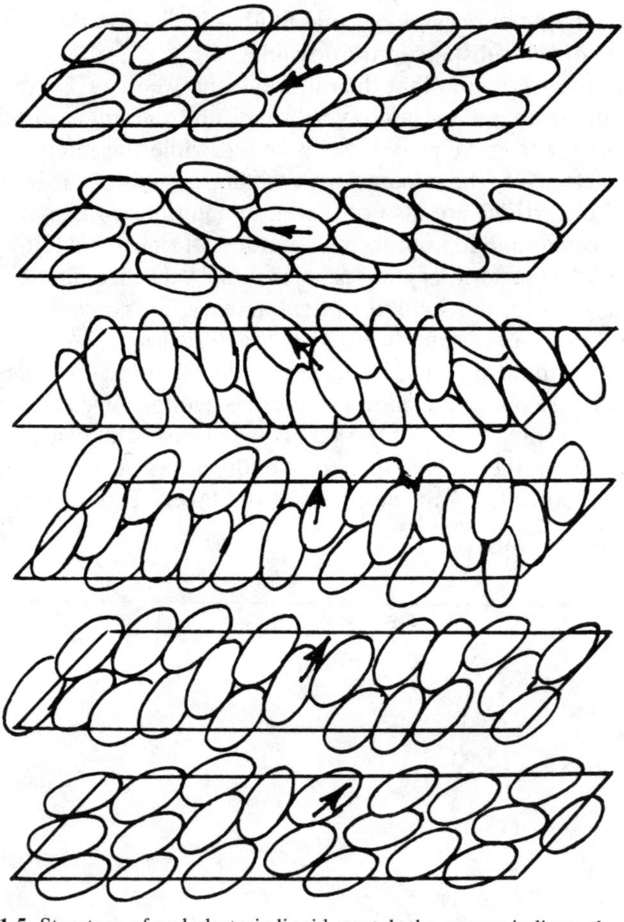

Figure 1.5. Structure of a cholesteric liquid crystal; the arrows indicate the preferred orientation of the long molecular axes.

of a layer in a smectic-C liquid crystal is less than the length of a molecule of the liquid crystal. In both cases the arrangement of the centers of the molecules within each layer corresponds to a two-dimensional liquid (the presence of short-range and absence of long-range order). In a smectic-B liquid crystal, a periodicity characteristic of a two-dimensional solid is observed in the positioning of the centers of the molecules in each layer formed by molecules whose long axes are perpendicular or inclined to the plane of the layer.

In the smectic chiral mesophase the angle of inclination of the molecules changes from layer to layer. The presence of a chiral admixture in a smectic-C liquid crystal causes the preferred orientation to rotate from layer to layer.

The characteristic molecular motions in SLC are translation in the plane of the layers (motion between layers are usually inhibited) and rotation around the long axis.

Much less is known at present concerning the more complicated forms of SLC (D, E, F, and G smectics). We will not be concerned

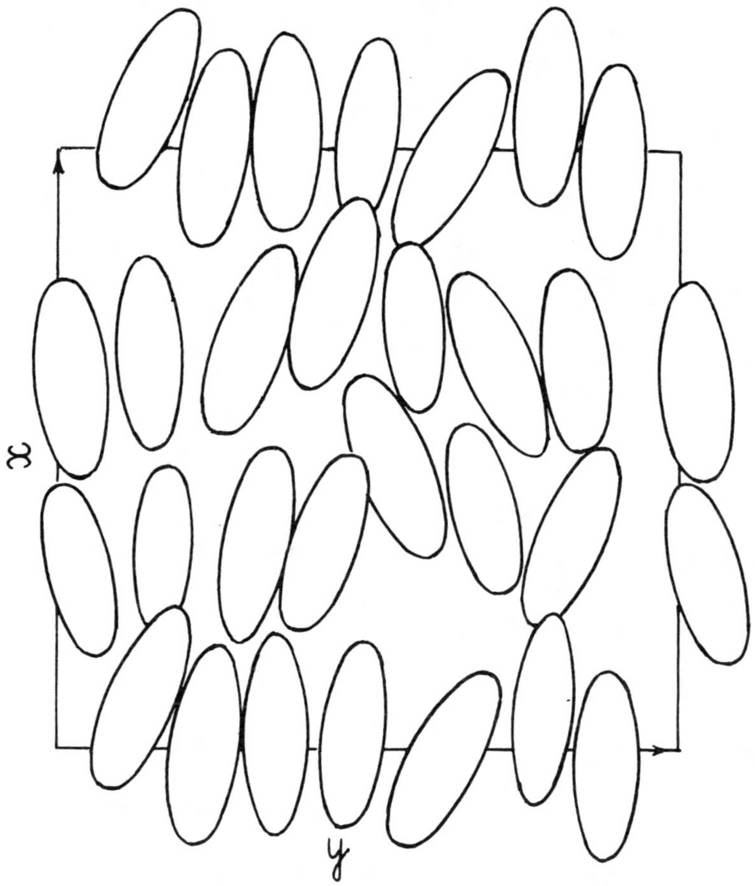

Figure 1.6. The structure of a smectic-A liquid crystal (flat layer, parallel to the director). The layered nature of the structure is clearly evident (the layers are parallel to the *y* axis). The figure is drawn according to the data in Ref. 22.

in detail in this book with these little-known substances, the structural peculiarities of which are discussed in Refs. 23 and 24.

Liquid crystals of a thermotropic type, which are characterized by the appearance or disappearance of mesomorphic state ranges in changing temperature, involve liquid crystals consisting of flat, in particular disc-like molecules, except the above mentioned types. Such disc-like liquid crystals are also characterized by orientational-ordered structures (column-like ones, etc.), thoroughly discussed in Part IV.

Liquid crystals of a lyotropic type are also of importance among various liquid crystals. Their transitions to mesomorphic states take place while the composition is being changed. But the study of this type of liquid crystals is beyond the range of this book.

CHAPTER II.

Orientational Order and the Molecular Dynamics of Nematic Liquid Crystals

We must then understand that order strives
against the unmitigated chaos lurking
along the convulsive backbone of the world.
Loren Eiseley, *Notes of An Alchemist*

2.1. DESCRIPTION OF THE ORIENTATION ORDERING OF MOLECULES

The determination of the characterstics of orientational ordering in liquid crystals is important because the related data permit distinguishing and identifying different mesomorphic phases. In addition, different theories of the liquid-crystalline state relate in one form or another the thermophysical properties of the liquid crystal to the orientational-ordering characteristics.

Special mathematical tools are used to analyze data from physical experiments that supply information on the ordering in liquid crystals (nuclear magnetic resonance, x-ray scattering, neutron scattering, and so on). These techniques also play an important role both in the development of the theory and in the study of liquid crystals with the help of computer experiments.

The unit vector **n** that defines the direction of preferred orientation of molecules in the case of NLC, SLC-A, and SLC-B coincides with the orientation of the optic axis of the medium and, in the case of SLC-A and -B, is perpendicular to the smectic planes. In CLC, the director rotates along the axis of the helix.

We shall examine briefly the basic characteristics of orientational ordering of the simplest case of uniaxial liquid crystals (NLC and the simplest types of SLC).[25] These same characteristics can be used to analyze ordering within CLC layers or more complicated modifications of SLC. The model of rigid, cylindrically symmetrical molecules[25] is usually applicable to such systems and orientational ordering can be described by single-particle density distribution functions of the paths traced out by the molecular axes on a unit sphere[26]

$$f(\cos\theta) = \langle \Delta N(\cos\theta)/(N\Delta\cos\theta)\rangle \tag{2.1}$$

or

$$D(\theta) = \langle \Delta N(\theta)/(N\sin d\theta)\rangle, \tag{2.2}$$

where ΔN is the number of particles for which the orientation of the long molecular axes corresponds to the interval ($\cos\theta$, $\cos\theta + \Delta\cos\theta$) for Eq. (2.1) and the interval (θ, $\theta + \Delta\theta$) for Eq. (2.2). The normalization conditions have the following form:

$$\int_{-1}^{+1} f(\cos\theta)\, d(\cos\theta) = 1;$$

$$\int_{0}^{2} D(N) \sin\theta\, d\theta = 1.$$

The distribution over the tilt angles relative to the director **n** can be found from the relation

$$N(\theta) = \langle \Delta N(\theta)/(N\Delta\theta)\rangle.$$

The average value of any characteristic $x(\cos\theta)$ of the liquid crystal, related to the orientational order, can be easily determined if the function $f(\cos\theta)$ is known:

$$\langle x(\cos\theta)\rangle = \int_{-1}^{1} x(\cos\theta) f(\cos\theta) d(\cos\theta).$$

With regard to the experimental determination of the orientational distribution functions, such a possibility does exist in principle (for example, with x-ray scattering). Unfortunately, in practice, it is presently possible to obtain reliable information only on some of the characteristics (for example, $\langle\cos^2\theta\rangle$).[27]

Let us examine the expansion of the function $f(\cos\theta)$ in a series of Legendre polynomials[25]

$$f(\cos\theta) = \sum_l \frac{2l+1}{2} \langle P_l(\cos\theta)\rangle P_l(\cos\theta), \tag{2.3}$$

where $P_l(\cos\,\theta)$ is a Legendre polynomial of degree l (l is even due to the equivalence of **n** and -**n**). The coefficients in the series are defined by the expressions

$$\langle P_l\,(\cos\theta)\rangle = \int_{-1}^{1} P_l(cos\theta) f(\cos\theta) d(\cos\theta),$$

and the first four coefficients are as follows:

$$\langle P_0\,(\cos\theta)\rangle = 1;$$

$$\langle P_2\,(\cos\theta)\rangle = (3\langle\cos^2\theta\rangle - 1)/2;$$

$$\langle P_4\,(\cos\theta)\rangle = (35\langle\cos^4\theta)\rangle - 30\langle\cos^2\theta\rangle + 3)/8;$$

$$\langle P_6\,(\cos\theta)\rangle = (231\langle\cos^6\theta\rangle - 315\langle\cos^4\theta\rangle + 105\langle\cos^2\theta\rangle - 5)/16.$$

Thus the experimental data give reliable information on the order parameter $\langle P_2\rangle$, first introduced by V. N. Tsvetkov.[28] The parameter $\langle P_4\rangle$ was determined for PAA and MBBA with the help of neutron scattering[27] and Raman scattering[29] experiments.

In a perfect crystal $\langle P_2\rangle = 1$ and in an isotropic liquid $\langle P_2\rangle = 0$. The values of $\langle P_2\rangle$ depend on the nature of the substance as well as on temperature and pressure; in addition, the quantities $\langle P_l\rangle$ are affected by an external field, which has orienting effect on the sample. In this case, as the external field is increased saturation is usually achieved, i.e., the extreme values of $\langle P_2\rangle$ are reached for the given conditions.

A knowledge of the low-kind order parameters $\langle P_2\rangle$, $\langle P_4\rangle$, and so on permits calculating the truncated distribution function f^M (cosθ), where M is the number of terms retained in the series. In some cases, however, the series (2.3) converges slowly so that it may be necessary to know a large number of coefficients $\langle P_l\rangle$. The important problem of the number of coefficients $\langle P_l\rangle$ necessary to describe the function $f(\cos\theta)$ can be solved using the results of molecular-dynamics computer experiments, which permit finding the complete function $f(\cos\theta)$ as well as any coefficient $\langle P_l\rangle$.[30] The characteristic form of the functions $f(\cos\theta)$, as well as that of $D(\theta)$ and $N(\theta)$, found for NLC by the molecular-dynamics method is shown in Figs. 2.1–2.3.

From the solution of the equations of motion of particles by the molecular-dynamics method, we obtain complete information about the spatial distribution of all of the particles. If the direction of the director **n** in space is not known, then the orientational order is studied with the help of the orientational-order tensor Q,[32] whose compo-

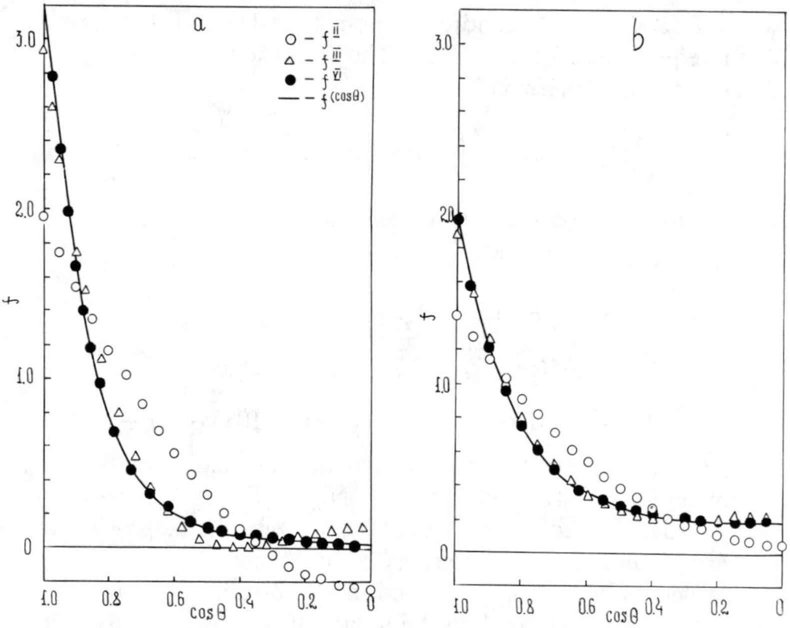

Figure 2.1. The complete $f(\cos\theta)$ and truncated $f^M(\cos\theta)$ orientational distribution functions, found by the method of molecular dynamics[30]: a) highly ordered state ($T^* = 0.75$; $\rho^* = 0.420$; $\sigma_\parallel/\sigma_\perp = 2.5$; $\langle P_2\rangle = 0.58$); b) state with low order ($\rho^* = 0.382$; $T^* = 0.70$; $\sigma_\parallel/\sigma_\perp = 2.5$; $\langle P_2\rangle = 0.36$).

nents have the form

$$Q_{\alpha\beta} = N^{-1}\left(\sum_{i=1}^{N} u_{\alpha_i}\, u_{\beta_i}\right) - \delta_{\alpha\beta}/3, \tag{2.4}$$

where u_{α_i} and u_{β_i} are projections of the unit vector $\mathbf{u}$ along the long axis of the molecule and along the coordinate axes; $\alpha, \beta = x, y, z$; and, $\delta_{\alpha\beta}$ is Kronecker's symbol. Expression (2.4) defines the components of a symmetric tensor of rank two with zero trace, which generally has five independent components. Such a tensor can be put into a diagonal form[33] for each instantaneous configuration studied in the molecular-dynamics experiments. As a result, it is possible to find the eigenvectors $\mathbf{n}_1$, $\mathbf{n}_2$ and $\mathbf{n}_3$ and the eigenvalues λ_1, λ_2, and λ_3. The maximum eigenvalue (λ_{max}) determines the order parameter $\langle P_2\rangle$:

$$\langle P_2\rangle = \frac{3}{2}\langle\lambda_{max}\rangle,$$

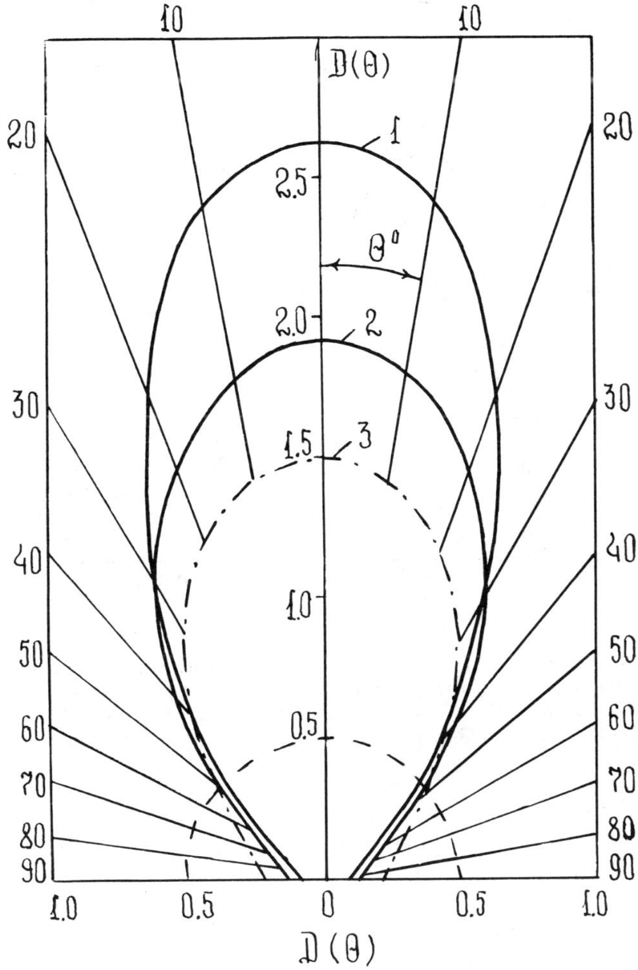

Figure 2.2. The orientational distribution functions $D(\theta)$ of NLC: 1,2-molecular dynamics "experiment" data[30] for PAA model at T = 400 K, $\langle P_2 \rangle$ = 0,54 and at T = 408 K, $\langle P_2 \rangle$ = 0,43 respectively (potential ϕ_2, $\sigma_{\|}/\sigma_{\perp}$ = 3,0; σ_0 = 5,01 Å, ε_0/k = 520 K); 3-X-ray experiment for PAA orientated by magnetic field 185,5 Gs (T = 393 K, $\langle P_2 \rangle$ = 0,30)[31]; broken line-for isotropic liquid.

while the director corresponds to the vector **n** whose characteristic value is λ_{max}.

To simplify the molecular-dynamics calculations, it is useful to give the initial orientation of the particles so that the orientation of the director coincides with an axis of the coordinate system (for example,

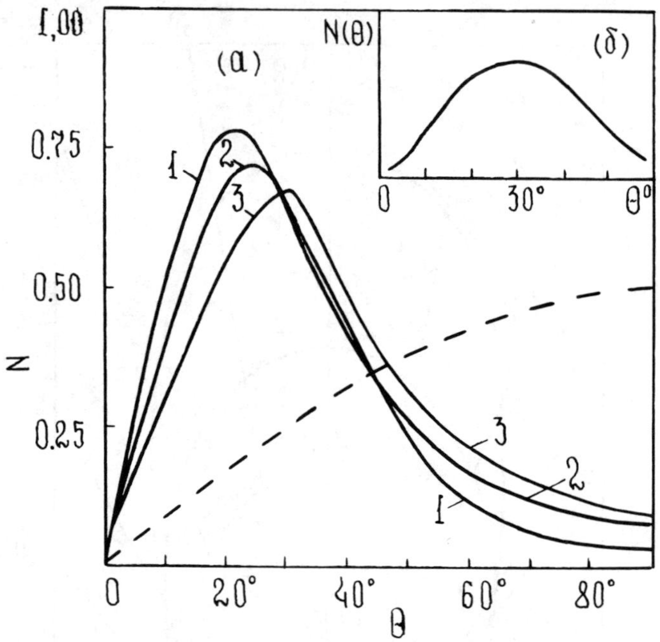

Figure 2.3. Distribution functions of particles over tilt angles with respect to the director for NLC. a) molecular-dynamics experiment[30]: 1) $T = 395$ K, $\langle P_2 \rangle = 0.58$; 2) $T = 400$ K, $\langle P_2 \rangle = 0.54$; 3) $T = 408$ K, $\langle P_2 \rangle = 0.43$; b) x-ray experiment for PAA, oriented by a magnetic field of 8 kG at $T = 391$ K; --- is for an isotropic liquid.

the x axis). Then, the quantity $\cos\theta$ is equal to the projection of $\mathbf{u}$ on the x axis. For the order parameter $\langle P_2 \rangle$, we obtain

$$\langle P_2 \rangle = \langle [3N^{-1} \sum_{i=1}^{N} u_{ix}^2 - 1]/2 \rangle. \tag{2.5}$$

The higher-kind *order* parameters can also be calculated in a simple manner, for example,

$$\langle P_4(\cos\theta) \rangle = \langle [35N^{-1} \sum_{i=1}^{N} (\mathbf{u}_i \cdot \mathbf{n})^4 - 30N^{-1} \sum_{i=1}^{N} (\mathbf{u}_i \cdot \mathbf{n})^2 + 3]/8 \rangle \tag{2.6}$$

Figure 2.4 shows the characteristic dependences of the order parameters of NLC.

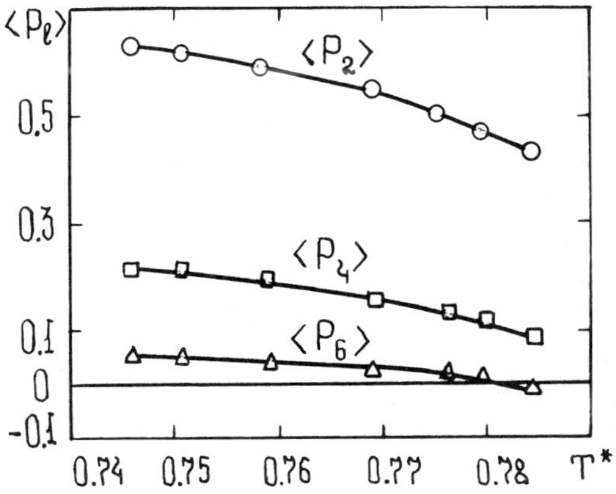

Figure 2.4. The order parameters $\langle P_1 \rangle$, obtained for NLC by molecular dynamics method[22,38] ($\sigma_\parallel/\sigma_\perp = 3.0$; $\rho^* = 0.348$).

The molecular-dynamics modeling results presented in this figure indicate that all order parameters are positive or zero (within the limits of error of the calculations). Figure 2.5 shows the characteristic experimental values of $\langle P_2 \rangle$ and $\langle P_4 \rangle$ for the materials 5CB, found with the help of resonance Raman measurements.[34] It is noteworthy that negative values of $\langle P_4 \rangle$ are observed in a number of cases. Since

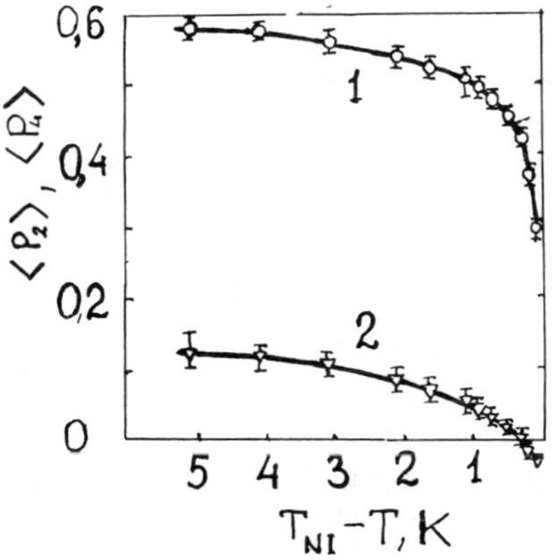

Figure 2.5. Measured values of the order parameters $\langle P_2 \rangle$ (1) and $\langle P_4 \rangle$ (2) of 5CB, obtained with the help of resonance Raman measurements.[34]

some authors believe that the appearance of negative values of $\langle P_4 \rangle$ is attributable to deficiencies of the experimental procedure and the analysis of the experimental results, this question remains controversial. In particular, taking into account the anisotropy of the local field when optical methods are used (Raman scattering of light, optical probing) increases somewhat both $\langle P_2 \rangle$ and $\langle P_4 \rangle$.[35,36]

At the same time, the use of recently carried out two-photon dichroism studies of molecular-orientational-distribution in nematic liquid crystal 5CB, together with a specially worked out model of anisotropic local field, which takes into account both short- and long-range orientational order,[37] led to values of $\langle P_4 \rangle$, less than those found in accordance with the Maier–Saupe theory. It is found that near the transition to an isotropic liquid $\langle P_4 \rangle < 0$. Authors[37] explain this feature by the high local (short-range) order, which is due to the high packing density of molecules 5CB.

2.2. THE CHARACTERISTIC DYNAMIC BEHAVIOR OF MOLECULES OF LIQUID CRYSTALS

It is of great interest to study the dynamic properties of liquid-crystalline systems in which a relative mobility of molecules is observed at their long-range orientational order. A knowledge of these properties provides information about the translational and rotational motion of molecules and permits explanation and interpretation of many phenomena occurring in liquid crystals as well as the results of experimental physical studies (neutron scattering, nuclear magnetic resonance, electron paramagnetic resonance, Rayleigh and Raman light scattering, dielectric relaxation, and infrared absorption).[39] Information on particle motion enables the identification of the mechanism of transport processes, which are important in practical applications of liquid crystals. Unfortunately, the dynamic properties of molecules in liquid crystals have been hardly studied. The experimental values of transport properties (self-diffusion and viscosity coefficients, etc.), obtained so far, are not very accurate. The proposed molecular theories of transport properties (in particular, self-diffusion) are based on a number of approximations and their use requires experimental data. Thus, further investigations are necessary in this area.

The use of time correlation functions (TCF, employed effectively in the nonequilibrium statistical theory of dense isotropic media[40,41]) is promising for the study of the dynamic behavior of molecules and the transport properties of liquid crystals. For example, the translational particle motion can be studied with the TCF of linear velocities

$$\psi_i^*(t) = \langle v_i(0)v_i(t)\rangle / \langle v_i(0)v_i(0)\rangle, \tag{2.7}$$

where $v_i(0)$ and $v_i(t)$ are the projections of the center of mass particle velocities on the i-th axis of a Cartesian coordinate system initially ($t = 0$) and at time t; the brackets indicate averaging (over particles and the initial reference time). The division by $v_1(0)v_i(0)$ is performed in order to go over to the reduced TCF.

Tiem correlation functions of the linear velocity components parallel and perpendicular to the direction of the director **n** are of interest for the study of bulk samples of liquid crystals. In the study of liquid-

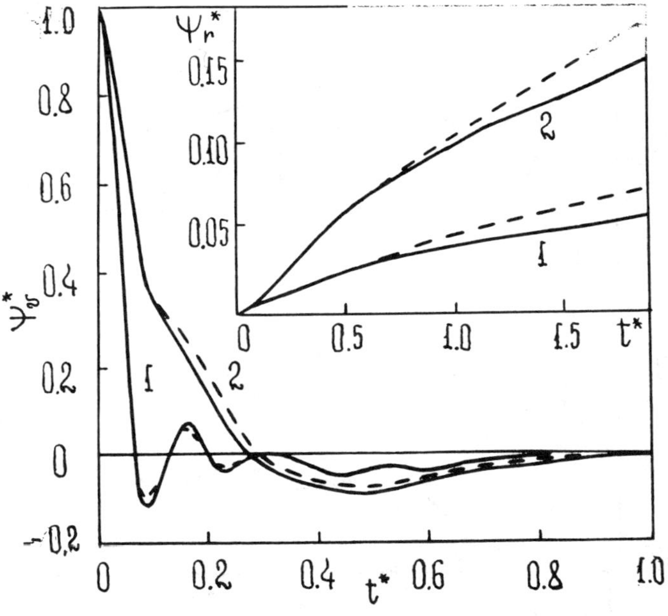

Figure 2.6. The TCF of linear velocities (ψ_v^*) and mean-square displacements (ψ_r^*) of molecules in NLC, obtained by the molecular-dynamics method[38]: 1) perpendicular to the director; 2) parallel to the director; —— is for the ϕ_2 potential, $T^* = 0.77$; - - - is for the ϕ_3 potential, $T^* = 0.76$.

crystalline films it is also important to know the TCF of velocities parallel and perpendicular to the planes of the bounding surfaces.

Information on the translational motion of molecules can also be obtained by determining the mean-square displacements

$$\psi_i(t) = \langle [r_i(t) - r_i(0)]^2 \rangle,$$

where $r_1(t)$ is the projection of the radius vector of the center of mass of a particle on the i-th axis of a Cartesian coordinate system (Fig. 2.6).

The TCF of linear velocities can be obtained from the results of neutron-scattering experiments. Very detailed information on TCF and on the mean-square displacements can be obtained for molecular-dynamics studies of liquid crystals (see Sec. 3.5). Figure 2.6 shows the TCF of linear velocities characteristic of NLC.

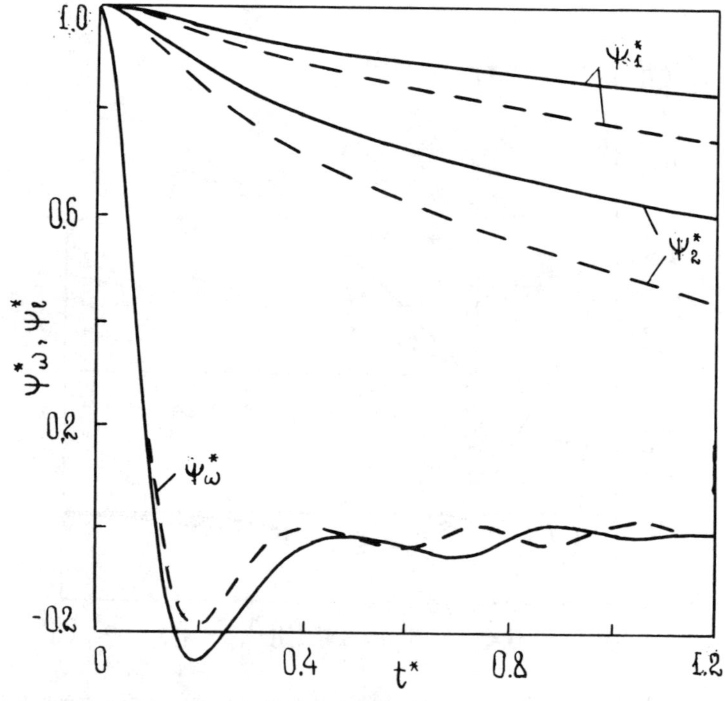

Figure 2.7. Time correlation functions of the angular velocities ψ^*_ω and reorientational TCF ψ^*_l, obtained by the molecular-dynamics method[262]: 1) ψ_ω; 2) ψ^*_1, 3) ψ^*_2. —— is for the ϕ_2 potential and – – – is for the ϕ_3 potential.

The rotational motions of molecules in liquid crystals can be studied with the help of the TCF of angular velocities

$$\psi^*_\omega(t) = \langle \boldsymbol{\omega}(0)\boldsymbol{\omega}(t)\rangle / \langle \boldsymbol{\omega}(0)\boldsymbol{\omega}(0)\rangle, \tag{2.8}$$

and the vector ($l = 1$) and tensor ($l = 2$) reorientational TCF $\psi_l(t)$:

$$\psi_l(t) = \langle P_l[\mathbf{u}(0) \cdot \mathbf{u}(t)]\rangle, \tag{2.9}$$

where $\boldsymbol{\omega}$ is the angular velocity vector of the particle, $\mathbf{u}$ is a unit vector oriented along the long axis of the molecule, and P_l is a Legendre polynomial of degree l.

Experimental data on the TCF of $\psi_1(t)$ can be obtained with the help of Rayleigh and Raman scattering and infrared absorption experiments. Such data, however, are practically nonexistent for liquid crystals though the following quantity has been estimated in an indirect manner[40]:

$$\tau_\omega = \int_0^\infty \psi_\omega(t)dt.$$

The molecular-dynamics method is one of the most promising methods for obtaining detailed information on the TCF of rotational velocities and on the reorientational TCF (Fig. 2.7).

CHAPTER III.

The Computational-Theoretical Study of the Thermophysical Properties and Structure of Nematic Liquid Crystals

It must be so, for miracles are ceased
And therefore we must needs admit the means
How things are perfected.
William Shakespeare, *Henry V* (Act I, Scene I)

3.1. BORN'S THEORY AND TSVETKOV'S MODEL

Nematic liquid crystals have been studied most completely with the use of computational-theoretical methods. Many theories have been developed for NLC, since these liquid crystals have the simplest structure and have wide applications. In addition, it is often possible to generalize successful theoretical approaches used in the study of NLC to other liquid crystals.

One of the most important problems which any theory of the liquid-crystalline state of matter must solve is the problem of establishing the relationship between the properties of the molecules and the structure of the substance, the macroscopic properties of the substance, and the nature of the particle motion. Starting with such information it is possible to determine the molecular properties responsible for the transition into the liquid-crystalline state.

In this review, we are concerned with the molecular-statistical theories of liquid crystals. We are not concerned with continuum models and theories (for example, the theory of elasticity), which are described in Refs. 13 and 42.

It is assumed in Born's theory[43] that the orientational molecular order in NLC is a result of the interaction of permanent molecular dipoles. Using a method similar to that applied by Langevin (theory of paramagnetism) and Debye (theory of polarization), Born analyzed the spontaneous orientation of dipolar molecules in the internal electric field formed by their surroundings. It was found that in systems with sufficiently polar molecules orientational order arises at temperatures below some value T_c (the latter can be compared with the NLC–IL transition temperature). Using the measured value of this temperature, the molecular weight, and the density of *p*-azoxyanisole, it is possible to find the magnitude of the dipole moment. The value obtained ($\mu = 3.6$ D) corresponds in order of magnitude to the dipole moment of some NLC, but it is much greater than the dipole moment of *p*-azoxyanisole (2.5 D).

Born's theory, however, contradicts a number of established experimental facts. As noted, there is no correlation between the ability of substances to form liquid crystals and the values of their molecular dipole moments. The molecules of a number of NLC are nonpolar, although many highly polar substances do not form a liquid-crystalline state. According to Born's theory, the IL–NLC phase transition must be a second-order transition, while in practice first-order phase transitions are observed. In addition, Born's theory cannot explain the absence of the piezo- and pyro-effects in NLC.[15] Born's theory also suffers from the general shortcomings of mean-field theories (particularly the neglect of interactions of a separate molecule with nearest-neighbor molecules which results in an incorrect description of short-range order). In this connection Born's theory is not satisfactory, although it can be useful in solving special problems involving estimates of the effect of permanent molecular dipole moments on some properties when it is important (jump in the average statistical dielectric constant accompanying the IL–NLC transition,[44,45] pretransitional phenomena in NLC[15]).

A semiempirical model of NLC was developed by V. N. Tsvetkov,[28] who proposed a quadratic dependence of the interaction between a molecule and its surroundings on the angle between the direction of the molecular axis and the director **n**. Viewing orientational order as resulting from the combination of the disordering effect of thermal motion and the elastic force (which encourages orientational order) and using some experimental values of the elastic moduli, V. N. Tsvetkov obtained an expression which describes the change of the order parameter $\langle P_2 \rangle$ with temperature. This dependence agrees qualitatively with the experimental data. Since the elastic energy is not related directly to intermolecular interaction, however, V. N. Tsvetkov's approach is based on a combination of molecular-statistical and continuous representations.

Born's and V. N. Tsvetkov's theories prepared the foundation for the development of the well-known Maier–Saupe theory.

3.2. THE MAIER–SAUPE THEORY AND ITS MODIFICATIONS

The Maier–Saupe theory is based on the molecular-field approximation, similar to that proposed by Weiss for ferromagnetic substances. This approach was first contemplated by Grandjean[46] (see, also, Ref. 13) and Maier and Saupe later performed detailed studies and obtained quantitative results which could be compared to experimental data.[47]

In its initial form the Maier–Saupe theory was based on a number of assumptions, the most important of which are as follow:

1. The molecular-field approximation and the assumption that the environment of a chosen molecule is spherically symmetrical are used;
2. Only the dispersion interaction forces between molecules in the ground states (second-order terms in the perturbation theory) are considered; these forces correspond to the lowest-order dipole term in the dispersion interaction.

Instead of calculating the interaction energy between the chosen molecule and the remaining molecules exactly, in the Maier–Saupe theory the action of the surrounding molecules on the chosen molecule is estimated; in so doing, it is assumed that the internal field is constant in time. In making this estimate, all angles describing the actual orientation of the chosen molecule relative to surrounding molecules are replaced by some average angle relative to the director. At the same time it is assumed that the surroundings are spherically symmetrical. These assumptions are only admissible for particles far away from the chosen molecule. At small distances, where repulsion plays a leading role, there is a marked correlation between the translational and rotational motions of neighboring molecules. As the distribution functions of the mass centers of the particles are anisotropic, these assumptions are not valid. For this reason the Maier–Saupe theory cannot describe the phenomena and properties associated with short-range order.

Such a calculation of the interaction energy between a molecule and its surroundings leads to the result

$$W = -\frac{U}{\tilde{V}^2}\langle P_2\rangle P_2(\cos\theta), \tag{3.1}$$

where $\tilde{V}$ is the molar volume of the NLC and U depends on the molecular properties of the substance.

Using the distribution function

$$f = C\exp(-W/kT), \tag{3.2}$$

where k is Boltzmann's constant, it is possible to find $\langle\cos^2\theta\rangle$ and the order parameter $\langle P_2\rangle$ and it is possible to determine the temperature at which the IL–NLC transition occurs. This temperature is approximately equal to 0.22 U/kV^2 and the value of the order parameter $\langle P_2\rangle$ at the transition point is equal to 0.429, which agrees better with experiment (Fig. 3.1) than Onsager's theory (0.84).

To make it easier to take into account short-range order, an additional formal assumption was used, namely, that there exist clusters of molecules whose axes are parallel.[47,48] It is then possible to calculate the changes in thermodynamic functions associated with nematic ordering

$$\Delta E = -\frac{NU}{2mV^2}\langle P_2\rangle^2, \tag{3.3}$$

$$\Delta S = \frac{N}{M}\left[-\frac{U}{2TV^2}\langle P_2\rangle(2\langle P_2\rangle + 1) + k\ln\mathcal{T}\right], \tag{3.4}$$

$$\Delta F = \frac{N}{M}\left[\frac{1}{2}\frac{U}{V^2}\langle P_2\rangle(\langle P_2\rangle + 1) - kT\ln\mathcal{T}\right], \tag{3.5}$$

$$\Delta\phi = \frac{N}{M}kT\left[\frac{1}{2}\frac{U\langle P_2\rangle^2}{kTV^2}\left(\frac{\Delta V}{V} + \frac{\langle P_2\rangle + 1}{\langle P_2\rangle}\right) - \ln\mathcal{T}\right], \tag{3.6}$$

$$\Delta V_n = \frac{kV^3Tn}{U\langle P_2\rangle_n^2}\left[\frac{U\langle P_2\rangle_n(\langle P_2\rangle + 1)}{kTn\,V_n^2} - 2\ln\mathcal{T}\right]. \tag{3.7}$$

Here,

$$\mathcal{T} = \int_0^{\pi/2}\exp\left(\frac{3}{2}\frac{U\langle P_2\rangle}{kTV^2}\cos^2\theta\right)\sin\theta d\theta,$$

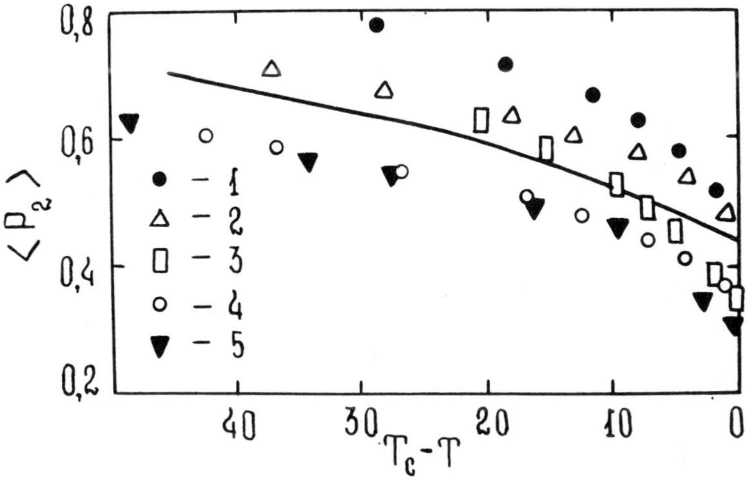

Figure 3.1. The order parameter $\langle P_2 \rangle$ as a function of $T_c - T$ for different NLC: 1) 4-butoxyphenyl ester of 4′-hexyloxybenzoic acid (BEHA); 2) *p*-azoxyphenetole (PAP); 3) methoxybenzylidene-butyl-aniline (MBBA); 4) *p*-azoxyanisole (PAA); 5) 4-butyl-4-methoxyazoxybenzene (BMAOB).

E is the energy, S is the entropy, F is the Gibbs free energy, ϕ is the Helmholtz free energy, N is Avogadro's number, m is the number of molecules in a cluster, and the index n corresponds to quantities evaluated at the IL–NLC transition point. According to the Maier–Saupe theory the change in entropy $\Delta S/Nk$ accompanying the phase transition is about 0.417, which agrees much better with experiment ($\simeq$0.3) than Onsager's theory ($\Delta S/Nk \simeq 8$).

We shall now study a more general and rigorous variant of the molecular-field theory (proposed in Refs. 49 and 50) which in a particular case reduces to the Maier–Saupe theory. The starting expression for the interaction energy of the molecules is a relation in which the parts of the potential dependent on the distance r between the centers of mass of the particles and on angles θ and ϕ describing the orientations of the long axes of the molecules are separated from one another:

$$V_{12} = 4\pi \sum_{l_1,l_2,m} U_{l_1,l_2,m}(r) Y_{l,m}(\theta_1, \varphi_1) Y^*_{l_2,m}(\theta_2, \varphi_2), \qquad (3.8)$$

where $Y_{l,m}(\theta, \phi)$ are the spherical harmonics and $U_{l_1,l_2,m}$ are the coefficients in the series.

Since it is difficult to use the potential (3.8) directly, the molecular-field approximation is introduced. For this, a three-fold averaging of V_{12} is performed: over all orientations of the vector **r**; over all orientations of molecule 2 and over all values of the distance r. We obtain this result:

$$V_1(\cos\theta) = U_2\langle P_2\rangle P_2(\cos\theta) + U_4\langle P_4\rangle P_4(\cos\theta) + U_6\langle P_6\rangle P_6(\cos\theta). \tag{3.9}$$

In expression (3.9) the term corresponding to $l = 0$ (which can be viewed as an additive constant) is dropped. Retaining only the first term in (3.9), we arrive at an expression characteristic of the usual variant of the Maier–Saupe theory.

The average values of the order parameters $\langle P_l\rangle$ entering into the starting potential V_1 can be found from relations of the form

$$\langle P_l\rangle = \frac{\int_0^1 P_l(\cos\theta)\exp[-\beta V_1(\cos\theta)]d(\cos\theta)}{\int_0^1 \exp[\beta V_1(\cos\theta)]d(\cos\theta)}. \tag{3.10}$$

The free energy per particle

$$F = kT\ln \int_0^1 \exp\left[-\frac{V_1(\cos\theta)}{kT}\right] d(\cos\theta) - \frac{1}{2}\sum_l U_l\,\langle P_l\rangle^2$$

has a minimum, corresponding to the stable phase.

Serious difficulties arise in determining the potential parameters U_l, which depend on the nature of the intermolecular interactions. At present, simple dependences of U_l on the volume with the constants determined from the experimental data are postulated.[51] In the usual variant of the Maier–Saupe theory, a simple particular case with an inverse quadratic dependence on V is used.

The temperature dependence of U_l remains unclear. According to Maier and Saupe, if this dependence is neglected then the order parameter $\langle P_2\rangle$ is a universal function of T/T_c for all NLC.

The advantages of the Maier–Saupe theory are simplicity and clarity, which have led to its popularity. This theory predicts a first-order phase transition, which agrees with experiment. The temperature dependence of the order parameter $\langle P_2\rangle$ is qualitatively correctly predicted. A useful estimate is obtained for $\langle P_2\rangle$ at the IL–NLC transition point (see above; the experimental data for MBBA, BEHA, and

BMAOB range from 0.32 to 0.45). Using the transition moments entering into the parameter U, it is possible to determine the role of the molecular optical anisotropy in the formation of liquid crystals.

We call attention to the paper by Luckhurst *et al.*,[52] who attempt to explain the fact that in spite of its shortcomings the Maier–Saupe theory is more successful than theories based on models of rigid elongated molecules (the Onsager method, lattice theories, and the scaled-particle theory). This explanation is based on the fact that short-range forces (neglected in the Maier–Saupe theory) lead to the formation of molecular clusters. This increases the role of long-range interactions. The parameter U in the potential is thus determined not by the interaction of separate molecules, but rather by the interaction of clusters. Since clusters remain even after the NLC makes the transition into an isotropic liquid, the entropy change accompanying the phase transition is much smaller than the value computed for rigid particles. Luckhurst *et al.* point out that the existence of clusters must reflect the nature of the distribution functions (the presence of short-range order at distances of several molecular dimensions). Based on these considerations, Luckhurst *et al.*[52] note that both attractive and repulsive forces are involved in establishing nematic ordering. Since, however, the effect of repulsive forces is manifested in the formation of clusters existing both before and after the IL–NLC transition, a theory based precisely on attractive interactions will be comparatively successful.

Nevertheless, the Maier–Saupe theory has serious shortcomings. The most important shortcoming is the fact that steric repulsion, which plays an important role in establishing nematic order, is ignored. This is indicated by computer experiments on rigid and realistic molecular models,[22,38] by experimental data on the temperature dependence of $\langle P_2 \rangle$ and other properties,[53–55] as well as by theoretical work.[56–58] The assumption of spherical symmetry of the distribution of the centers of mass of molecules (or groups of molecules) is also a serious shortcoming. Both assumptions exclude the possibility of correctly representing the short-range order and the related properties of NLC. Attempts to account for the effects of short-range order in the simplest possible manner, by replacing a single molecule by a spherical cluster of m identically oriented molecules and using the corresponding equations (3.3)–(3.7), leads to agreement between the computed and experimental data with $m \simeq 2$ (for *p*-azoxyanisole, $m = 1.64$). There is no physical meaning for such a value of m, so m is regarded as a kind of adjustable parameter.

The universal nature of $\langle P_2 \rangle$ (T) derived from theory does not correspond to experiment (Fig. 3.1). Independent experiments can determine the temperature dependence of the potential parameter U. It is found in so doing that this dependence is very critical. For this reason, Maier and Saupe's assumption that U is independent of temperature is apparently incorrect. In this connection, the agreement with experiment of the function $\langle P_2 \rangle$ (T) obtained from the Maier–Saupe theory could be accidental, which was pointed out by de Gennes.[13]

The constant U in the potential used in the Maier–Saupe theory (expression (3.1)) was calculated by Priest.[59] It was found that the dipole–dipole dispersion interaction forces are not sufficient to establish nematic order. From here follows the necessity for including higher-order terms and the repulsive forces in the expansion of the potential function V_1.

The Maier–Saupe theory also gives rise to serious objections, because the use of the molecular-field approximation while ignoring the effects of short-range order can lead to a false orientational phase transition. Priest[59] studied in detail a variant of the theory that uses expansion (3.9) while retaining the first few terms in the series and showed that the phase transition does not appear. An exact solution of the problem for a model very similar to the Maier–Saupe model and a study using the cluster variational method also indicate the absence of a first-order transition.[60,61]

The shortcomings of the Maier–Saupe theory noted above led to the appearance of numerous modifications and refinements. In Refs. 49 and 50, the number of terms retained in the expansion of $V_1(\cos\theta)$ was increased to two. Although this approach reproduced the experimental results with acceptable accuracy, the use of additional adjustable parameters greatly decreases its theoretical value. In addition, as noted by Wojtowicz,[51] the use of the inadequately justified dependence $U_l \simeq V^{-\gamma}$ led to the absurd value $\gamma = 10$ for MBBA.

The effect of the higher-order terms in the expansion of $V_1(\cos\theta)$ was also examined by Chandrasekhar and Madhusudana[56–58,62] who took into account anisotropic repulsive interactions. The dependence $U_l \simeq V^{-3}$ $(l = 2$ and $4)$ was used. The role of short-range order was estimated by introducing a clusterization parameter so that the theory contains three parameters.

An inconsistency in the derivation of the basis equation of the Chandrasekhar–Madhusudana method for the free energy and the entropy which led to a nonsingle-value result was recently noted by

V. K. Semenchenko and B. Chochtayev.[63] An inaccuracy in the derivation of basis relationships of the Chandrasekhar–Madhusudana method was also pointed out by Humphries *et al.*,[49] who regard the satisfactory agreement of the results obtained by this method with experimental data merely the results of the use of three adjustable parameters.

Deviations in the actual shape of molecules from the cylindrical shape were taken into account approximately (see Ref. 64) and this improved the agreement of the dependence $\langle P_2 \rangle$ (T) for constant volume with experiment.

A simple molecular-statistical computation method,[65] which permits obtaining an analytical solution, was developed based on the Maier–Saupe theory. Since this approach leads to uncomplicated working expressions for the most important thermodynamic and structural characteristics (for example, the temperature dependence of the order parameter), this approach substantially simplifies the calculations. The method in Ref. 65 was also used for other models (the Maier–Lubensky model for a nematic–smectic-A transition was studied as an example[65]).

In the last few years the work using different variants and modifications of molecular field theory of NLC continued. The approximation to self-consistent molecular field, the freezing of a system of hard ellipsoids, as well as features of isotropic–plastic and isotropic–nematic transitions[66] were studied. Calculations of free energy and plotting of phase diagrams of NLC with the use of anisotropic modification of the Lennard–Jones potential were carried out in the approximation to the mean-field theory.[67] The theoretical study of anisotropic and isotropic phases composed of uniaxial, biaxial or flexible molecules[68–70] is of a certain interest. The influence of degree of biaxiality, rigidity of molecules and lengths of alkyl end chains on properties of NLC were studied in detail by this method.

The theory of the mean molecular field turned out to be effective enough to study the surface tension of nematics. Paper 71 considers the theoretical model of rod-like molecules which interact by the attraction and repulsion of hard cores. The competition of these forces explains normal molecular orientation with respect to a free surface, which is observed in 8CB and 5CB, as well as cases of inclined orientation of molecules of MBBA and EBBA. Temperature dependences of surface tension were obtained, and in accordance with the experiment, the jump of this quantity at the transition temperature of NLC–isotropic liquid was found.

3.3. ONSAGER'S METHOD, FLORY'S MODEL, AND LATTICE MODELS

Onsager[72] examined the problem of determining the properties of a system of long rigid rods. The free energy per particle f of such a system can be represented in the form

$$F - F_0 + kT \int f_n \ln(4\pi fc)d\Omega + \frac{1}{2} c \iint f_n f_m \, \beta_1 \, (\mathbf{nm}) d\Omega d\Omega + 0(C^2), \quad (3.11)$$

where F_0 is an additive constant, c is the average concentration of rods, $f_\mathbf{n}$ is the angular distribution function of rods (the quantity $cf_\mathbf{n}d\Omega$ is the number of rods per unit volume having orientation corresponding to the unit vector $\mathbf{n}$ in the solid angle element $d\Omega$), and β_1 $(\mathbf{nm})$ is the volume excluded by one rod oriented along $\mathbf{n}$ for another rod oriented along $\mathbf{m}$:

$$\beta_1 = 2\, l^2 \, d \, |\sin\gamma|,$$

where l is the rod length, d is the rod diameter, γ is the angle between $\mathbf{n}$ and $\mathbf{m}$, and it is assumed that $l >> d$. The second term on the right side of (3.11) is related to the decrease in the entropy as a result of the orientational molecular order, while the third term describes the effect of the exclused volume. If only the term proportional to c is retained on the right side of (3.11), then the correlations between different rods will not be taken into account in this approximation (self-consistent field approximation).

The distribution function $f_\mathbf{n}$ can be determined with the use of the self-consistent equation obtained by minimizing the free energy with respect to variations of $f_\mathbf{n}$, satisfying the normalization condition

$$\int_{f\mathbf{n}} d\Omega = I$$

Solving this equation by a variational method, Onsager came to the conclusion that both an isotropic and an anisotropic solution exist (the latter corresponds to the nematic phase). The volume fraction of the rods

$$x = c \frac{\pi}{4} l d^2$$

near the transition point is equal to 4.5 d/l in the nematic phase and 3.3 d/l in the isotropic phase (independent of the temperature of the system). The order parameter $\langle P_2 \rangle$ near the transition is equal to approximately 0.84, i.e., much higher than that predicted by the Maier–Saupe theory (0.429) and observed experimentally (about 0.3).

The problem of rigid rods was also studied by Flory using a self-consistent approximation.[73] Flory's model is based on a representation of the rods as a system of points forming a lattice, in which same points can be regarded as belonging to the different rods depending on the direction of the rods (the description of the angular distribution function in this case is quite inexact). According to Flory the volume rod concentrations in the nematic and isotropic phases near a transition are equal to about 12.5 d/l and 8 d/l (for $l/d > 10$), respectively. The magnitude of the order parameter $\langle P_2 \rangle$ is even higher than for the Onsager solution and corresponds even less well to available experimental data.

An improved variant of the lattice model was substantiated and developed by Flory and Ronca.[74,75] In this model,[74] a system of solid rods, placed in a solvent, is studied. The equilibrium distribution of the orientations of the rod-shaped particles, forming a nematic phase, was obtained and directly applied in calculations; in addition, in the asymptotic limit of a high degree of order, the thermodynamic properties of this model approximately correspond to the data obtained from the early model. The errors in the latter were established by comparing the results with those obtained from theories based on the virial expansion. It is interesting that in the first approximation the many-body forces (with the exception of repulsive forces accompanying direct contact) can be ignored.[74]

Further refinements also took into account the orientation-dependent attractive forces between particles.[75] Here the energy of attraction was regarded not as the result of the interaction of entire molecules, but rather in the form of interactions of pairs of segments, which are in contact. It was established that steric effects resulting from the asymmetry of the molecular shape play the most important role. The region of the nematic–isotropic–liquid transitions and the relationship between this region and the characteristics of the molecular model were determined. It is significant that according to the results obtained both the change in the entropy and the orientational part of the heat capacity C_p vary monotonically with a NLC–IL transition, while the divergence of C_p occurs at a much higher temperature, when the metastable anisotropic state becomes unstable. Thus there arises the possibility of relating, by means of comparison of the theoretical and experimental data, the conditions for a NLC–IL transition and the structure of the molecules for the case of nematics which can be approximated by solid rods.[75] Theoretical models of this type may turn out in the future to be very useful not only for studying pure NLC, but also for studying solutions and metastable liquid-crystal states.

The self-consistent field approximation can also be used to study lattice models, in which the number of allowed orientations of each molecule is restricted. Such models greatly simplify the real situation, where a continuous distribution of molecular orientations is characteristic, and the study of such systems is greatly simplified. Similar to the results of Onsager and Flory, at low densities an isotropic phase is observed in the lattice models, while at high densities, certain orientations turn out to be dominant.[76,77] There are cases, however, where it is possible to observe configurations with "disordered domains" corresponding to systems in which the long-range order typical for nematics is absent.[13,78] This important circumstance and the significant distortion of the important properties of fluctuations in NLC are a result of replacing the continuous rotation group by a discrete point group and greatly restrict the applicability of lattice models in the theory of the liquid-crystalline state.

Renormalization group study of a lattice model of the isotropic–nematic transition[79] has led to the discrepancy with experimental results and with mean field approximation. This approach has led to the conclusion that the studied phase transition is the transition of the second kind. Apparently this is also connected with defects of a lattice model.

Bogolyubov–Tyablikov-like approximation (used recently for magnetics) was applied for the lattice model of nematic liquid crystals.[80] In accordance with this theory, the order parameter is defined by the spectrum of elementary excitations of a system depending on temperature. In the infinite ineraction radius limit results of this theory, predicting phase transition of the first kind NLC–isotropic liquid, agrees with the Maier–Saupe theory. However, in the case of short-range forces the jump of the order parameter at the NLC–isotropic liquid transition becomes less than in the mean-field theory.

3.4. LANDAU–DE GENNES MODEL AND BETHE'S APPROXIMATION

Even after a substance makes the transition from the nematic into the isotropic phase, local order remains in the substance over characteristic distances which depend on the temperature and pressure. In other words, nematic swarms (or clusters) are observed in the isotropic phase.

Thus, in spite of the jump-like disappearance of long-range order at the transition temperature, definite short-range order remains in the isotropic phase. This is indicated by experimental data on magnetic birefringence, electric birefringence (Kerr effect), and light scattering.[13,16,17] Foex[81] pointed out the analogy to the behavior of a ferromagnetic substance at a temperature above the Curie point. In using the idea of swarms, however, it is necessary to consider the continuous nature of the change in the order parameter. It is useful to follow the ideas of the theory of the continuum for such a description. De Gennes[82] proposed a description of the IL–NLC transition based on L. D. Landau's well-known theory of phase transitions.[83]

The excess free energy of the system in the absence of an external field can be expanded in powers of the order parameter tensor Q:

$$\Delta F = \frac{1}{2} A(p,T) Q_{\alpha\beta} Q_{\beta\alpha} - \frac{1}{3} B(p,T) Q_{\alpha\beta} Q_{\beta\alpha} Q_{\gamma\alpha} + \frac{1}{4} C(p,T) Q_{\alpha\beta} Q_{\beta\alpha} Q_{\gamma\delta} Q_{\delta\gamma} + \ldots \quad (3.12)$$

where $A(p,T)$, $B(p,T)$, and $C(p,T)$ are expansion coefficients. Expression (3.12) predicts a phase transition at a higher temperature than that at which coefficient A vanishes. The following form of the temperature dependence of this coefficient is proposed:

$$A = A_0 (T - T^*) \quad (3.13)$$

Since for uniaxial liquid crystals an expression of the form

$$Q_{\alpha\beta} = \langle P_2 \rangle \left(n_\alpha n_\beta - \frac{1}{3} \delta_{\alpha\beta} \right),$$

is justified,

$$\Delta F = \frac{1}{3} A \langle P_2 \rangle^2 - \frac{2}{27} B \langle P_2 \rangle^3 + \frac{1}{9} C \langle P_2 \rangle^4 \quad (3.14)$$

The dependence of ΔF on the order parameter $\langle P_2 \rangle$ is shown in Fig. 3.2 for different temperatures. We can see that at the temperature

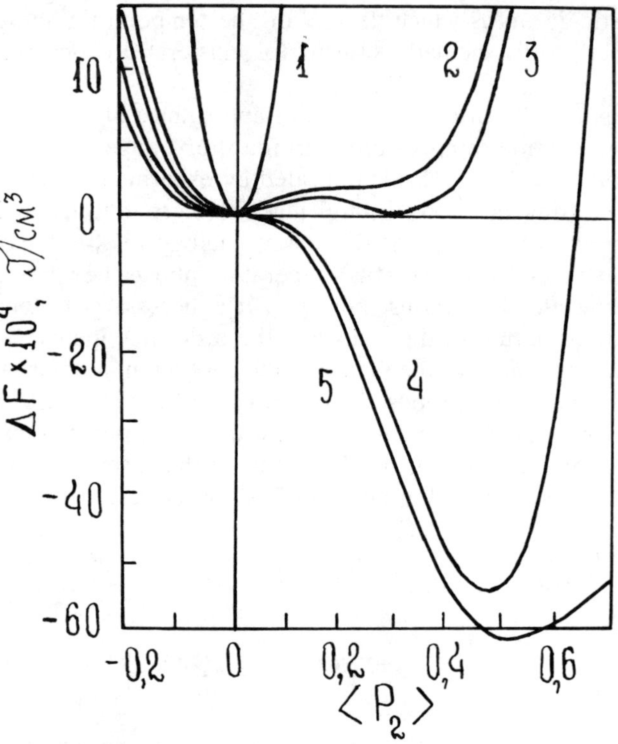

Figure 3.2. The dependence of the excess energy ΔF on the order parameter $\langle P_2 \rangle$ for MBBA[45]: 1) isotropic liquid (328 K); 2) stable limit ($T^{**} = 319$ K); 3) IL-NLC phase transition ($T_c = 319$ K); 4) stable limit ($T^* = 318.5$ K); 5) NLC.

$T = T_c$ there is a first-order phase transition accompanied by a discontinuous jump in $\langle P_2 \rangle$. The temperature T^*, which is somewhat lower, corresponds to the limit of metastable states of the isotropic phase (in principle it is possible to supercool the isotropic liquid to T^*). The isotropic phase, however, becomes unstable at $T = T^*$ (the coefficient A changes sign). On the other hand, $T = T^{**}$ determines the limit of metastable states of the nematic phase when the latter is heated. The temperatures T^* and T^{**} can be viewed as critical points of the isotropic and ordered phases.[16]

The equilibrium values of $\langle P_2 \rangle$ can be obtained as the roots of the equation

$$A\langle P_2 \rangle - \frac{1}{3} B\langle \mathrm{P}_2 \rangle^2 + \frac{2}{3} C\langle P_2 \rangle^3 = 0. \tag{3.15}$$

The corresponding solutions are:

$$\langle P_2 \rangle = 0 \text{ (isotropic phase)};$$

$$P_2 = (B/4C)\left[1 + \left(1 - 24\,\frac{AC}{B^2}\right)^{1/2}\right] \text{ (nematic phase); and,}$$

the third solution corresponds to maximum free energy. In general, both solutions can be physically significant; the solution corresponding to the lowest free energy corresponds to the stable phase. At the temperature T_c the free energies of both phases are equal and, in this case,

$$T_c = T^* + \frac{1}{27}\left(\frac{B_2}{A_0 C}\right), \qquad \frac{AC}{B^2} = \frac{1}{27}. \tag{3.16}$$

At temperatures above T_c the isotropic phase is stable; at lower temperatures the nematic phase is stable. The nematic phase becomes unstable at the temperature T^{**}:

$$T^{**} = T^* + \frac{1}{24}\left(\frac{B_2}{A_0 C}\right), \qquad \frac{AC}{B_2} > \frac{1}{24}. \tag{3.17}$$

It is especially significant that as a result of the presence of a term with an odd power of $\langle P_2 \rangle$ in expression (3.14) the phase transition must be a first-order transition (which is clear from Fig. 3.2). Jumps in the density, in latent heat, and so on are small, however, so that these phase transitions are weak first-order transitions.[13] Fundamental problems concerning the relation of the symmetry of these states to the order of the phase transition were discussed by L. D. Landau and E. M. Lifshitz.[83]

The magnitude of the order parameter of the nematic phase at the phase transition point is

$$\langle P_2 \rangle_c = \frac{B}{3C}. \tag{3.18}$$

The change in enthalpy (latent heat of the transition) is

$$\Delta H = \left(\frac{A_0 B^2}{27C^2}\right) T_c. \tag{3.19}$$

Though the Landau–de Gennes model can be used to describe the IL–NLC transition and even to study NLC near the transition temperature, it should be kept in mind that the starting expansion is valid only for small $\langle P_2 \rangle$. At the same time, the order parameter $\langle P_2 \rangle$ of typical NLC is quite large (0.3–0.4) even near the NLC–IL transition. Thus this model can be used with confidence only for a slightly ordered isotropic phase.

The Landau–de Gennes model, which describes the effects of short-range order, supplements to a certain extent the Maier–Saupe theory. It is interesting to compare the results of both approaches when both approaches can be applied. Using the expansion and integrating the equation obtained for ΔF in the Maier–Saupe theory (3.2), we obtain

$$\Delta F = NkT\left[\frac{U\langle P_2\rangle^2}{2kT^2V^2}(T - T^*) - 0.0762\,\frac{U^3\langle P_2\rangle^3}{8k^3T^3V^6} + 0.0122\,\frac{U^4\,\langle P_2\rangle^4}{16\,k^4T^4V^8} + \ldots\right], \tag{3.20}$$

where $T^* = U/5kV_c^2$. The form of expression (3.20) coincides with the expansion of the free energy in the Landau–de Gennes model (3.14). For the phase transition points, however, the Landau–de Gennes model does not lead to satisfactory quantitative results. Thus at the transition point $A/kT_cV_c^2 = 4.54$ and $T^*/T_c = 0.908$; for *p*-azoxyanisole, $T_c = 408$ K and $T_c - T^* \simeq 40$ K, whereas the experimental value of this difference is close to 1 K. Attempts to refine the Maier–Saupe theory by taking short-range order into account were made in Refs. 84–87.

The Landau–deGennes theory was also successfully used for the study of substrate-induced orientational order in the isotropic phase of liquid crystals.[88] The investigation by this method of nematic ordering of the isotropic phase near the wall restricting the sample at $T > T_{IL-NLC}$ allowed a conclusion to be drawn about the existence of a number of various conditions of wall ordering versus the orientational twisting moment created by the wall. This theory also allowed the free energy density of nonequilibrium nematics to be calculated.[89] New, interesting results were received concerning nematics which underwent mixed deformations of longitudinal and cross bend or of torsion and longitudinal bend.[89]

The Landau–deGennes theory allowed finding the effect of orientational melting of end fragments of flexible chains of molecules, which are uniaxial with NLC at phase transition.[90] The application of Landau's theory made it possible to determine the position of the isotropic–nematic critical point of the biphenyles (that is, the point in which phase transition IL–NLC becomes the transition of the second kind, arising upon the action of the external electric or magnetic field[91]).

The use of Bethe's approximation,[92] first proposed for describing short-range order in binary alloys, leads to better quantitative results. It is assumed in this method that each molecule is surrounded by neighbors (their number is $Z > 3$) and that interactions between neighbors can be neglected in the intermolecular interactions. Various forms have been proposed for the pair potential E_{ij} (i denotes the central molecule and j denotes a neighbor) as a function of the spherical coordinates θ_i, φ_i, θ_j, and φ_j and for the potential $V(\theta_j)$ describing the interaction of the neighbor molecule with the outer sphere. One of the best methods[17] is based on the expressions

$$E_{ij}(\theta_{0j}) = -\mathrm{B}^* P_2(\cos\theta_{0j}),$$

$$V(\theta_j) = -\mathrm{B}P_2(\cos\theta_j) - cP_4(\cos\theta_j),$$

where the quantities B and C are expressed in terms of B*, so that all properties of the system are determined by this parameter. It is also possible to find the temperature dependences of the order parameters. All order parameters change in a jump-like manner at the NLC–IL transition temperature. However, if the long-range order parameters ($\langle P_2\rangle$ and $\langle P_4\rangle$) vanish at the transition, then the short-range order parameter remains nonzero even in the isotropic phase.

Calculations using this method lead to the values $\langle P_2\rangle_c \simeq 0.4$ and $T_c - T^* = 25.3$ (with $Z = 8$). This result is closer, but still considerably different from, the experimental value.

3.5. STUDY OF NEMATIC LIQUID CRYSTALS BY MEANS OF COMPUTER EXPERIMENTS

The rapid development of computer technology and the development of methods of computational mathematics have encouraged the use of numerical methods (the Monte Carlo and molecular-dynamics methods), first successfully used to study the simplest model systems (hard spheres, Lennard–Jones liquids, etc.) in statistical physics, and they are used effectively to study the structure and thermophysical characteristics of complex molecular liquids and liquid crystals.

The Monte Carlo method, first proposed and applied to the study of liquids (see Refs. 93–95), is based on averaging some function of state F over a set of random configurations forming a Markov chain with fixed transition probabilities. This averaging replaces the usual method of statistical physics involving the calculation of a configurational integral or another integral forms, which determine the average values of functions of the coordinates. A sufficiently long Markov chain converges to the canonical Gibbs ensemble, since the states of this chain form a well-defined ergodic class. The probabilities of transitions between states of the system are chosen so that when such a computer experiment is performed over a sufficiently long time, different states appear with frequences proportional to the Boltzmann factor $\exp(-U/kT)$.

Since it is impossible to realize the Monte Carlo method for systems with a number of particles of the order of Avogadro's number, periodic boundary conditions are generally used, which greatly decrease the error arising from surface effects. These conditions are achieved by surrounding the specimen under study (usually cube-shaped) on all sides by identical specimens so that the configurations and velocities of particles contained in the main specimen are exactly repeated in the surrounding specimens. The entire space is thus separated into equal volumes with translational periodicity. There is a basis for assuming[96] that use of periodic boundary conditions leads to quite reliable results if the number of particles in the main specimen is not too small. Practical calculations show that several tens or hundreds of particles in the main specimen are usually sufficient. The foundations of the Monte Carlo method as applied to problems in statistical physics are presented in detail in Refs. 97 and 98.

Lasher[99] studied an approximate model of a liquid crystal by the Monte Carlo method. In this work the centers of the molecules were assumed to be fixed at the sites of a cubic lattice and only the interaction of molecules with nearest neighbors was considered. The possible orientation of the long axes of molecules corresponded to one of 12 orientations. In this lattice variant of the Maier–Saupe model a first-order orientational phase transition is observed and the order parameter at the transition point is equal to 0.82, which is much greater than the experimentally observed value. Proceeding from a discrete set of possible molecular orientations to a continuous set[100,101] greatly improved the agreement with experimental data: $\langle P_2 \rangle = 0.33$. Nevertheless, it is clear that many important features of NLC cannot be studied because of the lattice nature of the model.

A two-dimensional system of rigid ellipses was studied by Vieillard-Baron using the Monte Carlo method.[102] The main specimen contained 170 particles for which the ratio of the axes was equal to six. It was possible to establish the presence of two first-order phase transitions for this system: an order–disorder transition for the centers of gravity of the ellipses and an orientational phase transition (at a lower density). Thus the influence of the anisotropy of particle shapes on possible phase transitions in two-dimensional systems was clarified. Vieillard-Baron[103] later extended this model to three-dimensional systems (spherocylinders with a length-to-diameter ratio of 3). Unfortunately, at high densities the metastable states arising in the studied systems made thermodynamic equilibrium very difficult to achieve. As a result, it was impossible to investigate the model at nematic phase densities. Nevertheless it was established that the isotropic liquid phase in this model remains stable up to much higher densities than those expected from the scaled-particle theory.[104]

A system of rigid spherocylinders (with a length-to-diameter ratio of 2 and 3) was also studied by the Monte Carlo method (see Ref. 105). The results agreed with the data provided in Ref. 103 within the limits of accuracy of the calculations but agreed poorly with the results found in Ref. 106.

The effects of the packing density and of a deviation in the shape of molecules from the spherical shape on the limit of stability of the solid phase were studied in detail by Few and Rigby[106] who investigated a system of elongated rigid spherocylinders with a length-to-diameter ratio of 2 using the Monte Carlo method.

Thus the Monte Carlo method has mainly so far been used to study quite idealized models that differ considerably from real substances (rigid particle and lattice systems). Some results are quite interesting, but they are hardly related directly to real liquid-crystalline objects. In addition, the Monte Carlo method does not permit determining the characteristics of the dynamic behavior of molecules and the transport properties.

Many more results have been obtained using another computer modeling method, the method of molecular dynamics. This approach is based on integrating the system of classical equations of motion of particles contained in the main specimen. This integration can be performed using a finite-difference scheme with an appropriate time step. As in the Monte Carlo method, periodic boundary conditions are used. Initially all particles are usually situated at lattice sites (cubic face-centered, hexagonal close-packing, etc.), while the velocities are chosen to have equal magnitudes but random directions. The system passes into an equilibrium state (an equilibrium Maxwell distribution is reached) as a result of the time evolution of the system (after the nonequilibrium period is completed).

Data from a molecular-dynamics computer experiment in the form of a sequential set of coordinates and velocities yield exhaustive information on the dynamic behavior and properties of substances. Thus the thermodynamic properties can be defined as the time average of functions of dynamic variables which are the corresponding microscopic analogs

$$\langle f \rangle = \lim \frac{1}{\tau} \int_0^{\tau} f(t)dt. \qquad (3.21)$$

Starting from the response theory, on the other hand, the transport properties can be found in terms of the corresponding time correlation functions or in terms of the mean-square displacements.[107] Of course in an actual computer experiment, averages of type (3.21) are replaced by finite sums, giving rise to statistical errors that can be reliably estimated.

The basic assumptions of the molecular-dynamics method reduce to choosing an interaction potential, the number of particles in the main specimen, and an averaging time (number of integration steps), whose role can be estimated. For this reason the molecular-dynamics method can be viewed as being directly based on the first principles of mechanics and can be considered completely reliable. Therefore it is now possible to check the validity of approximate statistical theories, find acceptable forms of the pair potential with the help of control experiments, and solve other important problems. A discussion of the molecular-dynamics method applied to simple liquids is given in Refs. 96, 108, and 109.

Kushick and Berne[110] used the molecular-dynamics method to study model systems consisting of considerably nonspherical particles. They studied a two-dimensional system of ellipses with axial ratios of 2.5 and 3.0 and established the region of existence of the nematic phase. They also studied one state of a three-dimensional system of ellipsoids of revolution (the axial ratio is equal to 3.5) interacting via a modified anisotropic Lennard–Jones potential (12-6). The experimental and computed values of the ratios of the components of the self-diffusion tensor, however, were in great disagreement.

A detailed study of liquid crystals was performed in a series of papers,[22,30,38] and we shall examine those results in greater detail.

Considering the shape of most molecules in liquid crystals (see Fig. 1.1) and the fact that the molecules rotate comparatively freely around their long axes as noted by many authors,[1,13,15] it seemed expedient to use the model of elongated ellipsoids of revolution. Anisotropic repulsion and attractive dispersion forces (usually the most important in establishing nematic order) were taken into account in this molecular model. Since the exact form of the interaction potential for molecules in NLC is not known, molecular-dynamics experiments were performed for four forms of the potential:

1) A soft repulsive potential that describes the interaction of anisotropic particles with the help of ellipsoidal Gaussian functions[111]

$$\phi_1(\mathbf{u}_i,\mathbf{u}_j,r = \epsilon_2\epsilon\ (\mathbf{u}_i,\mathbf{u}_j)\exp\left[\frac{r^2}{2\sigma_\perp^2\ \sigma^2\ (\mathbf{u}_i,\mathbf{u}_j,\mathbf{r})}\right]; \tag{3.22}$$

2) A modified (12-6) Lennard–Jones potential which allows for attraction and repulsion of nonspherical particles:

$$\phi_2(\mathbf{u}_i,\mathbf{u}_j,\mathbf{r}) = 4\delta,\epsilon\left\{\left[\frac{\sigma_\perp\sigma(\mathbf{u}_i,\mathbf{u}_j,\hat{\mathbf{r}})}{r}\right]^{12} - \left[\frac{\sigma_\perp\sigma(\mathbf{u}_i,\mathbf{u}_j,\hat{\mathbf{r}})}{r}\right]^6\right\}; \tag{3.23}$$

3) A repulsive potential:

$$\phi_3\ (\mathbf{u}_i,\mathbf{u}_j,\mathbf{r}) = \phi_2(\mathbf{u}_i,\mathbf{u}_j,\mathbf{r}) + \epsilon_0\epsilon; \tag{3.24}$$

4) An interaction potential that is specially constructed so that the dispersion interaction energy of *p*-azoxyanisole molecules corresponds to the energy of attraction of ellipsoidal particles according to the Lifshitz–MacLachlan theory[112] for the four most important orientations of two interactng molecules:

$$\phi_4 = \frac{4\delta_0\epsilon\ (\mathbf{u}_i,\mathbf{u}_j)}{\sigma^2\ (\mathbf{u}_i,\mathbf{u}_j,\mathbf{r})}\left\{\left[\frac{\sigma_\perp\ \sigma(\mathbf{u}_i,\mathbf{u}_j,\hat{\mathbf{r}})}{r}\right]^{12} - \left[\frac{\sigma_\perp\sigma(\mathbf{u}_i,\mathbf{u}_j,\hat{\mathbf{r}})}{r}\right]^6\right\}. \tag{3.25}$$

The following orientational dependences of ϵ and σ were used in expressions (3.22), (3.23), (3.24), and (3.25)[111]

$$\epsilon(\mathbf{u}_i,\mathbf{u}_j) = [1 - \chi^2(\mathbf{u}_i,\mathbf{u}_j)^2]^{-1/2},$$

$$\sigma(\mathbf{u}_i,\mathbf{u}_j,\hat{\mathbf{r}}) = \left\{1 - \frac{1}{2}\chi\left[\frac{(\hat{\mathbf{r}}\cdot\mathbf{u}_i + \hat{\mathbf{r}}\cdot\mathbf{u}_j)^2}{1 + \chi(\mathbf{u}_i\cdot\mathbf{u}_j)} + \frac{(\hat{\mathbf{r}}\cdot\mathbf{u}_i - \hat{\mathbf{r}}\cdot\mathbf{u}_j)^2}{1 - \chi(\mathbf{u}_i\cdot\mathbf{u}_i)}\right]\right\}^{-1/2}, \quad (3.26)$$

$$\chi = \frac{\sigma^2 - \sigma_\perp^2}{\sigma_\parallel^2 + \sigma_\perp^2}$$

Here, $\sigma_\parallel$ and $\sigma_\perp$ are the axes of the ellipsoid, r is the distance between the centers of mass of the particles, $\mathbf{u}_i$ and $\mathbf{u}_j$ are unit vectors oriented along the long axes of the ellipsoids, $\hat{\mathbf{r}}$ is a unit vector oriented along a line connecting the centers of the ellipsoids, ϵ_0 is a force parameter, and σ is the anisotropy parameter. The calculations were performed for ratios of the molecular axes equal to 2.5, 3.0, and 3.5.

The results obtained with the help of the potential ϕ_1 were used to clarify the possibility of modeling NLC. This potential is, however, not sufficiently realistic at short distances and most of the calculations were therefore performed with the use of the potentials ϕ_2 and ϕ_4. The range of distances for which the interaction of particles was taken into account was limited for the ϕ_2 and ϕ_4 potentials by the quantity $r/\sigma_\perp = 2.5\sigma$, while it was limited by the quantity $r/\sigma_\perp = 2^{1/6}\sigma$ (at large distances, ϕ was set equal to zero) for the potential. The computer experiments enabled determination of the parameters of the ϕ_2 potential: for *p*-azoxyanisole (PAA) $\sigma_0 = 5.01$ Å, $\epsilon_0/k = 510$ K, and $\gamma = \sigma_\parallel/\sigma_\perp = 3.0$; for methoxybenzylidene-butylaniline (MBBA) $\sigma_0 = 5.30$ Å, $\epsilon_0/k = 401$ K, and $\gamma = 3.0$.

Comparison of results obtained with the ϕ_2 and ϕ_3 interaction potentials made it possible to estimate the effect of repulsive and attractive forces on the results of calculations of the most important NLC properties.

The depth of the potential well of ϕ_2 depends comparatively little on the relative orientation of the interacting particles (Fig. 3.3); only the distance of maximum attraction varies significantly with orientation, which is characteristic for interacting force centers arranged on the surface of a molecule. On the other hand, for the ϕ_4 potential both the depth of the potential well and the distance of maximum attraction vary considerably as a function of orientation. This cor-

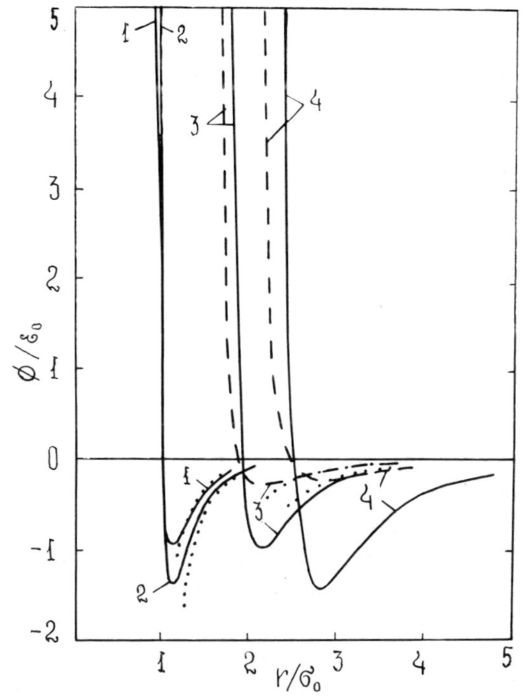

Figure 3.3. The ϕ_2 and ϕ_4 interaction potentials for ellipsoidal particles: 1) |· orientation; 2) || orientation; 3) |– orientation; 4) – – orientation; the continuous curves are for the ϕ_2 potential and the dashed curves are for the ϕ_4 potential; the circles correspond to the Lifshitz-McLachlan theory.[112]

responds to an arrangement of interaction force centers along the particle axes (or a uniform distribution of these centers over the volume) and agrees with the Lifshitz–MacLachlan theory of attraction of ellipsoidal objects.

The Hamiltonian of a system of N particles, with each particle having five degrees of freedom, has the following form:

$$H = \frac{1}{2m}\sum_{i=1}^{N}(p_{ix}^2 + p_{iy}^2 + p_{iz}^2) + \frac{1}{2J}\sum_{i=1}^{N}(p_{i\theta}^2 + p_{i\varphi}^2/\sin^2\theta_i) + \sum_{i=1}^{N}\sum_{j=1}^{N}\phi_{ij}, \tag{3.27}$$

where m is the mass of a particle, J is the moment of inertia relative to the short axis, and p_i are the momenta. In Eq. (3.27) the Cartesian coordinates of the centers of mass and the polar angles θ and φ, which define the orientation of the long axis, are used as generalized coordinates. During the course of the molecular-dynamics experiment, however, it was found useful to describe the rotational particle motion with the help of the Cartesian coordinates of a unit vector oriented along the long axis of the ellipsoid.

The centers of mass of the ellipsoidal particles were initially distributed in an ordered manner with complete orientational order ($\langle P_2 \rangle = 1$). The equations of motion in the form presented above were integrated using the scheme proposed by Beeman,[113] which requires much less computer time than the usual Rahman scheme. The main specimen (a cube or a rectangular parallelepiped) contained 108 or 168 particles. The criteria for a correct solution of the system of equations of motion were the constancy of the total energy of the system and the closeness of the total momentum to zero. Attainment of equilibrium in the system was signified by the achievement of constant average values of the functions of dynamic variables, a Maxwellian distribution of particle velocities, a uniform distribution of kinetic energy over the degress of freedom of the particles. The required time for the system to reach the equilibrium state was about 1000 steps; the functions of dynamic variables were averaged over an interval of 1000–1500 steps after reaching equilibrium (a time step is equal to $\Delta t^* = 0.0046$; for PAA $\Delta t = 2 \cdot 10^{-14}$ s).

The molecular-dynamics method permits solving one of the most important problems in the theory of the liquid-crystalline state, namely the determination of the characteristics of orientational ordering (and therefore, the identification of the NLC). This method permits determining simply all order parameters $\langle P_l \rangle$ of the nematic phase. Typical temperature dependences of $\langle P_2 \rangle$, $\langle P_4 \rangle$, and $\langle P_6 \rangle$ are shown in Fig. 2.4. Figure 3.4 shows the temperature dependence of $\langle P_2 \rangle$ for *p*-azoxyanisole at atmospheric pressure. The results of the molecular-dynamics experiment (the axial ratio is equal to 3) agree with experimental data within the range of error. The molecular-dynamics model with $\gamma = 2.5$ (as well as the Maier–Saupe theory[47]) leads to a flatter dependence $\langle P_2 \rangle(T)$. It should be noted that the quantities $\langle P_4 \rangle$ computed by the molecular-dynamics method using the same values of the potential parameters ($\epsilon_0/k = 520$ K, $\sigma_0 = 5.01$Å, and $\gamma = 3.0$) also agree within the range of error with experimental data obtained by neutron scattering.[27] The results obtained from the Maier–Saupe theory[47] are much higher than the experimental results over the entire temperature range of existence of the NLC. The use of

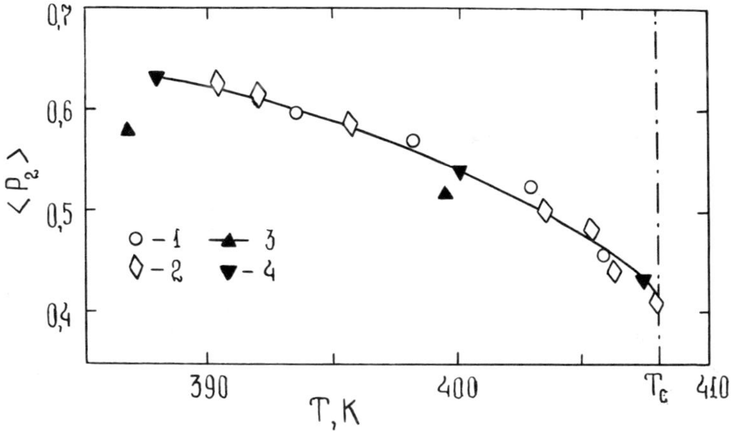

Figure 3.4. The order parameter $\langle P_2 \rangle$ of PAA at atmospheric pressure as a function of temperature (T_c = 408.5 K): 1) x-ray scattering experiments[114]; 2) NMR experiments[115]; 3) molecular dynamics calculations, γ = 2.5; 4) same, γ = 3.0.[30]

modified variants of the mean-field theory[49,50] could lead to better agreement with experiment, but this would require using several adjustable parameters, which often assume physically meaningless values.[51]

The temperature range in which the nematic mesophase of PAA exists increases significantly if the system has a constant density. Figure 3.5 compares results of a molecular-dynamics experiment[30] performed with a constant density of ρ^* = 0.420 to experimental data for $\langle P_2 \rangle$ and $\langle P_4 \rangle$ for PAA (v = 221 cm^3/mole). The agreement is very good and, in addition, the molecular-dynamics results are nearly independent of the number of particles in the main specimen (108 or 168 particles). The theory of Chandrasekhar and Madhusudana[57,58,62] will also lead to good agreement, but only for appropriate values of the adjustable parameters. It is interesting to note that an insignificant orientational order remains for $T > T_c$ also. This order can be attributed to the restricted number of particles in the main specimen as well as to the short-range order remaining in the system.

Figure 3.6 shows the temperature dependences of $\langle P_2 \rangle$ and $\langle P_4 \rangle$ found by the molecular-dynamics method for MBBA. The computed values of the parameter $\langle P_2 \rangle$ correspond well to experimental values.[29,116] The experimental values of the parameter $\langle P_4 \rangle$[29] are, how-

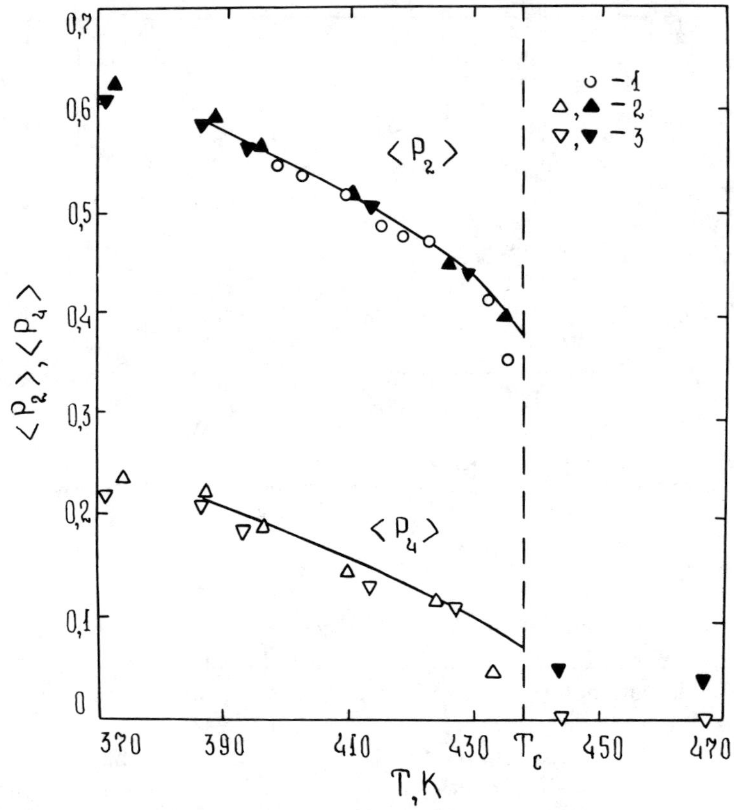

Figure 3.5. The temperature dependences of $\langle P_2 \rangle$ and $\langle P_4 \rangle$ for PAA at constant density: 1) experimental data (NMR) for $\rho = 1.169$ g/cm^3 (Ref. 54); 2) molecular-dynamics calculation,[30] $N = 108$; 3) same, $N = 168$; the continuous line shows the Chandrasekhar–Madhusudana theory.[57,58,62]

ever, much lower than computed values. This disagreement is difficult to explain, but the inadequacy of the molecular model used for MBBA and/or the inaccuracy of the experimental data[29] for $\langle P_4 \rangle$ could be possible reasons. These data indicate negative values of $\langle P_4 \rangle$, which are difficult to explain by an analysis of the corresponding orientational distribution functions (see below). The results of the Maier–Saupe theory[47] differ significantly from experimental results for both $\langle P_2 \rangle$ and $\langle P_4 \rangle$. A more flexible variant of the molecular-field theory (essentially a semiempirical approximation) with two adjustable parameters improves the agreement with the experimental curve of $\langle P_2 \rangle$ (T), but one of the parameters assumes an absurd value.

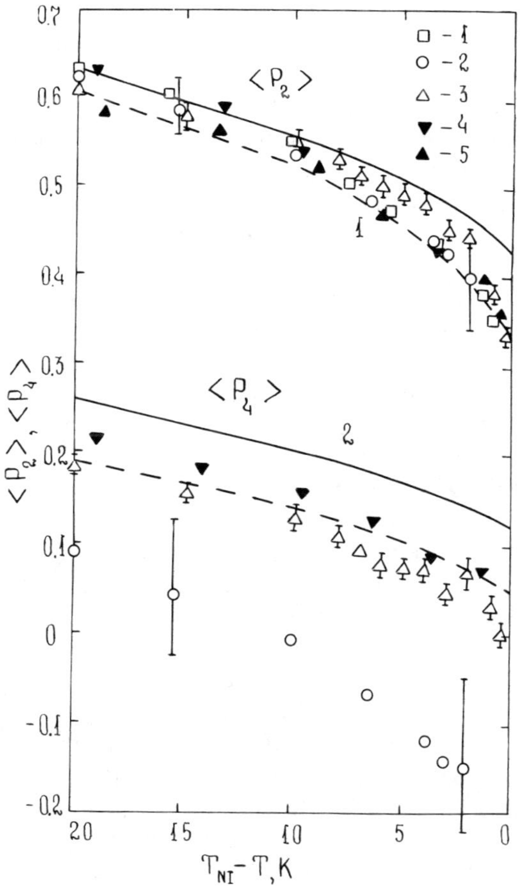

Figure 3.6. The order parameters $\langle P_2 \rangle$ and $\langle P_4 \rangle$ of MBBA (T_{NI} = 318 K) at different temperatures: 1-experimental data (measurement of dichroism of dyes[116]); 2-experimental data (measurement of Raman scattering[29]); 3-experimental data (measurement of resonance Raman scattering[117]); 4-molecular dynamics calculation ($\gamma = 3.0$; $\sigma_0 = 5{,}30$ Å; $\epsilon/k = 401$ K[22,38]); 5-the same, $\gamma = 2{,}5$;[22,38] the continuous lines correspond to results of the Maier–Saupe theory[47]; the broken lines were obtained by using the theory in Ref. 52.

Let us examine the results obtained in determining the more general characteristics of orientational order in NLC, namely, the orientational distribution functions (Figs. 2.1–2.3). The qualitative nature of the functions $f(\cos\theta)$ and $D(\theta)$ is the same for systems with the intermolecular potentials ϕ_2 and ϕ_3. A change in the state parameters (for example, a decrease in orientational order) decreases the maximum and slope of the curve (the distribution function approaches the distribution function of an isotropic liquid). It should be kept in mind, however, that the order parameter $\langle P_2 \rangle$ gives limited information on the ordering of the system so that it is necessary to know the higher-order coefficients.

There are serious difficulties in solving the problem of the experimental determination of higher-order $\langle P_l \rangle$. It is therefore of interest to establish the number of order parameters $\langle P_l \rangle$ that are sufficient to describe the orientational ordering completely (i.e., necessary for a satisfactory representation of the function $f(\cos\theta)$). This problem can be solved successfully with the help of the molecular-dynamics method, by which any $\langle P_l \rangle$ and the full function $f(\cos\theta)$ can be found.[30]

Figure 2.1 shows the full distribution functions $f(\cos\theta)$, as well as the truncated functions f^{II}, f^{III}, and f^{IV} (computed using the appropriate number of terms in the series (2.3)). It is evident that with high orientational order ($\langle P_2 \rangle = 0.58$) the parameters $\langle P_2 \rangle$, $\langle P_4 \rangle$, and $\langle P_6 \rangle$ must be included; for states with low order ($\langle P_2 \rangle = 0.36$), however, $\langle P_2 \rangle$ and $\langle P_4 \rangle$ will give a good representation of the function $f(\cos\theta)$. This result agrees with the results found in Ref. 27, which are based on the experimental study of PAA using coherent neutron scattering. The fact that the low-order parameters $\langle P_2 \rangle$ and $\langle P_4 \rangle$ are sufficient to describe states with low order explains to a certain extent the relative success of the modified molecular-field theories which include these terms in the expansion of the potential.

The form of the experimental functions $f_{\mathrm{exp}}{}^{\mathrm{III}}$ for MBBA (Fig. 3.7) cannot be interpreted in a reasonable manner based on the considerations presented above. Therefore, if the experimental data in Ref. 29 are correct, then the molecular-dynamics model used for MBBA is not completely adequate.

The molecular-dynamics results for the distribution function $D(\theta)$ for PAA are compared with the experimental data obtained using x-ray scattering by V. M. Chaikovskii and I. G. Chistyakov[118,119] and by V. V. Andrushkevich[31] in Fig. 2.2. While the computed and experimental curves agree qualitatively, there are definite quantitative disagreements: the molecular-dynamics model yields higher maxima in $D(\theta)$ and higher values of the order parameter $\langle P_2 \rangle$. These disagreements can be explained by the inadequate orienting action of a magnetic field of 185.5 G.[118,119] With powerful orienting fields the characteristics of microscopic regions of the NLC (obtained in the computer experiment) and the macroscopic characteristics (obtained

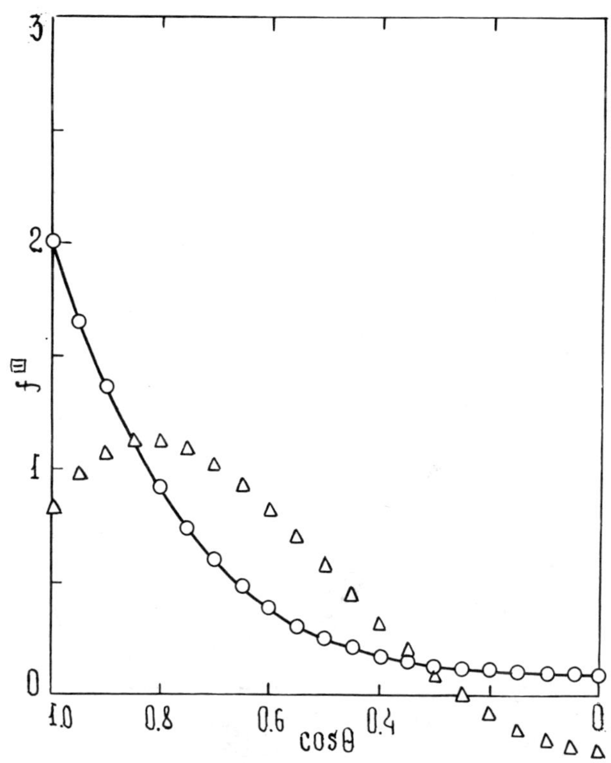

Figure 3.7. The function f^{III} for PAA (○) and for MBBA (△) with $T^c - T = 2$ K. The continuous line is a result of a molecular-dynamics calculation.[30]

in the x-ray experiment) must coincide or differ very little. It is well known, however, that a magnetic field continues to increase the orientation of separate microregions up to intensities of the order of 20 kG.[120] The inadequacy of the molecular-dynamics model used could be manifested to a certain extent also.

The distribution function $N(\theta)$ obtained by the molecular-dynamics method is compared with the experimental function[26] in Fig. 2.3. The qualitative agreement is accompanied by quantitative disagreements. The reasons for these disagreements are apparently the same as for the function $D(\theta)$.

The dominant role of short-range repulsive forces in establishing orientational order is clearly demonstrated by molecular-dynamics experiments performed using the potentials θ_2 and θ_3 (Fig. 3.8). It is evident that the contribution of attractive forces is insignificant. This

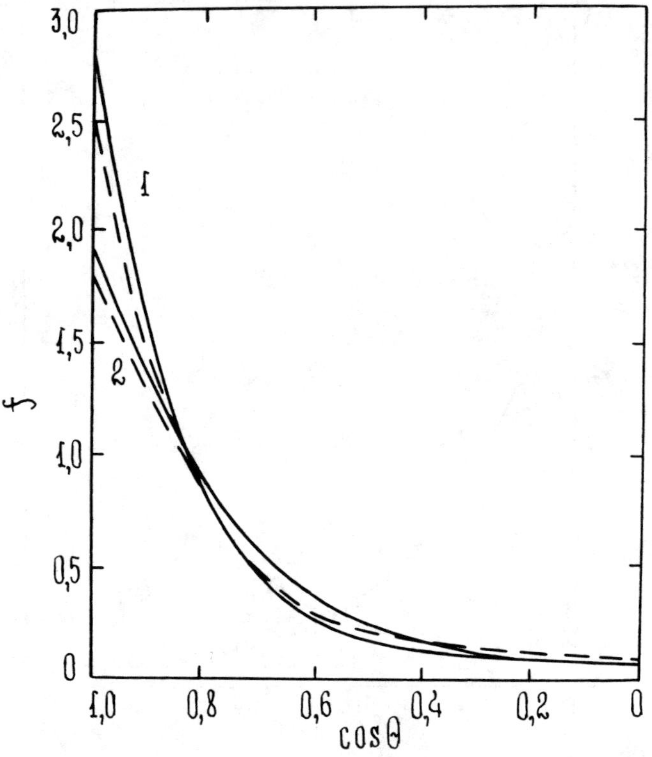

Figure 3.8. The orientational distribution function $f(\cos\theta)$ for the complete ϕ_2 (——) and repulsive ϕ_3 (---) interaction potentials:[30] 1) $\gamma = 3.0$, $\rho^* = 0.350$, and $T^* = 0.77$; 2) $\gamma = 3.0$, $\rho^* = 0.344$, and $T^* = 0.78$.

is confirmed by the results in Ref. 121 obtained from the study of orientational pair correlations in MBBA using Rayleigh light scattering and by the investigation of the thermodynamic properties of PAA performed by S. M. Stishov et al.[122]

The application of the molecular-dynamics method to the study of the dynamic behavior of molecules in liquid crystals and the related transport properties opens up quite extensive possibilities. The technique of time correlation functions (TCF) and Kubo's linear response theory can be used for this purpose. Figure 2.6 presents the characteristic dependence of the TCF of the linear velocities of molecules in NLC obtained by the molecular-dynamics method.[38] It is evident that in the nematic phase the forms of the TCF perpendicular and parallel to the director are different: $\psi_v^\perp$ decreases much more rapidly and oscillates for quite a long time, while $\psi_v^{\parallel}$ has a single minimum in the negative region. This indicates considerably freer motion along the axis of nematic order. A similar result can be obtained from an analysis of data on the mean-square displacements (see Fig. 2.6).

The TCF of linear velocities obtained for the potentials θ_2 and θ_3 correspond qualitatively to each other. Some quantitative disagreements are observed at intermediate and long times; in addition, the effect of attractive forces is not great and somewhat decreases the mobility of particles.

The molecular-dynamics method permits studying the features of the rotational motion and reorientation of molecules, which plays an important role in physical phenomena observed in NLC (rotational diffusion, dielectric relaxation, infrared absorption, light and neutron scattering, NMR, EPR, etc.). In the systems which have been studied,[38] the TCF of the angular velocities have a deep negative region (Fig. 2.7). The most characteristic feature of the reorientational TCF is the very slow decrease in ψ_l with time. This indicates the difficulty of reorientation of elongated particles in dense systems.

As in the case of TCF of linear velocities, the use of the potentials θ_2 and θ_3 leads to qualitatively similar functions ψ_ω^* and ψ_l^*. Attractive forces lead to more pronounced oscillations in the function ψ_ω^* and retard the decrease in ψ_l^*, which corresponds to the representation of attractive forces as an additional ordering factor for molecules in NLC.

The molecular dynamics method developed and widely used for investigation of NLC[123–125] also makes it possible for thin films contained between rigid walls to be studied. Detailed investigations of such nematic layers and definition of many characteristics (order parameter, distribution functions, time correlation functions, mean-square displacements and diffusion coefficients in different directions with respect to director and to directions of restricting surfaces) were carried out in Refs. 126, 127. Many interesting features of NLC films were found, as well as the influence of restricting surfaces on their properties.

In the last few years, the lattice variant of the Maier–Saupe model has been studied by the methods of computer simulation.[128–130] The interest in this model stems from the fact that it is comparatively simple while at the same time it gives a qualitatively correct description of the important features of the nematic-phase–isotropic-liquid phase transition. Because of the simplicity of this model, the calculations can be performed for a large number of particles of the main sample (104 particles and more). Moreover, this model has been studied many times in the past by different approximate theoretical methods, which is important for making comparisons and assessments of the results.

Some works in this direction were concerned with an examination of different improvements and modifications of the model, as well as refinement and checking of preceding results. Thus Luckhurst and Romano[128] investigated the effect of the translational degree of freedom on the properties of the lattice model. Using the Monte Carlo method, they studied a system of 256 cylindrical symmetrical particles, which interacted via a potential which consisted of a scalar part of the Lennard-Jones form, and a Maier–Saupe anisotropic part. In contrast to preceding studies, the particle locations were not restricted to the lattice sites, and their interaction was not restricted to nearest neighbors only. The results of the calculation of the thermodynamic properties agreed satisfactorily with the experimental data for para-azoxy-cyanisole. It was noted that the mean-field theory agrees better with the data obtained from computer "experiments," performed for this model, than with the lattice models with "imposed" translational order and truncated anisotropic pair potential.

The lattice model with a Maier–Saupe interaction potential turned out to be very convenient for studying a number of other questions regarding the physics of liquid crystals. In Ref. 129, the Monte Carlo method was used to study the effect of an external field on a lattice system of particles, interacting with one another via a potential of the form $P_2(\cos\beta_{ij})$. The interaction potential of the i-th particle with an external field had the following form:

$$U_i/\epsilon = -xP_2(\cos\beta_i),$$

where β_i is the angle between the direction of the field and the symmetry axis of the particle and x is the interaction force parameter. The modeling established that the orientation of the director can be fixed by applying a moderate external field, which increases only insignificantly the orientational order parameter of the system. The results obtained for a system of 1000 particles in the main sample suggest that when the number of particles is increased up to 8000 and more the orientation of the director can also be fixed at a lower field intensity, which does not change the magnitude of the order parameter.

In order to study the effect of the form of the intermolecular interaction potential on the computational results and in order to take into account the anisotropic dispersion forces more accurately, Humphries, Luckhurst, and Romano[131] studied by the Monte Carlo method the behavior of a system of 1000 linear particles, located at the sites of a simple cubic lattice. In contrast to the Maier–Saupe potential, the pair interaction potential depended not only on the relative orientation of the particles, but also on the spatial arrangement. It is noteworthy that the results of the calculations differed little from the well-known data for the simpler Maier–Saupe potential. It was also noted that the orientation of the director was virtually unchanged in the modeling process. This result is important for the development of the computer-modeling methods and agrees with the results of calculations performed by the molecular-dynamics method.[22,30,38]

The fundamentally important question of the dependence of the results of computer "experiments" on the number of particles in the main sample was examined by Luckhurst and Simpson.[129] They performed a careful study of the region of the phase transition in the lattice model with a Maier–Saupe interaction potential for the case of 8000 particles in the main sample. The results of the calculations of the internal energy, heat capacity, and order parameter agree satisfactorily with the data from previous "experiments" for 1000 particles, but the features of the first-order phase transition are much more distinct in the case of the system with the larger number of particles.

In Ref. 130, the molecular-dynamics method was used to study a model system of 1000 cylindrically symmetrical particles, located at the sites of a cubic lattice and interacting via a pair potential of the form

$$U_{ij} = -\epsilon_{ij} P_2(\cos \beta_{ij}),$$

where ϵ_{ij} is a positive coefficient, which differs from zero only for neighboring particles. Both the static and dynamic properties of the system were calculated for three temperatures, corresponding to the states of the nematic phase and the isotropic liquid. Different types of correlation functions, characterizing the structure of the system and the rotational motion of the particles, were studied in detail. The form of these functions is qualitatively analogous to the form which we obtained previously for dense systems of ellipsoidal particles.[30,38]

Further investigations of this trend[132] were based on taking into fuller account the character of inter-molecular interaction described by the equation:

$$U = \xi P_2(\cos \theta) + \gamma P_4(\cos \theta),$$

where θ is the angle between the axes of symmetry of the interacting particles, ξ and λ-force parameters . The application of the Monte Carlo method made it possible for internal energy and heat capacity for such a system to be calculated, revealing the availability of the NLC–IL transition. Computer simulation studies of the surface properties of the nematic model were made in another work.[133] The values of the surface tension found were compared with values received on the basis of mean molecular field theory. A two-dimensional hard-core nematic formed by infinite thin rigid rods was studied by Monte Carlo method in paper 134. It was shown that the transition between isotropic liquid and nematic phase occurs at the expense of the formation of free disclinations. Specific heat, as well as the first five virial coefficients and coefficients of expansion of free energy were calculated.

Computer simulation methods were repeatedly used for special studying phase transitions. Among these works it is necessary to note the investigation of the phase transition of solid crystal–nematic by the Monte-Carlo method[135] as well as the study of the nematic–smectic A transition.[136] In the latter paper a three-dimensional lattice de Gennes model was simulated and many features of this transition observed in experiments were reproduced.

More delicate features of molecular structure of a mesogenic substance were taken into account in a Monte-Carlo simulation of a system consisting of rod-like rigid cores with flexible tails.[137] Parameters of orientational ordering of rod-like rigid cores and of first segments of flexible tails were calculated. Unfortunately, this study was possible only for a two-dimensional lattice system.

Computer simulation studies of the anisotropic viscosity of liquid crystals with nonequilibrium molecular dynamics method[138] are of great interest. This method was successful for the study of systems of ellipsoids with an interaction of the Lennard–Jones type and agreed with experimental data.

The method of molecular dynamics investigation of lattice models of NLC made it possible to study the influence of external electric fields on a thin NLC film, which at the same time undergoes the influence of restricting walls.[139,140] Unlike the Monte-Carlo method one can study the process of influence of field in time (that is, transitions to equilibrium after switching-on or switching-off the field). This fact provided unique possibilities for theoretical study of dynamic features of Fredericksz's effect which are, as is well known, due to competitive influence of fields on NLC film.[139,140]

In conclusion, we cannot help but note the large amount of information and great clarity provided by molecular-dynamics models, which permits their use as a means for illustrating the basic concepts and ideas concerning the nature and behavior of liquid-crystalline systems (see Figs. 1.3–1.6).

3.6. OTHER METHODS OF CALCULATION OF STRUCTURE AND EQUILIBRIUM PROPERTIES

Aside from the key above-mentioned trends of calculation-theoretical investigations of structure and equilibrium (thermodynamical) properties of nematic liquid crystals, other approaches have been developed as well. Let us consider briefly some of those that have attracted attention in the last years.

A series of works of Soviet authors was devoted to molecular-statistical consideration of a system of rigid rectangular asymmetrical parallelepipeds with discrete distribution of orientations and with continuous distribution of centers of masses. This system was used as a simple model of NLC and their solutions.[141–143] It was shown that such a model enables one to describe thermodynamic properties and ordering of the liquid crystal phase. The expressions for pressure and for chemical potentials were found.[141] This approach proved to be convenient for the description of ranges of existence of isotropic and liquid crystal phases, both for a pure nematogenic substance and for systems of nematic–nonmesogenics and nematic–nematic. Free energy of nematic systems of rigid molecules was found on the basis of the Van der Waals model.[144] It has become possible to show that taking into account biaxiality of molecules provides a good agreement of order parameters and thermodynamic characteristics of nematic–isotropic transition for PAA with experimental data.

A special group of investigations is presented by works devoted to the development of a multipole model and to using multipole expansion in studying liquid crystals,[145,146] while at the same time problems of account of flexibility of molecules and their conformations and the contribution of steric interactions are solved. The prospects of this trend were noted, as well as the possibility to attain by means of it a more profound comprehension of the connections between molecular characteristics and macroscopic properties of NLC.

Special models were developed for the explanation of reentrant behavior in liquid crystals when in decreasing the temperature below the range of the existence of smectic A phase the range of nematic mesophase is observed again.[147] The mechanism of this phenomenon is explained by competition between strong dipole–dipole interaction of head parts of molecules and Van der Waals interaction of molecular "tails." Corresponding theoretical phase diagrams discussed on the basis of available experimental data were drawn.

Reentrant transitions in systems of flexible molecules were explained by the interaction of orientational and conformational degrees of freedom and by an abrupt conformational disordering of alkyl and alkoxyl chains on reaching a definite temperature in cyanobiphenyls and in compounds similar to them.[148]

A simple semiempirical formula describing the change of the temperature of transition NLC–isotropic liquid of substances belonging to different homologous series was derived in Reference 149. It was shown that this formula is suitable for quite a number of nematogenic substances and provides a good agreement with experimental data. Theoretical investigation of the kinetics of the isotropic–nematic phase transition in liquid crystals in adiabatic conditions was given in Reference 150. In this paper the Ginzburg–Landau theory of nucleation of new phase was used. Calculating the size distribution of the nematic drops appeared to be successful. The calculation of the evaporation heat of PAA in the vicinity of the transition point for NLC–isotropic liquid on the basis of acoustic data was done on paper.[151]

A number of theoretical works devoted to the study of the behavior of thin nematic films as well as to the investigation of the surface tension of NLC are of interest. Nematic ordering induced by the interaction with the wall described by the external field was studied in Reference 152. The phase diagram was drawn by using this investigation. The calculation of the surface tension of NLC in terms of correlation functions was performed by Roumanian investigators.[153,154] Using the calculated expression of J. Murakami[155] under the assumption of pair additivity of the molecular interaction enables one to calculate the surface tension of MBBA. However, the discrepancy between calculated and experimental data was significant.[153] The tensor coefficient of the surface tension as well as its translational and orientational components were studied in Reference 154.

Transient dynamics of orientational fluctuations of the director for a layer of NLC in the Freederickcz transition was studied in Reference 156. In that paper, the relaxation Ginzburg–Landau equation was used with the random Gauss noise for the angle describing the deformation of torsion for a nematic phase. Different stages of the transition process were analyzed and quantitative estimations were given.

3.7. MOLECULAR THEORIES OF TRANSPORT PROPERTIES

The theories of NLC examined above (with the exception of the molecular-dynamics method) do not permit studying the dynamic properties of molecules in NLC and calculating the transport properties. As in the case of simple liquids, the theory of transport properties is much more complicated and much less well developed than the theory of equilibrium properties.

One of the first attempts at constructing a theory of diffusion in NLC was undertaken by Franklin,[157] who generalized the well-known Kirkwood theory to the case of anisotropic molecules and obtained the following relations for the components of the self-diffusion tensor:

$$D_{\parallel} = kT[1/\mu\xi + (2 + \langle P_2\rangle)/(6\pi\, \mu^2\varphi) \times (4\alpha_1 + 3\alpha_2/2 + 13\alpha_3/2 + 6\alpha_4 + 10\alpha_5)], \tag{3.28}$$

$$D_{\perp} = kT[1/\mu\xi + (5 - \langle P_2\rangle)/(12\pi\mu^2\varphi) \times (-11\alpha_2/4 + 9\alpha_3/4 + 6\alpha_4 - 3\alpha_5/2)]. \tag{3.29}$$

Here, μ and φ are parameters which are determined by the geometry of the molecules, ξ is a friction coefficient, and α_i ($i = 1, 2, \ldots, 5$) are the viscosity coefficients of the NLC according to Ericksen and Leslie. The use of experimental values of the quantities entering on the right side of (3.28) and (3.29) as well as experimental data on the self-diffusion coefficient (required for determining the friction coefficient ξ) permits describing the temperature dependences of $D_{\parallel}$ and $D_{\perp}$ with satisfactory accuracy. Franklin's method is, however, phenomenological. In addition, as in the theory of simple liquids, the derivation of expressions (3.28) and (3.29) requires a number of assumptions, while the concept of a friction coefficient does not have a clear physical meaning.

A theoretical expression for the coefficient of rotational diffusion of molecules in NLC was obtained by Chu *et al.*[158] This result is based on a solution of a Vlasov-type kinetic equation in which molecular collisions are taken into account in an approximation of the collision integral by a model Bathnager–Gross–Krook function. The use of a single-particle Maier–Saupe potential (3.1) in this case, however, severaly restricts the generality and accuracy of this method.

The method of time correlation functions (TCF) and Kubo's linear-response theory[107] permit determining the transport properties of NLC. At present, however, there are no general and rigorous methods for determining the TCF (though recently the molecular-dynamics method has been successfully used for this purpose). For this reason, the real TCF is sometimes replaced by a specially chosen functional dependence. Chu and Moroi[159] calculated the average self-diffusion coefficient D_0 of NLC using this method. Using the idea of molecule clusters which are aligned parallel to one another they were able to derive the equations which relate the averaged coefficient of self-diffusion D_0 with the components $D_{\parallel}$ and $D_{\perp}$ and the order parameter $\langle P_2 \rangle$. Although this approach leads to satisfactory agreement with experiment, it is based on quite a rough approximation of the pair distribution function, on assumptions concerning the dimensions of the clusters, and on the form of the TCF.

In recent years, great attention was paid to revision, specification, and generalization of existing ideas about the character of particle motions in NLC and transport mechanism. A detailed review of individual and collective molecular motions and their agreement with experimental data obtained by the NMP method was given in paper 160. The authors of this work stated the present mathematical technique for description of molecular dynamics of nematics, smectics and cholesterics (method of correlation functions, projection operator technique, investigations and solutions of the diffusion equation for liquid crystals). The model of random travelling as an approach to the solution of the problem of diffusion of molecules in the presence of a potential pit was considered in Reference 161. The results of the study and discussion of solitary waves in nematics[162,163] found in shear flow experiments of nematics[164] offer an interesting modern section connected with travellings in NLC.

The development of equations of motion and other important characteristics were considered on the basis of irreversible thermodynamics (Onsager's linear theory) in paper 165. By using the variation principle, the relaxation time τ for the NLC layer contained between parallel walls in the presence of strong cohesion with them was calculated. It appeared that $\tau \sim \delta^2$ (δ is the cell thickness) remains a finite value for thin cells unlike the theory of Ericksen–Leslie–Parodi confirming $\tau \sim \delta^{-2}$.

A paper[166] devoted to computer simulation studies of the anisotropic viscosity of NLC was mentioned above. Another approximating method of estimation of viscosity of liquid crystals and their mixtures was proposed by a Japanese scientist who found the correlation between viscosity and parameters of effective intermolecular interaction potential, taking into account dispersion, dipole–dipole, and dipole–inductive dipole interactions.[166]

CHAPTER IV.

The Thermophysical Properties of Nematic Liquid Crystals and Comparison of Experimental Data and Computational Results

Gray, dear young fellow, is all theorizing,
And green, life's golden tree.
Johann Wolfgang Von Goethe, *Faust*

4.1. REGION OF EXISTENCE OF NEMATIC LIQUID CRYSTALS

The results of modern experiments and theoretical investigations point out the importance of anisotropic repulsive forces in establishing nematic order. It is therefore interesting to establish a theoretical relationship between the shape of the molecule, the density, the temperature, and the degree of order in model systems describing NLC. This problem was solved for rigid spherocylinders using the scaled-particle method,[104] for rigid ellipsoids using the integral-equation method,[167] and for ellipsoids with power-law repulsion using the molecular-dynamics method.[22,38]

Some results of the molecular-dynamics investigation carried out for a constant temperature $T^* = 0.76$ are presented in Fig. 4.1. It is evident that if the ratio of the axes of ellipsoidal molecules is less than 2, then even at high densities nematic order does not arise. Similar results are obtained by the scaled-particle and integral-equation methods though these approaches predict lower densities of NLC–IL transition.

For real NLC the region of existence is usually characterized by a pressure–temperature dependence; this region is situated between

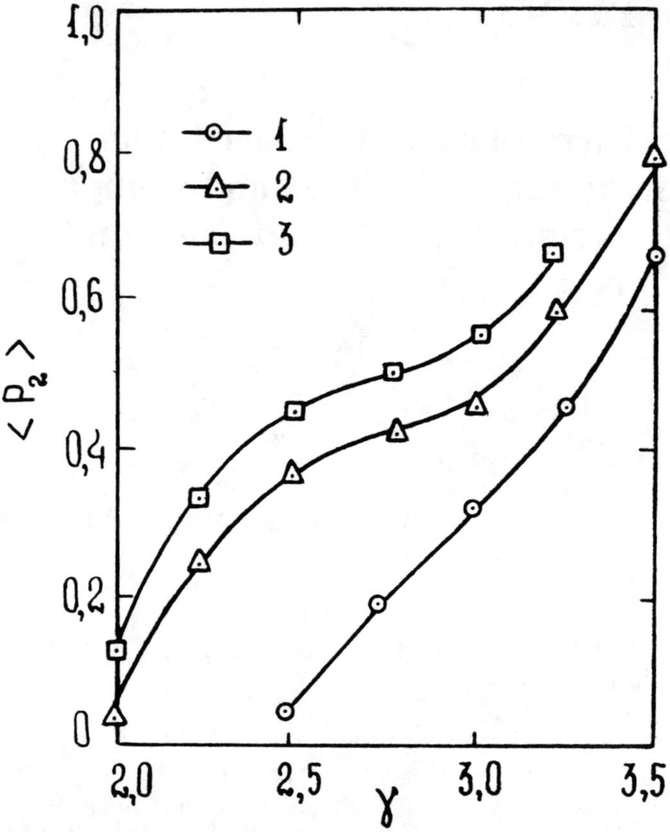

Figure 4.1. Dependence of ordering parameter $\langle P_2\rangle$ on $\gamma = \sigma_{\parallel}/\sigma_{\perp}$ and on packing coefficient $\eta = vN/V$ for a system of ellipsoidal particles at $T^* = 0.76$ (potential ϕ_3)[22,30]: 1—$\eta = 0.50$; 2—0.53; 3—0.55.

the curves of the NLC–solid and NLC–IL transitions (Fig. 4.2a). Figure 4.2b shows in reduced coordinates the computed P–T curves according to Chandrasekhar's data.[168] Good agreement is observed, which ensures a correct description of the P–T diagram with a proper choice of parameters. A detailed analysis of the stability of the nematic phase can be performed with the help of the dependence of the free energy versus the order parameter (see Fig. 3.2).

The P–T dependences of PAA and MBBA were studied experimentally up to a pressure of 700 MN/m^2 by N. A. Tikhomirova, L. K. Vistin', and V. N. Nosov.[169] The low-pressure data presented in

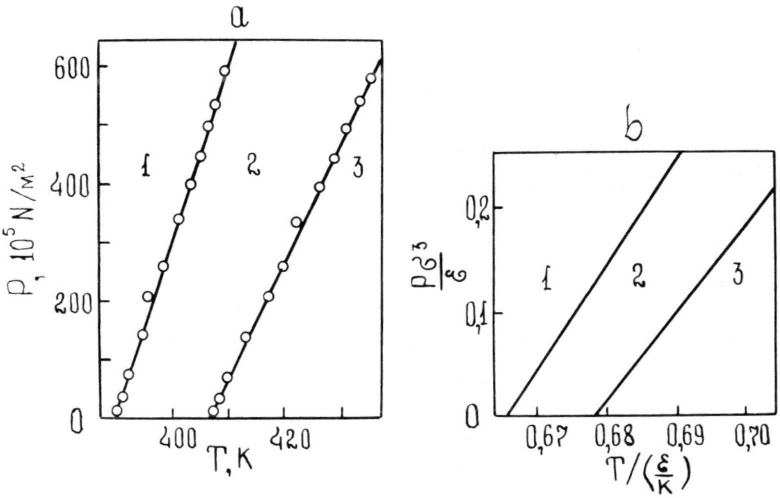

Figure 4.2. Comparison of experimental (a) and computed (b) pressure-temperature curves for PAA[17]: 1—solid phase; 2—NLC; 3—IL.

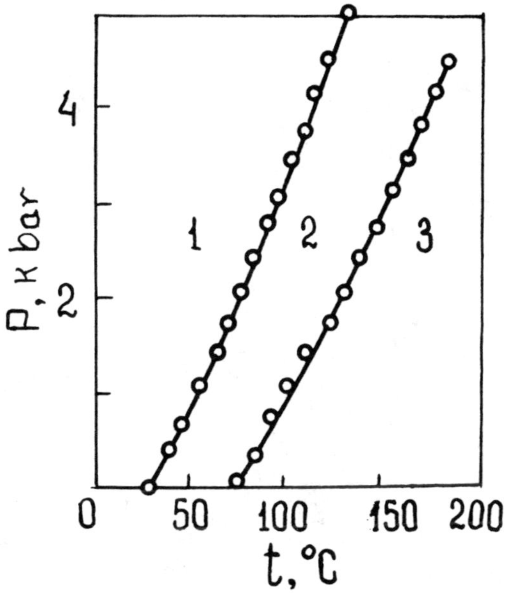

Figure 4.3. Phase diagram of *N*-(*p*-ethoxybenzylidene)-*p*-*n*-butylaniline (EBBA)[174]: 1—solid phase, 2—NLC, 3—IL.

Ref. 169 agree satisfactorily with the results of previous work.[170–172] Keyes, et al.[173] studied in detail the P–T dependences of a number of liquid crystals (including MBBA).

Examples of some P–T phase diagrams, demonstrating the regions of existence of NLC are presented in Figs. 4.3–4.5.[174]

The data obtained during the last few years on temperatures of transitions and phase diagrams of pure NLC are given in their most complete form in papers 175–178. So, the data on temperatures of phase transitions of a large group of 4-alkoxy-*N*(fluoren-2-ylmethylene)-anilines (AFYA), 4-alkoxy-*N*-(9-oxofluoren-2-ylmethylene)-anilines (AOFYA) and 4-alkoxy-(2-naphthylmethylene)-anilines (ANMA) made it possible to judge liquid crystal properties of these substances, some of which demonstrate both nematic and smectic mesophases.[175,176] An infra-red spectroscopic study of *N*-[4-(4-octyl-

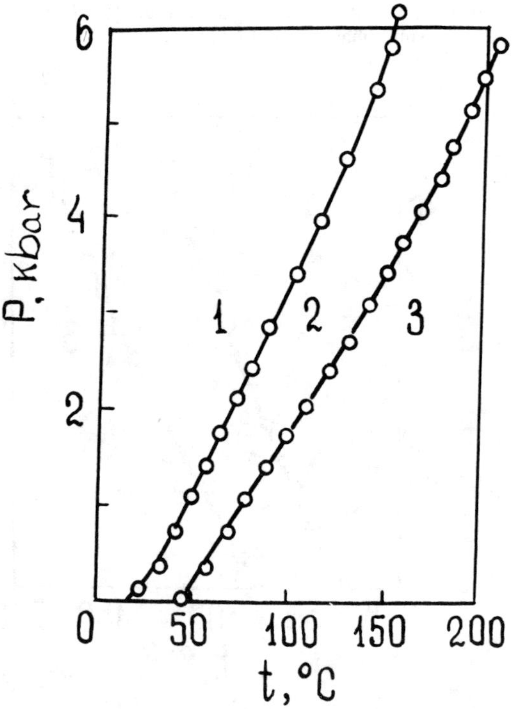

Figure 4.4 Phase diagram of *N*-(*p*-methoxybenzylidene)-*p*-*n*-butylaniline (MBBA):[175] 1—solid phase, 2—NLC, 3—IL.

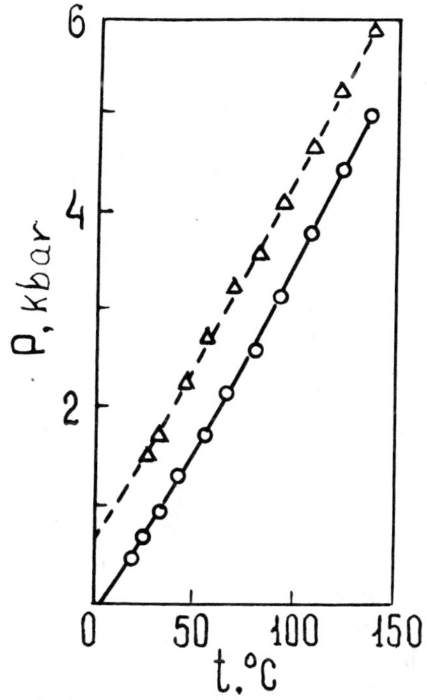

Figure 4.5. Phase diagram of *N*-(*p*-methylbenzylidine)-*p*-butylaniline (MBA)[174]: the solid line shows the NLC–IL transition (monotropic transition); the broken line shows the solid phase–IL transition.

oxybenzoyloxy) benzylidene]-4-cyanoaniline (OBBC) demonstrating smectic, nematic and reentrant-nematic phases was reported in paper 177.

Attention was also paid to the influence of pressure on phase transitions of isotropic liquid-nematic liquid crystals (as well as nematic–smectic and smectic–reentrant nematic liquid crystals) in 8OCB, CBOOA, CBNA etc.[178,179] Molecular conformation and orientational order in nCB (n = 5–8) liquid crystals was studied in phase transition range.[180] An evanescent-wave-ellipsometry study was used for investigation of wall-induced orientational order of a liquid crystal, 4′-*n*-pentyl-4-cyanobyphenile (5CB).[181]

A theoretical interpretation of temperatures of transitions of a nematic–isotropic liquid for a homologous series of substances was generalized in the above-mentioned paper[149] and the relation of orientational melting of chain fragments of molecules with the character of this phase transition was discussed in Ref. 182. The influence of pressure on local orientational fluctuations of NLC was investigated within the variation method.[183]

Complex experimental and theoretical investigation of orientational ordering of thin liquid crystal films near the range of transition to an isotropic liquid as well as the definition of boundaries of range of existence for metastable phases were made by Soviet investigators.[184] Another group of authors have studied the influence of pore sizes on temperatures of phase transitions of a liquid crystal substance 4-octyl-4′-ethoxytolane containing in porous matrix.[185] It was shown that with reducing the pore radius the temperature of the transition solid crystal-liquid crystal is decreased, but the temperature of the transition liquid crystal–isotropic liquid is increased in comparison with corresponding temperatures of a bulk sample.

The role of surface adsorption in the surface-induced alignment of nematic liquid crystals on evaporated SO_2 films was studied for an isotropic liquid that was in equilibrium with nematic phase.[186]

As it is known, external fields (electric and magnetic fields) also influence molecular ordering. This effect for NLC was studied quite thoroughly, for example, in the paper,[187] in which electric and magnetic field effects on nematic liquid crystals with high positive dielectric anisotropy were investigated. On the basis of data on temperature dependence of dielectric anisotropy of *N*-(*p*-ethoxybenzylidene)-*p*-*n*-cyanoaniline (EBCA) and *N*-(*p*-methoxybenzylidene)-*p*-*n*-cyanoaniline (MBCA) a conclusion was made regarding statistical ordering of an antiferroelectric character.

4.2. THERMAL PROPERTIES, SURFACE TENSION, AND DIAMAGNETIC SUSCEPTIBILITY

The pressure and temperature dependence of the volume of MBBA in the region of the NLC–IL phase transition were studied experimentally by Zawisza and Stecki.[188] The results of their study are presented in Fig. 4.6. The P–V curves contain sloping steps, indicating a first-order phase transition. As the temperature increases, the region of the transition shifts toward higher volumes and pressures. Apparently it is difficult to establish the temperature depend-

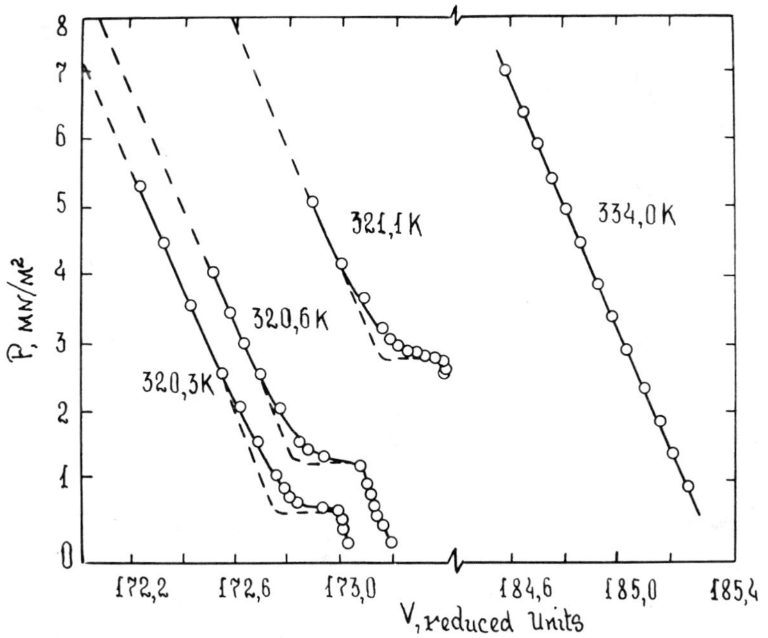

Figure 4.6. P-V curves of MBBA at constant temperatures.[188]

ence of the volume change accompanying the phase transition from the data in Ref. 188.

The P–V–T dependence of PAA was studied in detail in Ref. 189. Unfortunately the reliability of these data is doubtful. Thus the volume changes accompanying the NLC–IL transition found for pressures close to atmospheric turned out to be much higher than the values indicated in the literature,[47] and they do not have a reasonable interpretation.[122]

The change in the volume of PAA in the region of the NLC–IL transition up to high pressures was studied in detail (see Ref. 122 (Fig. 4.7)). It was established that this phase transition is a first-order transition over the entire range of pressures studied, and that the pressure dependence of the volume change is quite weak ($\Delta v/v_N \approx 3.5.10^{-3}$).

It is of interest to study the dependences of the volume and order parameter of two nematogens (5 CB and MBBA) on the temperature and pressure based on the data obtained from optical experiments,

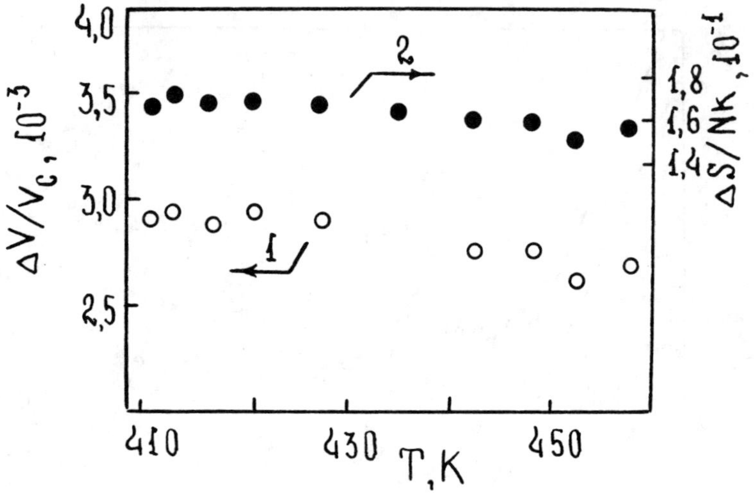

Figure 4.7. The temperature-dependences of $\Delta v/v_c$ (1) and $\Delta S/Nk$ (2) for PAA at high pressures.[122]

performed up to high pressures.[190] These results are presented in Figs. 4.8 and 4.9. It is noteworthy that none of the variants of the mean molecular-field theory of nematic ordering leads to satisfactory agreement with all data based on the experiments.

A detailed study of the thermal properties of MBBA and EBBA was performed in Ref. 191. Some results from this study are presented in Tables 4.1–4.3. Tait's equation

$$(V_o - V)/V_o = C \log (B + P)/(B + P_o),$$

was useful in describing the P–V–T data, but, because of the jump in the density accompanying the transition, the isotropic and nematic phases must be described separately. It is interesting that the compressibility of these substances k at atmospheric pressure is much smaller than the compressibility of n-decane and benzene. The compressibility at the transition point is approximately the same in both phases.[191] A comparison with more recent data[192] for MBBA indicates the satisfactory agreement, but, according to the data in Ref. 191, the jump in the density accompanying the transition is somewhat larger.

The temperature dependence of the volume of NLC has been studied in greatest detail at atmospheric pressure.[193] Figure 4.10 presents this dependence for PAA according to the data in Ref. 194, which

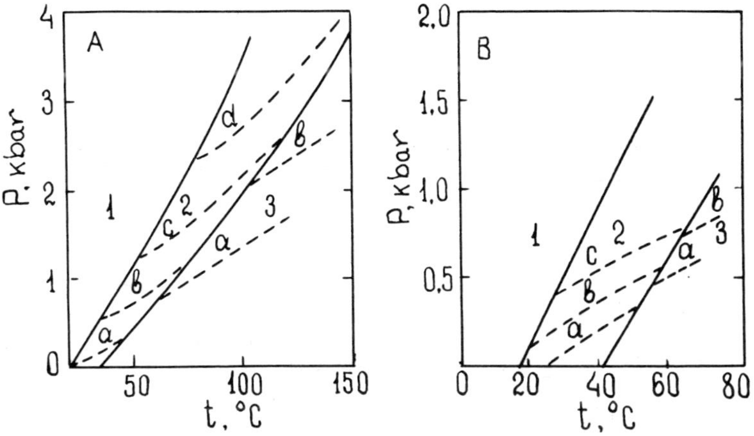

Figure 4.8. Isochores of 5 CB and MBBA[190]: 1) solid crystal; 2) nematic; 3) isotropic liquid; A) 5 CB: a) 248; b) 244; c) 240; d) 236 cm^3/mole; B) MBBA: a) 256; b) 254; c) 252 cm^3.mole.

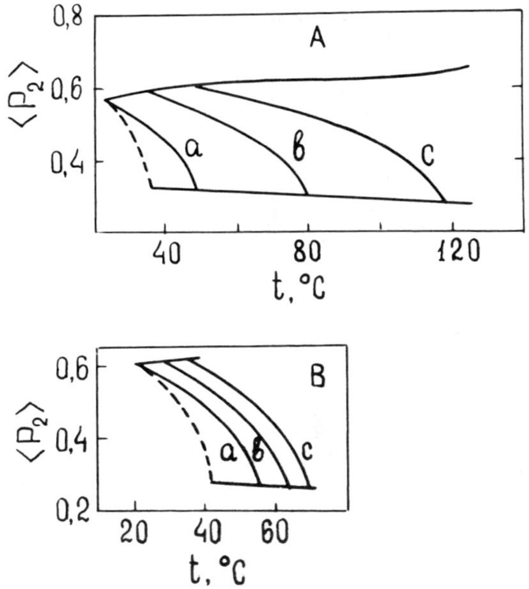

Figure 4.9. Curves of the parameter $\langle P_2 \rangle$ versus the temperature.[190] A) 5 CB: the solid lines are the isochores: a) 248; b) 244; c) 240 cm^3/mole; the broken curve is for atmospheric pressure; B) MBBA: the solid curves are the isochores a) 256; b) 254; c) 252 cm^3/mole; the broken curve is for atmospheric pressure.

Table 4.1
Density of MBBA (ρ, $g \cdot cm^{-3}$) in the Isotropic and Nematic Phases as a Function of the Temperature and Pressure[191]

	t, °C					
P, bar	40	50	60	70	80	90
1	1.0332	1.0228	1.0146	1.0064	0.9983	0.9901
100	1.0387	1.0284	1.0200	1.0117	1.0038	0.9957
200	1.0441	1.0359	1.0251	1.0171	1.0092	1.0014
300	1.0493	1.0414	1.0306	1.0224	1.0145	1.0068
400	1.0540	1.0464	1.0357	1.0277	1.0197	1.0120
500	1.0588	1.0515	1.0428	1.0328	1.0248	1.0171
600	1.0636	1.0562	1.0478	1.0376	1.0298	1.0221
700	solid	1.0607	1.0525	1.0442	1.0345	1.0270
800		1.0652	1.0569	1.0490	1.0391	1.0316
900		1.0693	1.0612	1.0535	1.0435	1.0361
1000		1.0730	1.0655	1.0577	1.0496	1.0405
1100		solid	1.0694	1.0617	1.0539	1.0448
1200			1.0734	1.0656	1.0580	1.0489
1300			1.0774	1.0694	1.0619	1.0548
1400			solid	1.0732	1.0658	1.0589
1500				1.0771	1.0698	1.0628
1600				1.0809	1.0735	1.0665
1700				1.0847	1.0772	1.0702
1800				solid	1.0807	1,0739
1900					1.0841	1.0775
2000					1.0875	1.0811
P_{IL-NLC}, bar	130	130	400	682	970	1270
ρ_{IL}		1.0299	1.0358	1.0415	1.0465	1.0516
ρ_{NLC}		1.0320	1.0379	1.0435	1.0483	1.0536
P_{NLC-SC}, bar	645	1018	1398	1793	2208	2667

Table 4.2

The Compressibility κ and the Coefficients C and B in Tait's Equation for Both Phases of MBBA[191]

	$K \cdot 10^6$, bar^{-1}			C	B, bars	
	1 bar	Transition Point		For All Temperatures		
t, °C	IL	IL	NLC		IL	NLC
60	54.55	45.88	43.96	IL:	2116.2	2087.4
70	55.94	41.77	40.13	0.2658	2063.7	2034.5
80	57.35	38.92	36.65	NLC:	2012.5	1977.9
90	59.10	36.60	34.38	0.2619	1947.0	1908.7

Table 4.3
Density of EBBA (g cm^{-3}) in the Isotropic and Nematic Phases as a Function of the Temperature and Pressure[191] *

P, bar	t, °C						
	50	70	90	100	110	120	130
1	1.0152	0.9990	0.9803	0.9726	0.9649	0.9572	0.9490
100	1.0207	1.0050	0.9865	0.9790	0.9715	0.9639	0.9558
200	1.0260	1.0108	0.9927	0.9853	0.9780	0.9706	0.9624
300	1.0311	1.0165	1.0015	0.9914	0.9843	0.9767	0.9689
400	1.0361	1.0219	1.0077	0.9975	0.9905	0.9830	0.9750
500	—	1.0271	1.0136	1.0034	0.9964	0.9891	0.9807
600	solid	1.0321	1.0192	1.0119	1.0024	0.9946	0.9867
700		1.0368	1.0245	1.0176	1.0079	1.0005	0.9924
800		1.0414	1.0296	1.0231	1.0134	1.0060	0.9980
900		1.0460	1.0344	1.0281	1.0210	1.0113	1.0034
1000		1.0505	1.0390	1.0331	1.0261	1.0164	1.0085
1100		1.0549	1.0433	1.0378	1.0311	1.0211	1.0134
1200			1.0475	1.0422	1.0360	1.0285	1.0183
1300			1.0516	1.0464	1.0403	1.0331	1.0228
1400			1.0556	1.0506	1.0446	1.0376	1.0274
1500			1.0595	1.0541	1.0484	1.0418	1.0345
1600			1.0633	1.0580	1.0521	1.0456	1.0385
1700			1.0670	1.0616	1.0559	1.0496	1.0425
1800			1.0705	1.0651	1.0595	1.0533	1.0462
1900			1.0740	1.0685	1.0630	1.0570	1.0500
2000				1.0719	1.0664	1.0606	1.0537
P_{IL-NLC}, bar			278	554	850	1150	1470
P_{NLC-SC}, bar	470	1180	1950	2370			

*Selected data.

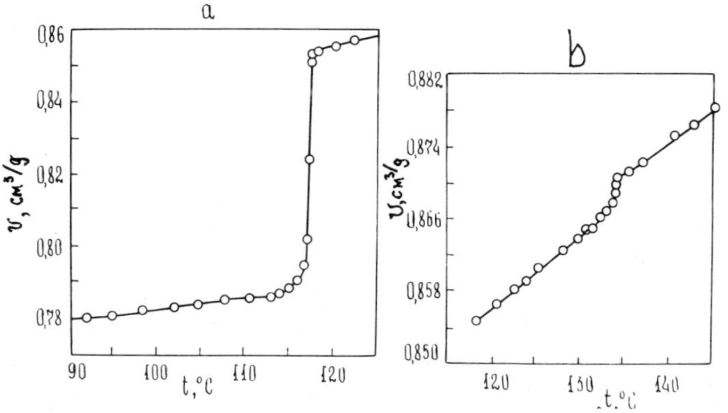

Figure 4.10. The temperature dependence of the specific volume of PAA in the region of solid phase–NLC (a) and NLC–IL (b) transitions.[199]

agree well with the results in Ref. 122. Table 4.4 compares calculations of the volume of PAA (performed in Ref. 195) with the experimental data. The basic working expression was obtained from the representation of the Gibbs free energy in terms of the order parameter and the orientational part of the distribution function.

The temperature dependences of the density of metoxybenzylidine butylaniline (MBBA) and hexyloxybenzylidine-amino-benzonitryl (HBAB) in the nematic phase and in the region of the transition into the isotropic liquid, as well as 4-cyanobenzylidene-4′-*n*-octyloxyaniline (CBOOA) in the NLC region and in the vicinity of transitions into the SLC-A and into the isotropic liquid are presented in Figs.

Table 4.4
Comparison of the Computed and Experimental Values of Volumes of PAA at Atmospheric Pressure (cm^3/g) (Ref. 195)

Temperature, K	Experiment	Calculation
393	0.8584	0.8582
395	0.8599	0.8597
397	0.8613	0.8612
399	0.8627	0.8628
401	0.8642	0.8644
403	0.8657	0.8660
405	0.8675	0.8678
406	0.8684	0.8686
407	0.8693	0.8696
408	0.8703	0.8705
408.5	0.8709	0.8710
408.8	0.8713	0.8713

4.11 and 4.12.[196] It is evident that there is a significant difference between the values of $\Delta\rho$, observed in the case of these two types of transitions. The densities of the two tolanes (MPT and PHT) and their equimolar mixtures in the vicinity of the NLC–IL transition was measured by French researchers[197] (Fig. 4.13), while the density of HBT and OBT was studied in India[198] (Fig. 4.14).

Table 4.5 presents the density of DPAB and EPAB for temperatures at which the nematic and isotropic phases exist.

It is evident from the data in the table that the density of EPAB is higher than that of DPAB in both the nematic and isotropic-liquid states. This difference is caused by the group $(CH_2)_3$. The volume of this group, which at a temperature of 30.96°C was equal to 48.60 $cm^3 \cdot mole^{-1}$ for the pair DPAB and EPAB and 49.56 $cm^3 \cdot mole^{-1}$ for

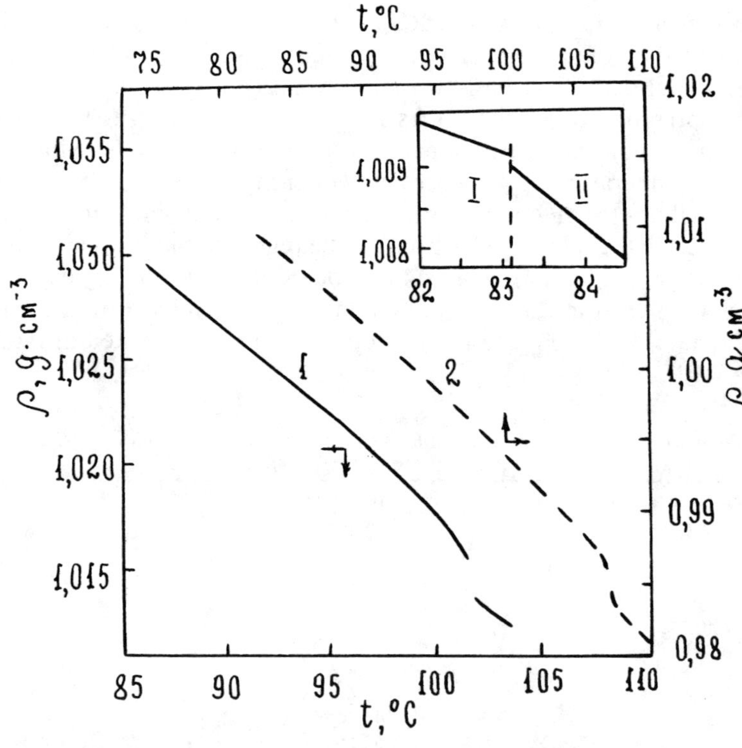

Figure 4.11. Temperature dependences of the density of HBAB (1) and CBOOA (2) [196]The change in the density of CBOOA in the vicinity of a SLC (I)–NLC (II) transition is shown in the inset.

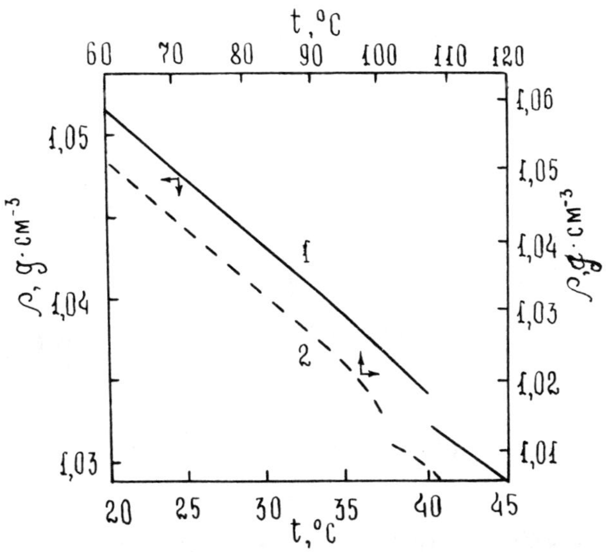

Figure 4.12. Temperature dependences of the density of MBBA ($\Delta\rho = 2.1 \cdot 10^{-3}$ $g{\cdot}cm^{-3}$) (1) and HBAB (2).[196]

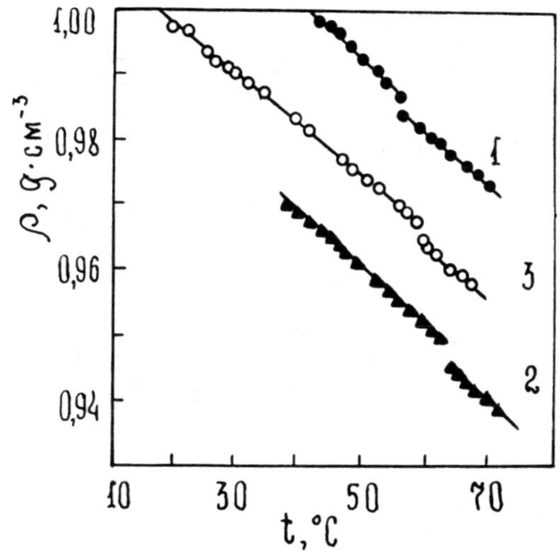

Figure 4.13. Temperature dependences of the density of the tolanes and their mixture in the region of the NLC-IL transition[197]: 1) *p*-methoxy-4-pentyl-4′-tolane (MPT), 2) propoxy-4-heptyl-4′-tolane (PHT), 3) equimolar mixture of these tolanes.

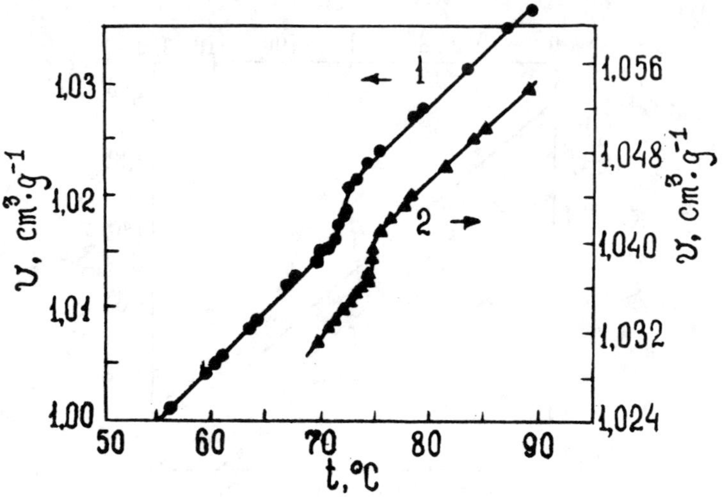

Figure 4.14. Temperature dependences of the specific volumes of (*p*,*n*-hexyloxy-benzylidene)-*p*-toluidine (HBT) (1) and (*p*,*n*-octyloxy-benzylidene)-*p*-toluidine (OBT) (2).[198]

the pair 4-oxtyl-4′-ethylazoxybenzene and EPAB, was calculated in Ref. 199. A somewhat larger difference is observed between the values found at one and the same reduced temperature

$$T^* = (T_{NI} - T)/T$$

for these pairs: 54.70 $cm^3 \cdot mole^{-1}$ and 50.45 $cm^3 \cdot mole^{-1}$.

The results of an experimental study of density and ultrasound velocity of *N*-(*n*-methoxybenzylidene)-*n*-aminobenzonitrile (PMBAB) and *N*-(*n*-ethoxybenzylidene)-*n*-aminobenzonitrile (PEBAB) made it possible to calculate many thermodynamic characteristics of these substances and to find out the availability of pretransition effects from

Table 4.5
Densities ($g \cdot cm^{-3}$) of 4,4′-dipentylazoxybenzene (DPAB) and 4-ethyl-4′-pentylazoxybenzene (EPAB) at Different Temperatures.[199]

t, °C	DPAB	EPAB
25	1.0190	1.0467
30	1.0153	1.0424
35	1.0116	1.0379
70	0.9788	1.0048
75	0.9748	1.0048

DPAB: t_{CN} = 24.5°C t_{NI} = 67.8–68.0°C
EPAB: t_{CN} = 5°C, t_{NI} = 38.9°C

both sides of phase transition NLC–isotropic liquid.[200] Similar measurements were also performed for *N*-(4-*n*-pentyloxybenzylidene)-4-*n*-phenylazoaniline (PBPAA).[201] The data on density and the expansion coefficient of 4-*n*-hexyloxy-phenyl-4-*n*-decyloxybenzoate (HPDB)[202] and *p*-*n*-nonyloxybenzoic acid (NOBA)[203] were obtained. Density and some other physical chemical properties of 1,4-*bis*-[4-*n*-octyloxybenzoyloxy]-2-*n*-alkylbenzenes were studied in paper 204.

Works devoted to simultaneous determination of pressure, volume and temperature (P–V–T-properties) of liquid crystal substance are of special interest. 4-Fluorophenyl-*trans*-4′-heptylcyclohexan carboxylate (HCCF)[205] and homologous series of di-alkylazoxybenzenes[206] are among the substances studied in such a way. On the basis of measurements it was concluded that carboxyl groups between rings of HCCF-molecules lead to the "buffer" effect playing the role of "softening potential."

It is interesting to note that a simple equation of state, based on the representation of molecules in NLC as particles with slightly nonspherical shape interacting via an anisotropic potential, gives a qualitatively correct description of the P–V–T diagram of PAA.[206] It is not possible, however, to give a correct quantitative prediction of both the thermal and caloric properties of the system near an NLC–IL transition (Table 4.6).

In spite of certain methodical difficulties, a number of measurements of the surface tension of NLC have now been performed. These data are briefly listed in Table 4.7,[209–218] where, in addition to the values of the surface tension, the measurement method is also indicated.

For liquid crystals, which are anisotropic diamagnetic substances, the diamagnetic susceptibility, which depends on the temperature and is determined by the magnetic properties of the molecules, is of considerable interest. The classical method, based on a measurement of the force acting on a diamagnet placed in a magnetic field, was used in the early work.[219] V. N. Tsvetkov and A. Sosnovskii measured the temperature dependences of the anisotropy of the diamagnetic susceptibility of a large group of NLC,[220] using the phenomenon of "entrainment" of the liquid crystal by a rotating field discovered by V. N. Tsvetkov.[221] The characteristic form of $\chi(T)$ is shown in Fig. 4.15. It is evident that in the interval of existence of the mesophase the susceptibility varies continuously, but a jump-like change is observed at the point of the transition into the isotropic liquid. The anisotropy of the susceptibility of the NLC is clearly manifested (Fig.

Table 4.6
Comparison of the Properties of PAA Near a NLC-IL Phase Transition and the Properties of Model Systems: A) Quasispherical Particles[207]; B) Slightly Nonspherical Particles[207]; C) Spherocylindrical Particles.[208]

Property	PAA (Ref. 207)	A	B	C
Transition temperature, K	408.8	408.1	409.4	410.4
Packing fraction	0.62	0.59	0.59	0.45
Order parameter $\langle P_2 \rangle$	0.40	0.44	0.47	0.54
Relative volume change	0.0035	0.0038	0.0102	0.040
Entropy change	0.17	0.54	0.59	0.89
$\frac{dT}{dP}$ at $P = 1$ bar at the transition point, K/kbar	48	19.7	48.6	175

Table 4.7
Surface Tension (δ, dyne · cm^{-1}) of Some Nematic Liquid Crystals[209–218]

Substance	t, °C	Method of Measurement	σ	Reference
P-azoxyanisole (PAA)	120	Maximum pressure of bubble	39.0	209
		The same	37.9	210
		Capillary	38.9	211
		Radius of curvature	38.8 ± 0.4	212
		Ring of du Nojy	39.6 ± 0.4	213
		Light scattering spectrum	38 ± 4	214
P-azoxyphenetole (PAP)	158	Maximum pressure of bubble	29.1	209
		Capillary	29.2	211
		Ring of du Nojy	29.8	213
Ethyl-*p*-azoxybenzoate	115	Maximum pressure of bubble	26.9	209
Anisaldazine	171	Maximum pressure of bubble	32.1	209
		Capillary	31.9	211
P-methoxy-*p*′-butylazoxybenzene	23	Volume of drop	37.3 ± 0.3	215
		Ring of du Nojy	38.8 ± 0.3	217
N-(*p*-methoxybenzylidene)-*p*′-butylanilinc (MBBA)	25		36	216
	35	Light scattering spectrum	38 ± 3	214
4-butyl-4′-ethoxy-α-chlorostilbene	23	Volume of drop	35.7 ± 0.5	215

4.15, 4.16). It was established that the magnetic anisotropy is attributable to the benzene rings and is proportional to their number.[220] The theoretical explanation of the anisotropy of the magnetic properties of the benzene ring was given by Pauling.[223] The magnetic susceptibility versus the magnetic-field intensity exhibits a peculiar dependence (as the intensity is increased, the susceptibility increases, but at some value it no longer depends on the intensity).[224] This behavior was explained based on the theory of swarms, under the assumption that the susceptibility is determined by the competing effect of the magnetic field and of the temperature on the swarm.[224]

These data are supplemented by the results of recently performed measurements of the surface tension of *p*-*n*-alkyl oxybenzoic acid *p*-cyanophenyl esters[225] and some of cyanobiphenyls (5CB, 6CB, 7CB and 8CB).[226] In paper 226 it was found that interfacial tension between nematic and isotropic phases of cyanobiphenyles changes from $(0.71 \pm 0.4) \cdot 10^{-2}$ erg/cm^2 (for 6CB) to $(1.82 \pm 0.7) \cdot 10^{-2}$ erg/cm^2 (for 7CB). The data obtained and recently found coefficients of reflection from interface for 7CB[227] enabled to define the thickness of

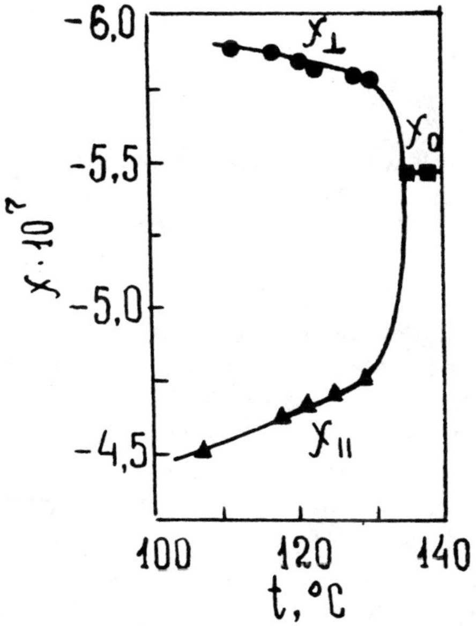

Figure 4.15. The specific diamagnetic susceptibility of PAA as a function of the temperature.[219,220]

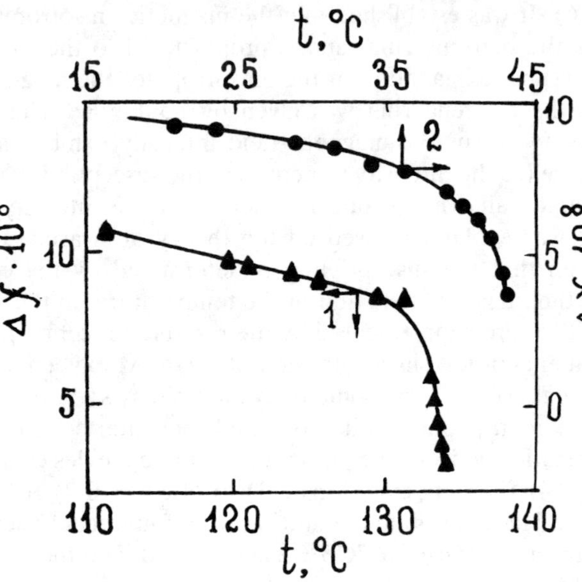

Figure 4.16. Temperature dependences of the anisotropy of the diamagnetic susceptibility of PAA (1) and MBBA (2).[222]

transition layer (270 ± 270)Å on the basis of experimental values which agrees well with the thickness predicted by the Landau-de Gennes theory (265 ± 290)Å.

Anisotropy of the surface tension of NLC was studied by Polish and Soviet authors.[228,229] Experimental techniques were developed and results obtained in this field were generalized. By using this method based on the Freederickcz transition in a magnetic field, the anisotropy of σ for MBBA was measured, which turned out to be close to the value $(2.1 \pm 0.5)\cdot 10^{-5}$ erg/cm^2.[229]

The results of comparison of the values of σ (5CB, 8CB, MBBA, EBBA) calculated by using the above-mentioned theoretical methods[230,231] with experimental data indicate that the statistical theory based on approximation of the mean field [230] leads to a satisfactory agreement. At the same time the calculation using the correlation functions leads to major discrepancies for MBBA, although the account of the real molecular correlations somewhat improves the agreement of calculated and experimental data.[231]

4.3. CALORIC PROPERTIES AND SOUND VELOCITY

The caloric properties of NLC, particularly the thermal effects of the solid–NLC and NLC–IL phase transitions, were studied in detail by Arnold,[170] Barall *et al.*,[232] N. M. Sakevich,[233] and Chow and Martire.[234] A theoretical estimate for PAA based on the Chandrasekhar–Madhusudana theory[57] resulted in the value 1230 J/mole, which is greater than the measured values (690 J/mole,[170] 740 J/mole,[232] 780 J/mole,[233] and 760 J/mole.[234] Chandrasekhar and Madhusudana[235] later refined their calculation and obtained the value 760 J/mole for PAA that agrees well with experimental data. The value 1500 J/mole found for *p*-azoxyphenetole coincided with the experimental value.[170]

Enthalpies and entropies of transitions NLC–isotropic liquid were found experimentally for a group of liquid crystalline compounds with lateral long-chain substituents (1,4-*bis*-4-*n*-octyloxybenzoyloxy-2-*n*-alkylbenzenes).[204] The authors role of forces of steric repulsion for stabilization of nematic phase.

The latent heat of the NLC–IL transition was calculated for MBBA by B. I. Ostrovskii *et al.*,[236] using quasistatic thermograms and experimental data on the heat capacity c_p which they obtained and which are equal to (0.5 ± 0.1) J/g.

The experimental results for the volumes of the coexisting phases (IL and NLC) and the pressure–temperature curve for PAA (obtained in Ref. 170) permitted calculating the change in the entropy ΔS, enthalpy ΔH, and internal energy ΔU (Figs. 4.7 and 4.17) using the Clausius–Clapeyron equation. The very slight temperature dependence established for these quantities precluded the observation of an isolated critical point even if the PAA molecule were more stable than is the real case.

A detailed experimental study of the thermodynamic properties of NLC at high pressures was performed for three substances: PAA, *n*-butylmetoxybenzene (MBAB), and etoxybenzylidene-*n*-butylaniline (EBBA).[237] The method of study consisted of measuring the temperature, pressure, and volume of the substance in the vicinity of the phase-transition curve. The measurements were performed, as a rule, in the isothermal state, but the thermal expansion at atmospheric pressure was also studied for each substance. Figure 4.18 shows the change in the molar volume of PAA as a function of temperature at atmospheric pressure near the melting point and near the NLC–IL transition and Fig. 4.19 shows the isotherm of PAA at 448.21 K. The

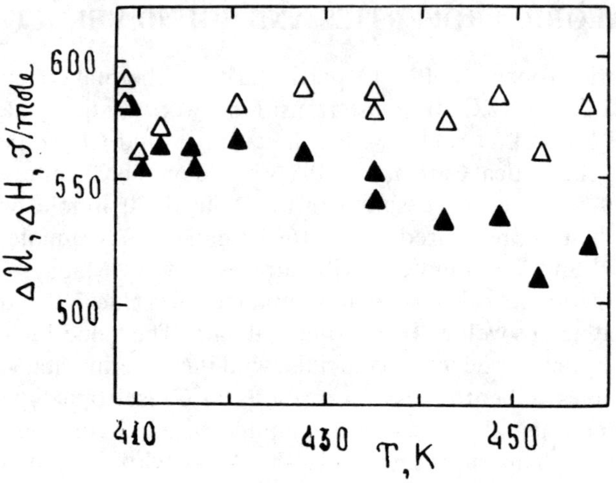

Figure 4.17. The temperature-dependence of ΔU (▲) and ΔH (△) for PAA at high pressures.[122]

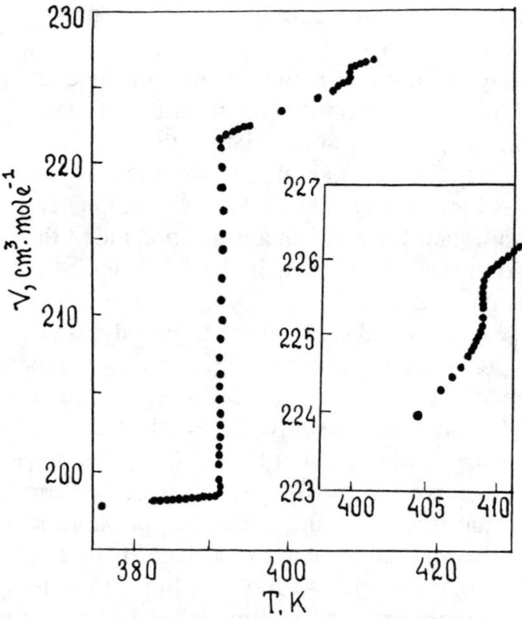

Figure 4.18. Temperature dependence of the molar volume of PAA in the region of melting and in the region of the NLC–IL transition.[237] The region of the NLC–IL transition in an enlarged scale is shown in the inset.

volume jumps ΔV accompanying the phase transitions were determined from these data and then the changes in the enthalpy ΔH, entropy ΔS, and internal energy ΔU were calculated using the Clausius–Clapeyron equation. The authors note that because of the small values of the volume jumps accompanying the NLC–IL transition and the difficulties in determining the onset of the phase transition against the background formed by pronounced pretransitional phenomena it was impossible to determine the values of the indicated thermodynamic quantities with an accuracy better than 5%. The results of the measurements and calculations[237] are presented in Tables 4.8 and 4.9 and in Figs. 4.18–4.20.

The entropy jumps ΔS/R and the relative volume jumps $\Delta V/V_N$ accompanying the NLC–IL transition decrease with temperature in all three substances (Table 4.9, Fig. 4.20), but these changes are small and they do not indicate the possibility of a critical point on the transition curve. For example, the change ΔS/R is small and does not

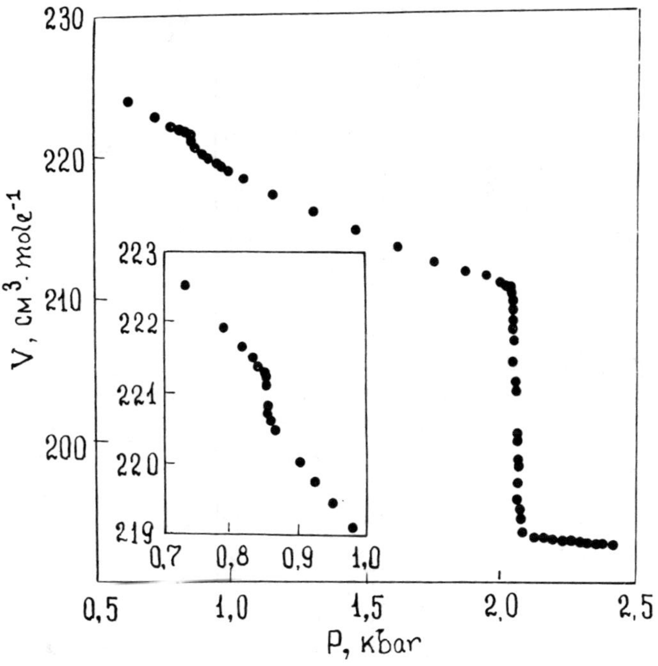

Figure 4.19. Isotherm of PAA at 448.21 K.[237] The region of the NLC–IL transition is shown in an enlarged scale in the inset.

Table 4.8
Thermodynamic Parameters of Melting of PAA

T, K	P, bar	ΔV, cm^3/mole	$\Delta S/R$	ΔU, kJ/mole	T, K	P, bar	ΔV, cm^3/mole	$\Delta S/R$	ΔU, kJ/mole
391.45	1	23.8	8.84	28.78	417.90	888	20.52	8.81	28.80
394.75	107	23.35	8.84	28.78	427.51	1240	19.43	8.75	28.70
405.55	443	21.95	8.83	28.81	448.21	2043	17.37	8.61	28.56
412.66	693	21.09	8.81	28.78	448.41	2056	17.37	8.62	28.58

Table 4.9
Thermodynamic Parameters of NLC-IL Phase Transitions[237]*

T, K	P, bar	V_N cm³/mole	ΔV, cm³/mole	$\Delta S/R$	ΔU J/mole
		PAA			
408.75	1	225.04	0.72	0.172	586
410.61	39	224.80	0.71	0.171	582
416.55	161	224.97	0.69	0.173	586
420.78	249	223.49	0.68	0.173	586
427.55	398	222.70	0.65	0.170	578
435.25	564	221.92	0.66	0.171	586
448.21	862	220.65	0.59	0.168	578
452.66	968	220.29	0.57	0.164	561
458.23	1104	219.80	0.54	0.160	548
		BMAB			
347.80	1	264.69	0.44	0.120	348
356.79	210	263.23	0.42	0.118	339
373.91	617	260.55	0.39	0.115	335
383.22	855	259.08	0.37	0.114	331
403.22	1379	256.3	0.34	0.110	322
413.14	1655	254.93	0.32	0.108	318
423.09	1935	253.61	0.31	0.106	314
433.05	2229	252.26	0.29	0.104	310
		EBBA			
352.41	1	283.50	0.60	0.182	532
369.91	471	279.23	0.53	0.173	507
380.74	765	277.11	0.49	0.167	494
387.80	965	275.70	0.47	0.165	486
394.15	1152	274.46	0.46	0.165	490
422.51	2044	269.42	0.40	0.159	477
438.91	2604	266.70	0.38	0.160	486
454.02	3148	264.34	0.34	0.150	461

*Selected data.

exceed 20%, which agrees with the insignificant change in the order parameter along the transition curve and the relation following from L.D. Landau's theory $S \simeq \langle P_2 \rangle^2$. It is significant that it follows from the trends in the changes of $\Delta V/V_N$ and $\Delta S/R$ that these quantities cannot vanish simultaneously (Fig. 4.21). As ΔV approaches zero, the relative volume jump $\Delta V/V_N$ approaches a small, but finite value $\simeq 0.2 \cdot 10^{-3}$, which is identical for all three substances.[237] This apparently indicates the absence of an appreciable contribution of flexible alkyl chains of the ends of molecules to the change in volume accompanying the phase transition.

Based on the studies performed, it was possible to show that

$$\Delta V/V_N \rightarrow \text{const}, \quad \Delta S/R \rightarrow \text{const},$$

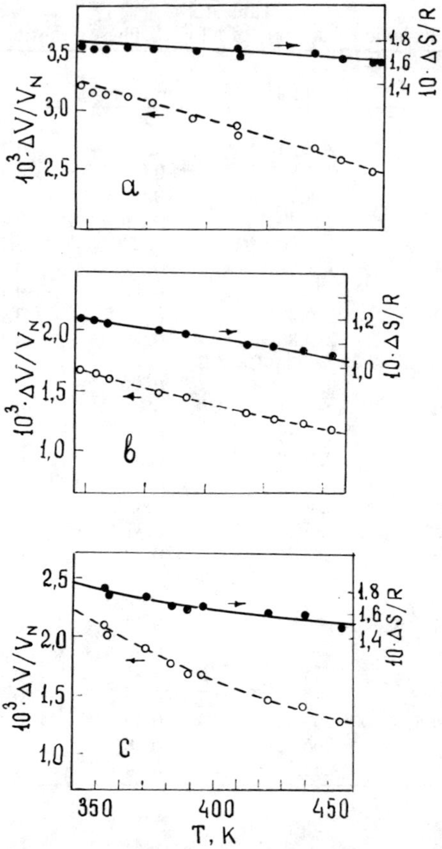

Figure 4.20. Temperature dependences of the relative changes of volume $\Delta V/V_N$ (broken line) and of the entropy $\Delta S/R$ (solid line) accompanying the NLC–IL transition[237]: a) PAA, b) BMAB, c) EBBA.

in a nematic phase transition and in the limit $T, p \to \infty$; in addition, the limiting values of $\Delta S/R$ for the different materials are somewhat different, whereas the values of $\Delta V/V_N$ are the same. It is interesting that there is a similarity between the behavior of the thermodynamic functions in a nematic phase transition and during crystallization of simple substances of the argon type. This analogy suggests that the energy of anisotropic repulsion makes the main contribution to the orientational energy,[237] which agrees well with the results of computer

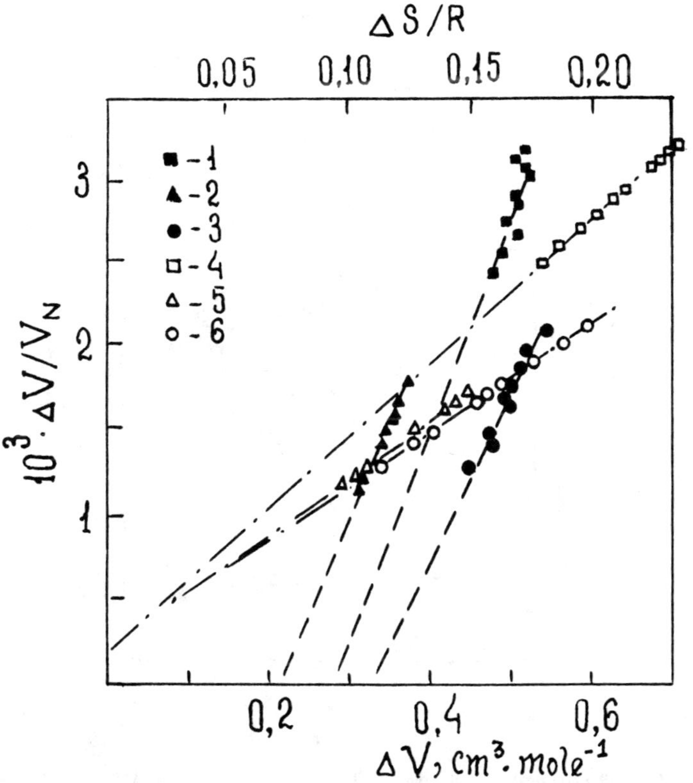

Figure 4.21. The dependences $\Delta V/V_N$ versus $\Delta S/R$:PAA (1), BMAB (2), EBBA (3) and the dependences $\Delta V/V_N$ versus ΔV :PAA (4), BMAB (5), EBBA (6) for accompanying the NLC–IL transition.[237]

modelling (see Sec. 3.5). At high temperatures, when the attractive interaction can be neglected, the parameters of the NLC–IL transition will be determined by the relations[238]:

$$V_{NI} \sim T^{n/3}, \quad \Delta V/V_N = \text{const}, \quad \Delta S/R = \text{const},$$

which agrees with the conclusion of finite asymptotic values for $\Delta S/R$ and $\Delta V/V_N$.

It we now consider intermediate temperatures and adopt the van der Waal's approximation for the energy of attraction, then the dependences acquire a different form[238]:

$$\Delta V/V_N = (\Delta V/V_N)_0 \exp(\epsilon/T), \ \Delta S/R \approx C + \gamma \, \Delta V/V_N$$

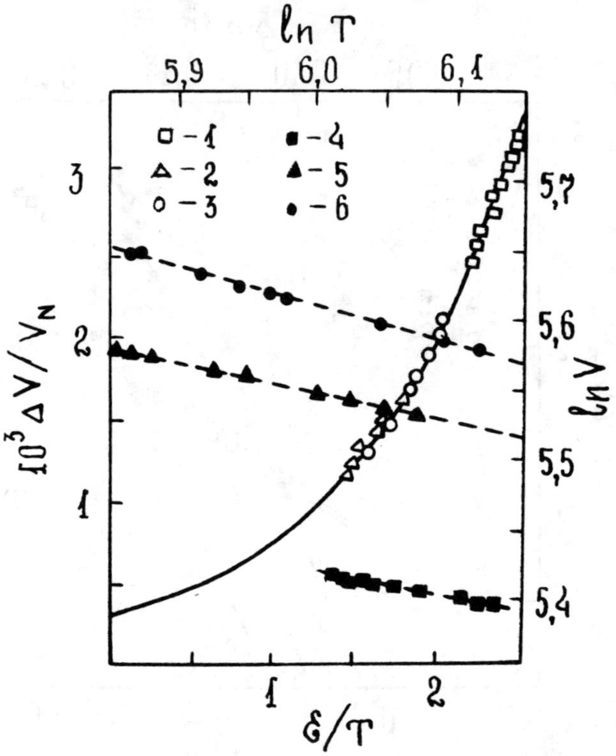

Figure 4.22. The dependence illustrating the expotential behavior $\Delta V/V_N = 0.27 \cdot 10^{-3}$ exp(E/T): solid line: 1) PAA, 2) BMAB, 3) EBBA and the dependences illustrating the power-law behavior $V_{NI} \sim T^{n/3}$: 4) PAA (n = 14.31), 5) BMAB (13.74), 6) EBBA (10.55).[237]

As is evident from the data in Fig. 4.22, the experimental temperature dependence of $\Delta V/V_N$ is described well by the exponential dependence presented above with the same value $(\Delta V/V_N)_0 \simeq 0.27 \cdot 10^{-3}$. With regard to the temperature dependence of the volume, this dependence is described well by a power-law function with $n \simeq 10^{-14}$. This indicates that the power-law dependence presented above for V_{NI} also holds in the intermediate temperature range. On the other hand, the estimates obtained in Ref. 237 correspond to possible values of n, representing the exponent, characterizing the law of interparticle repulsion.[239]

Figure 4.23 compares the calculations of the heat capacity c_p of PAA performed by Chandrasekhar and Madhusudana[57] with the experimental data obtained by Arnold.[169] It is evident that agreement is satisfactory.

A careful experimental investigation of the heat capacity of MBBA near the NLC–IL transition was performed by B. I. Ostrovskii *et al.*[236] (Fig. 4.24). They showed that an anomaly in the heat capacity of MBBA is manifested in a narrow temperature range $|T - T_c| < 1°C$ near the phase transition. A central region, corresponding to a δ-function divergence, was separated from this anomalous part. It was established that this region is associated with the liberation of the latent heat of the transition. The presence of a double peak in the heat capacity near the NLC–IL transition, found by Hirakawa and Kai,[240] who previously measured c_p for MBBA, was not confirmed in Ref. 236. An analysis using the phenomenological theory of L.D. Landau led to the conclusion that this theory cannot describe the anomaly in the heat capacity in the isotropic phase for $T - T_c < 1°C$.

Distinctive temperature dependences with singularities in the vicinity of NLC–IL transitions have been obtained for the adiabatic compressibility $\beta_S = v/w^2$ and the number $R_n = Mv(w)^{1/3}$, where v is the specific volume, w is the velocity of ultrasound, and M is the

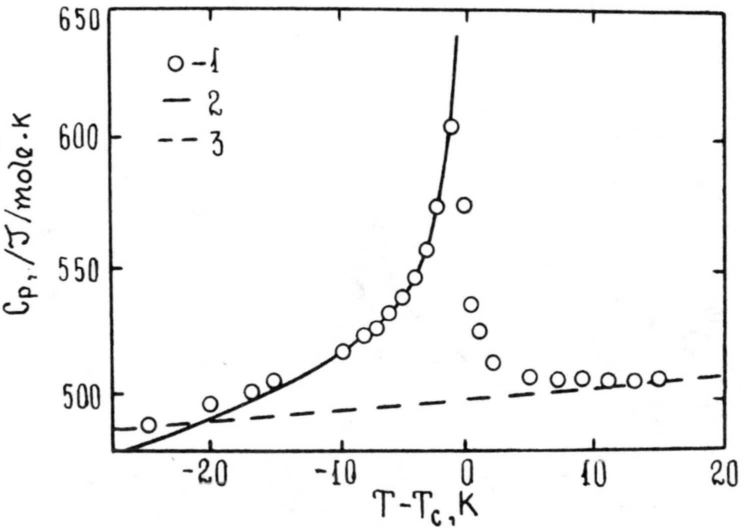

Figure 4.23. The temperature dependence of c_p for PAA (T_c = 408.5 K)[57]: 1) experimental data[170]; 2) computed data[57]; 3) contribution of the disordered system (extrapolated data for an isotropic liquid).

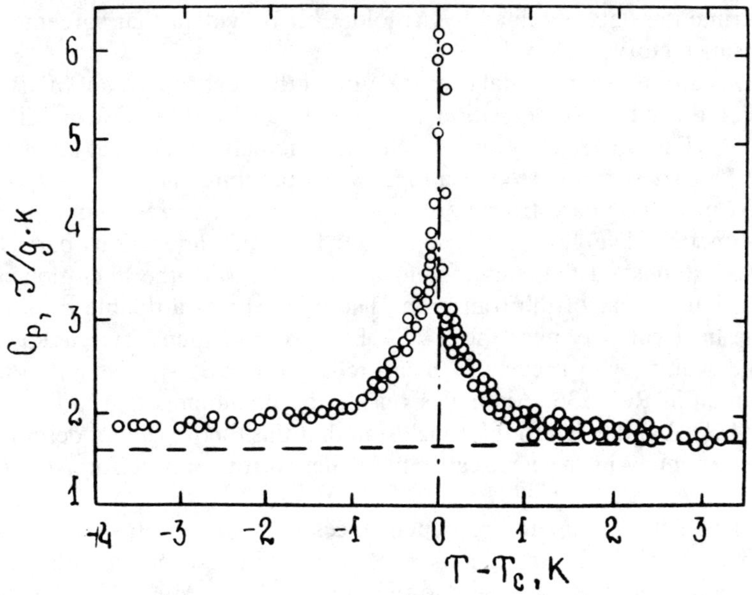

Figure 4.24. The temperature dependence of c_p for MBBA (T_c = 315.7 K).[236]

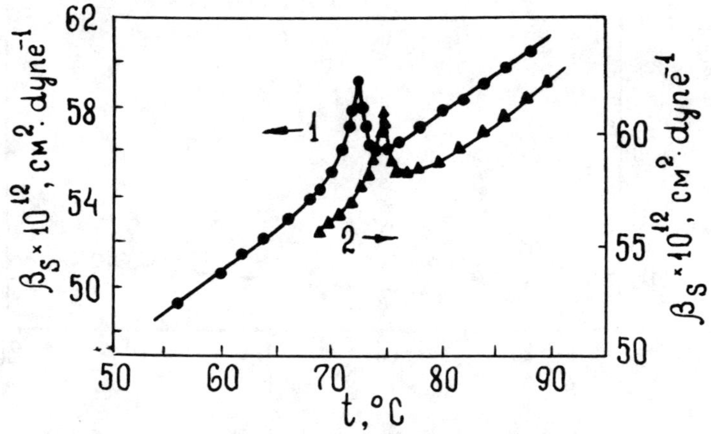

Figure 4.25. Temperature dependences of the adiabatic compressibility of HBT (1) and OBT (2).[241]

molecular weight.[241] These dependences are shown in Figs. 4.25 and 4.26 for NBT and OBT following the data in Ref. 241.

Figure 4.27 shows the correspondence between the computed (Chandrasekhar–Madhusudana theory[57]) and measured values of the isothermal compressibility of PAA. This correspondence can be considered satisfactory if the narrow region directly adjacent to the phase transition temperature is excluded.

Figure 4.28 shows the temperature dependence of the adiabatic compressibility of PAA in the vicinity of a phase transition according to the data in Ref. 242. The values of the adiabatic compressibility of PAA and *n*-azoxyphenetole are of the same order of magnitude as for the usual low-molecular liquids.[242,245]

Adiabatic compressibility, molar compressibility and molar sound velocity were calculated on the basis of measurements of ultrasound velocity for PMBAB and PEBAB.[200] The calculation of adiabatic compressibility was also performed for PBPAA in the similar way.[201] Anisotropy of the velocity of ultrasound and the coefficient of its absorption were specially studied to define characteristic times of relaxation processes.[246] The heat capacity and the expansion coefficient for HPDB in nematic and isotropic phases have been studied by German investigators.[202]

Let us examine the calculations, performed for PAA and MBBA by A. D. Bagmet and A. L. Tsykalo, of the thermodynamic properties of nematic liquid crystals using the molecular-dynamics method. The configurational component of the internal energy U, the excess (con-

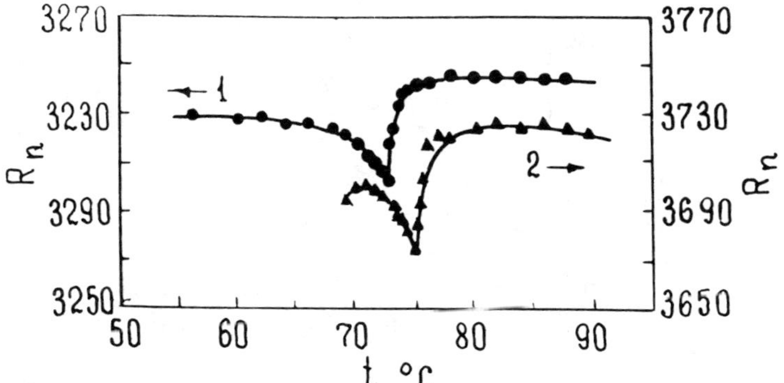

Figure 4.26. Temperature dependences of the Rao numbers of HBT (1) and OBT (2).[241]

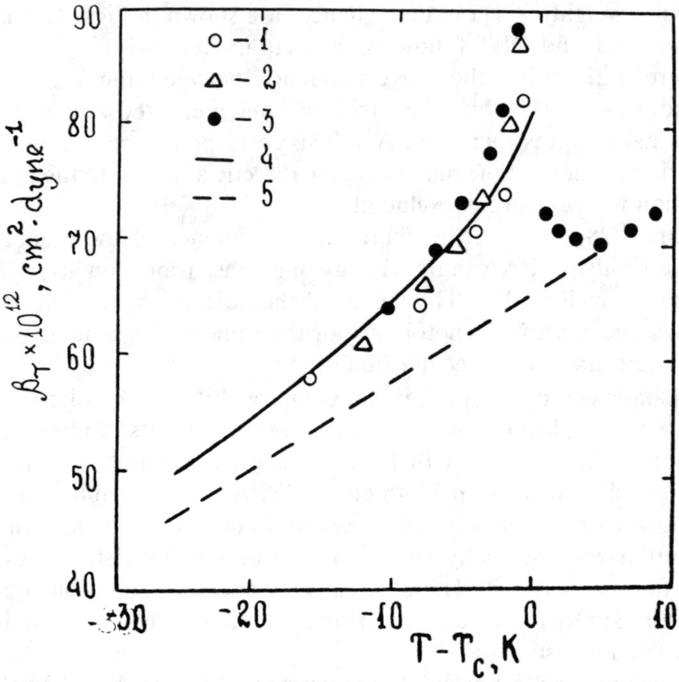

Figure 4.27. The isothermal compressibility of PAA as a function of temperature[57]: 1) experimental data[242]; 2) Ref. 243; 3) Ref. 244; 4) theoretical calculations[57]; 5) contribution of the disordered system (extrapolated isotropic-liquid data).

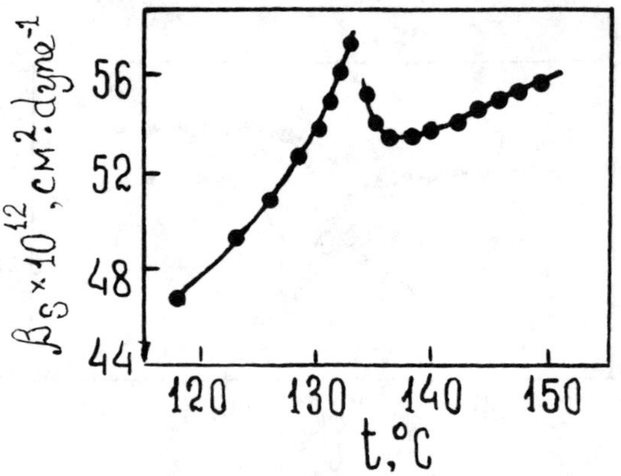

Figure 4.28. Temperature dependence of the adiabatic compressibility of PAA.[242]

figurational) heat capacity $\Delta c_v/c_v$, the adiabatic compressibility $\beta_s \equiv v^{-1}(\partial v/\partial P)_s$, the thermodynamic velocity of sound w, and the temperature coefficient of the pressure $\gamma_v = (\partial P/\partial T)_v$ were determined from data obtained from molecular-dynamics experiments with the help of the following expressions:

$$U = \langle \sum_{i \neq j}\sum \phi_{i,j} \rangle + \delta u; \tag{4.1}$$

$$\frac{\Delta C_v}{C_v} = \frac{5}{2} N \frac{\langle (\Delta K)^2 \rangle}{\langle K \rangle^2}; \tag{4.2}$$

$$\beta_s = \left[7P - \frac{28}{5}\frac{N}{v} kT - 8\frac{U}{v} - \frac{v\langle \Delta\pi^2 \rangle}{kT} \right]^{-1}; \tag{4.3}$$

$$W = \left[\frac{70}{\rho} - \frac{28}{5}\frac{kT}{m} - 8\frac{U}{mN} - \frac{Nm\,\langle \Delta\pi^2 \rangle}{\rho^2\,kT} \right]^{1/2} \tag{4.4}$$

$$\gamma_v = \frac{2}{5} C_v \left[v^{-1} - \frac{\langle \Delta k \Delta\pi \rangle}{\langle NkT \rangle^2} \right], \tag{4.5}$$

where δU is a correction for the truncation of the tail of the pair potential ϕ, ΔK denotes the fluctuations in the kinetic energy, and $\Delta\pi$ denotes the fluctuations in the microscopic analog of the pressure.

Some of the values obtained are tabulated in Table 4.10. Unfortunately, it is not yet possible to compare the computed and measured values of U because of the absence of experimental data. To calculate c_v, it is possible to use the value of Δc_v obtained from the $U(T)$ curve at constant volume in the molecular-dynamics experiment, as well as the ratio $\Delta c_v/c_v$ also calculated using expression (4.2), and data generated by the molecular-dynamics calculation. Then for PAA (potential (3.23), $\gamma = 3$), we obtain $c_v = 340$ J/mole · K ($\Delta c_v = 108$ J/mole · K and $\Delta c_v/c_v = 0.32$). According to data in Ref. 236, for a nearby state (1.160 g/cm^3 and 399 K), $c_v = 425$ J/mole · K. This agreement can be viewed as satisfactory. Worse agreement with experiment is obtained with the use of the potential (3.25): the computed value is $c_v = 296$ J/mole · K. The use of the potential (3.23) for MBBA gives the value $c_v = 520$ J/mol · K, which is close to the value $c_p = 530$ J/mol · K obtained experimentally at 308 K.[248] Considering the closeness of the values of the heat capacities c_v and c_p of the liquid-crystalline phase, this should be considered good agreement.

Table 4.10

Results of the Calculation of Thermodynamical Properties of PAA and MBBA by Molecular Dynamics Method and the Comparison with Experimental Data (ρ—density, T—temperature, U—configurational internal energy, ΔC_V—excess heat capacity, β_S—adiabatic compressibility, γ_V—thermal coefficient of pressure, W—thermodynamical velocity of sound)

Substance	Potential of Pair Interaction	ρ, g/cm³	T, K	$U \cdot 10^3$, J/mole	$\frac{\Delta C_V}{C_V}$	$\beta_S \cdot 10^{-12}$, cm · s²/g Calculation	$\beta_S \cdot 10^{-12}$, cm · s²/g Experiment	γ_V, bar · K^{-1} Calculation	γ_V, bar · K^{-1} Experiment	W, 10^5 cm/s Calculation	W, 10^5 cm/s Experiment
p-azoxyanisole (PAA)	$\phi_2(3.23)$	1.194	388	−51.3		58	46.7*	14	18††	1.2	1.35[236]
			400	−50.1	0.33	69	54.3**	15		1.1	1.26[236]
			412	−48.8	0.31						
	$\phi_4(3.25)$	1.169	371					12			
			394	−24.5	0.32			13	17†††		
			414	−22.7	0.29			11			
N-(*p*-methoxybenzylidene)-*p*-*n*-butilaniline (MBBA)	$\phi_2(3.23)$	1.044	299	−39.6		89	38.8***	10		1.1	1.54[13]
			308	−38.6	0.33	106	41.6†	11		1.0	1.56[247]
			314	−37.6	0.31						

*At 388 K and 1.173 g/cm³ [236]
**At 399 K and 1.160 g/cm³ [235]
***At 298 K and 1.088 g/cm³ [13]
†At 307 K and 0.990 g/cm³ [247]
††At 390 K and 1.282 g/cm³ [189]
†††At 390 K and 1.169 g/cm³ [54,55]

The values of the adiabatic compressibility found by the molecular-dynamics method correctly describe the temperature-dependence of β_s (Table 4.10). The deviations from the experimental values for PAA, however, reach 25%, and for MBBA the deviations are even larger.

The values of the temperature coefficient of the pressure γ_v, calculated by the molecular-dynamics method using expression (4.5), agree satisfactorily with the experimental data (Table 4.10). We note that the coefficient γ_v found by numerical differentiation of the isochore $P(T)$ obtained by the molecular-dynamics method also agrees well with experimental values and the values calculated using expression (4.5).

The temperature-dependence of the velocity of propagation of ultrasound, related in a well-known manner to the adiabatic compressibility, is of definite interest. The characteristic dependences for PAA at two frequencies are shown in Fig. 4.29.[249]

Table 4.10 presents the experimental values of the thermodynamic velocity of sound in PAA and MBBA as well as the values calculated using the molecular-dynamics method. Good agreement is observed

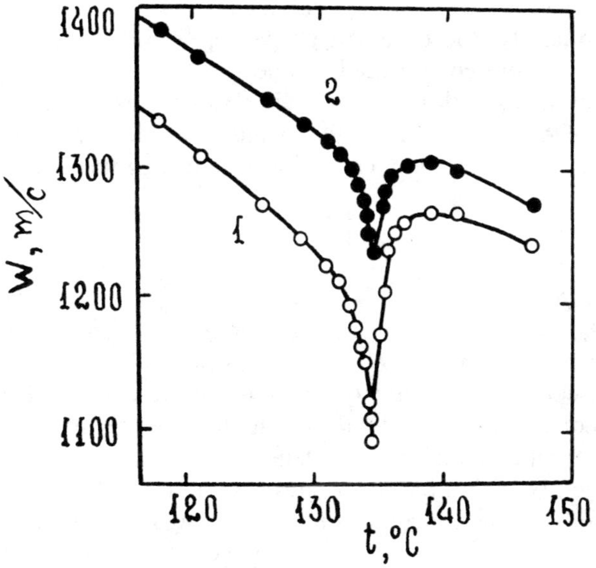

Figure 4.29. The temperature-dependence of the velocity of ultrasound with frequency 3 MHz (1) and 60.5 MHz (2) in PAA.[249]

with the experimental data for PAA (the deviations from the experimental values constitute 10–12%). For MBBA, the disagreement reached 36%.

We can thus conclude that with the use of an acceptable intermolecular pair-interaction potential the molecular-dynamics method gives a correct qualitative and, in many cases, quantitative prediction of the basic thermodynamic properties of nematic liquid crystals without resorting to adjustable empirical parameters.

Among other results of calculated determination of properties of NLC one should note the works of Soviet authors[250] which by using the method of correlative functions derived the relationship for the calculation of the latent heat of phase transition NLC–isotropic liquid and obtained a good agreement with experiment for PAA and MBBA. The calculation of the heat of evaporation for PAA was also carried out on the basis of acoustic data.[151]

4.4. TRANSPORT PROPERTIES

In contrast to ordinary isotropic liquids where the coefficient of self-diffusion, viscosity, or thermal conductivity is expressed directly as a scalar quantity, the transport properties of liquid crystals are expressed in a more complicated manner.

Translational self-diffusion in liquid crystals is usually described by a tensor of rank two. Due to the cylindrical symmetry of NLC, the coefficient of self-diffusion can be represented in the form

$$D = \begin{vmatrix} D_{\parallel} & 0 & 0 \\ 0 & D_{\perp} & 0 \\ 0 & 0 & D_{\perp} \end{vmatrix}$$

where $D_{\parallel}$ and $D_{\perp}$ are the tensor components corresponding to the coefficients of self-diffusion in directions parallel and perpendicular to the director **n**, respectively. It should be remembered that self-diffusion of a real NLC depends on whether the specimen is oriented by an external field and on the intensity and orientation of this field. At sufficiently high intensities, however, saturation is observed ($D_{\parallel}$ and $D_{\perp}$ for PAA no longer depend on the magnitude of the magnetic field induction at 1.5–2 kG,[249] though the orientation of microregions of the NLC continues to improve up to 20 kG.[120] The dependences of $D_{\parallel}$ and $D_{\perp}$ on the order parameter, which reflects the degree of orientation of the NLC, can be found using the following approximate expressions[110]:

$$\left.\begin{aligned} D_{\parallel} &= \tfrac{1}{3}\left[D_{\parallel}^{0}\left(2\langle P_2\rangle + 1\right) + 2D_{\perp}^{0}\left(1 - \langle P_2\rangle\right)\right] \\ D_{\perp} &= \tfrac{1}{3}\left[D_{\parallel}^{0}\left(1 - \langle P_2\rangle\right) + D_{\perp}^{0}\left(\langle P_2\rangle + 2\right)\right] \end{aligned}\right\} \tag{4.6}$$

Translational diffusion of NLC has been studied in detail experimentally, though there is sometimes poor agreement between data obtained by different researchers. The results obtained by Topler et al.[251] on self-diffusion of PAA using neutron scattering agree satisfactorily with the data obtained using radioactive markers by Yun and Fredrickson[252] (the disagreement is 10–20%). There are, however, serious quantitative and even qualitative (for the temperature dependence of D) disagreements with the results obtained by Otnes *et al.*[253] using neutron scattering. The coefficients of self-diffusion of MBBA,[249] TBBA,[254] and phenyl-4-benzoyloxybenzoate[249] have been studied.

Somewhat different values of the coefficients of self-diffusion of PAA, shown in Fig. 4.30, were obtained in Ref. 255. It is noteworthy that the relative change in the coefficient of self-diffusion accompanying saturation is equal to 16% in a longitudinal magnetic field and

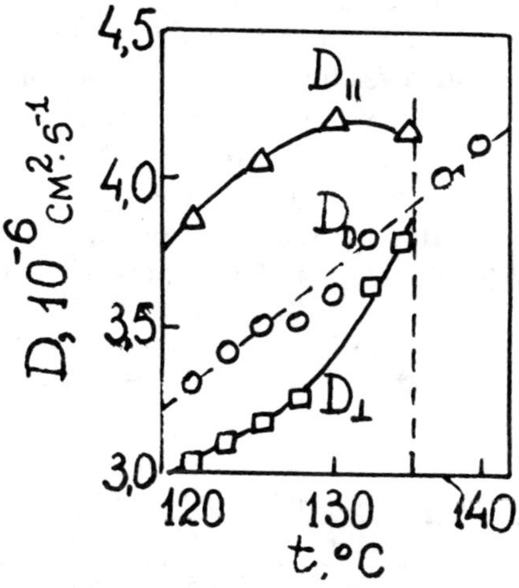

Figure 4.30. Temperature dependence of the coefficient of self-diffusion of PAA (the magnetic induction is equal to 3.5 kGs).[255]

Table 4.11
Coefficients of Self-Diffusion (10^{-7} cm^2 s^{-1}) of MBBA[249]

Method of Definition	t, °C	$D_{\parallel}$	$D_{\perp}$	D_0
Optical	24.5	2.7	1.3	
Optical	44	—	—	6.0
Diffusive-colouring	22.0	2.6	1.6	
Diffusive-trityl	25.0	5.0	3.5	

8% in a transverse field.[249] Tables 4.11 and 4.12 show the self-diffusion of MBBA and phenyl-4-benzyloxybenzoate.[248] A series of data for NLC are also presented in Ref. 256.

A more exact determination of the coefficients of self-diffusion of MBBA for a quite wide range of temperatures was apparently made by a new optical method.[257] These results are presented in Fig. 4.31. A detailed analysis of the separate contributions and the total apparent coefficients of self-diffusion for PAA and a number of other substances is presented in Refs. 258 and 259 (Figs. 4.32 and 4.33).

A new way based on the NMR method with cyclic switching of the field was proposed for the measurement of self-diffusion coefficients.[260] It was found that the characteristic values of the ratio $D_{\parallel}/D_{\perp}$ are between 1.1 and 1.7 for the NLC. The diffusion coefficients of MBBA and 5CB near the temperature of the phase transition NLC–isotropic liquid were studied by forced Rayleigh scattering.[261]

The theoretical interpretation of data on self-diffusion of NLC is usually based on Kirkwood's theory, modified by Franklin.[157] This approach, however, involves fundamental shortcomings, as noted in Sec. 3.7. For this reason, we shall briefly examine the results of a molecular-dynamics study of self-diffusion of NLC.

The components of the self-diffusion tensor $D_{\parallel}$ and $D_{\perp}$ can be determined by integrating the temporal correlation functions of the linear velocities $\psi_v^{\alpha}(t)$ discussed in Sec. 3.5:

$$D_\alpha = \langle v_\alpha^2 \rangle \int_0^\infty \psi_v(t)dt \qquad (\alpha = x,y,z), \tag{4.7}$$

Table 4.12
Coefficients of Self-Diffusion 10^{-7} cm^2 s^{-1} of Phenyl-4-benzoyloxybenzoate[249]

t, °C	$D_{\parallel}$	$D_{\perp}$	D_0	t, °C	$D_{\parallel}$	$D_{\perp}$	D_0
45.0	0.52	0.23	0.32	78.5	3.10	1.60	2.10
58.0	1.15	0.50	0.71	81.0	3.60	1.80	2.40
64.5	1.90	0.95	1.25	86.1	4.10	2.40	3.0
69.0	1.90	0.90	1.25				

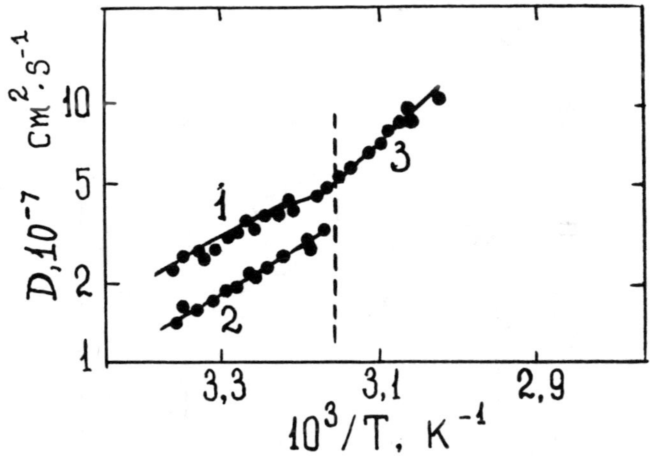

Figure 4.31. Temperature dependences of the coefficients of self-diffusion $D_{\parallel}$ (1), $D_{\perp}$ (2), and D_0 (3) of the nematic and isotropic phases of MBBA.[257]

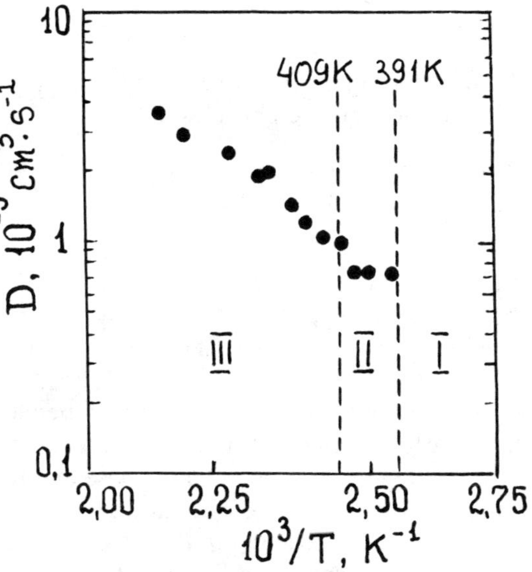

Figure 4.32. Temperature dependence of the apparent coefficient of diffusion of PAA (based on neutron scattering and NMR data):[258] I—solid crystal, II—NLC, III—IL.

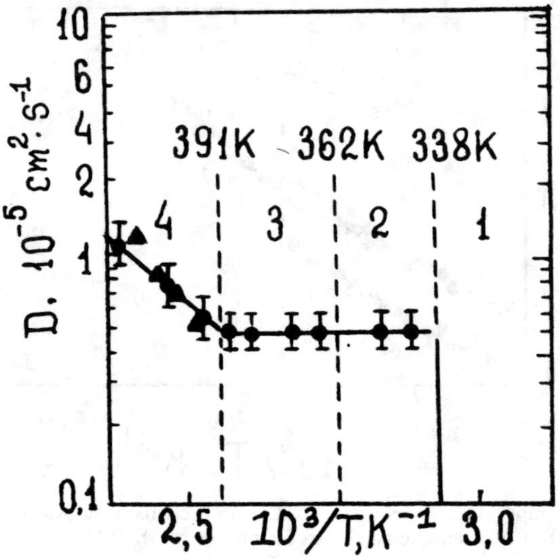

Figure 4.33. Temperature dependences of the apparent coefficients of self-diffusion based on neutron scattering and NMR data[258] for 4,4′-di-heptyloxy-azoxybenzene (PAH); 1—solid, 2—SLC, 3—NLC, 4—IL.

or from the slope of the linear part of the time-dependence of the mean-square displacement of the particles:

$$D_\alpha = \lim_{t\to\infty} \frac{\langle [r_\alpha(t) - r_\alpha(0)]^2 \rangle}{2t}. \tag{4.8}$$

In the molecular-dynamics experiments which have been performed,[30,38] the values of D_α obtained using expressions (4.7) and (4.8) coincided within limits of the statistical error in the calculation (10%).

Figure 4.34 shows the computed temperature-dependences of $D_\parallel$ and $D_\perp$ for the molecular-dynamics model (ϕ_4 potential, $\gamma = 2.5$, and $\rho^* = 0.420$)[262] in regions of the nematic phase and the isotropic liquid. The obtained curves describe qualitatively correctly the experimentally observed dependences for the studied NLC. An important result of the molecular-dynamics investigation is the establishment of a strong dependence of the ratio $D_\parallel/D_\perp$ on the orientational order parameter $\langle P_2 \rangle$. Thus when $\langle P_2 \rangle$ increases from 0.38 ($\gamma = 2.5$) to 0.81 ($\gamma = 3.5$) at constant temperature ($T^* = 0.76$) and packing

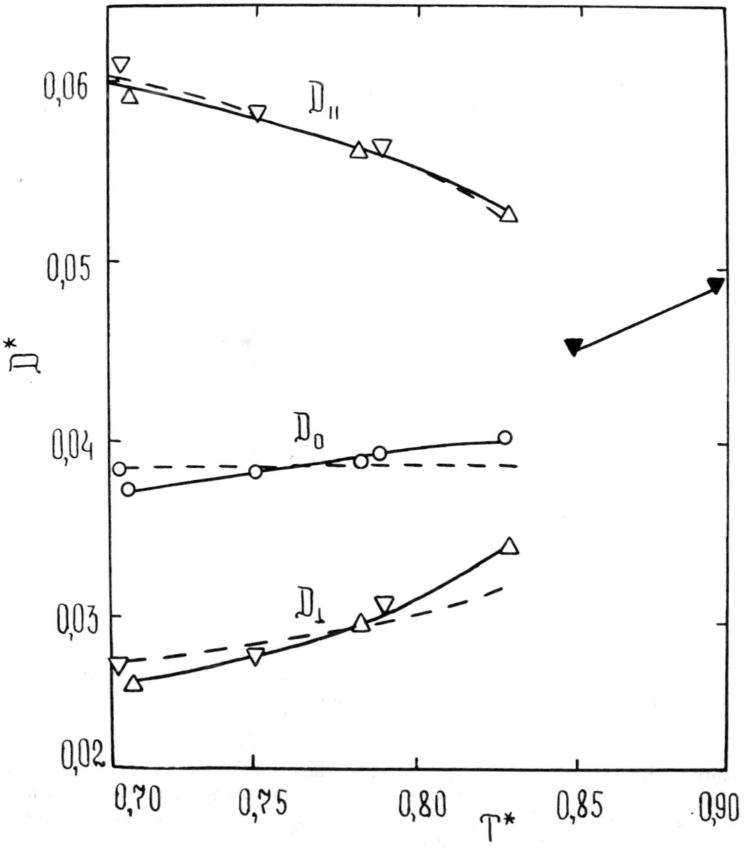

Figure 4.34. The temperature-dependences of the components of the self-diffusion tensor for a molecular-dynamics model in the region of existence of NLC and IL.[30] The dashed lines show the calculation using the approximate relation (4.6).

fraction ($\eta = 0.530$), $D_{\|}/D_{\perp}$ increases from 1.7 to 6.8. However, the averaged self-diffusion coefficient changes insignificantly in this case (by approximately 25%).

Figure 4.35 compares the experimental values of the components of the self-diffusion tensor of PAA[253] with the results of molecular-dynamics experiments.[30] The latter were obtained using the parameters of the ϕ_2 potential determined from data on the temperature-dependence of $\langle P_2 \rangle$ ($\gamma = 3.0$, $\sigma_0 = 5.01$ Å, and $\epsilon_0/k = 520$ K). It is evident that the model correctly describes the temperature dependences of the components of the self-diffusion tensor of PAA. The

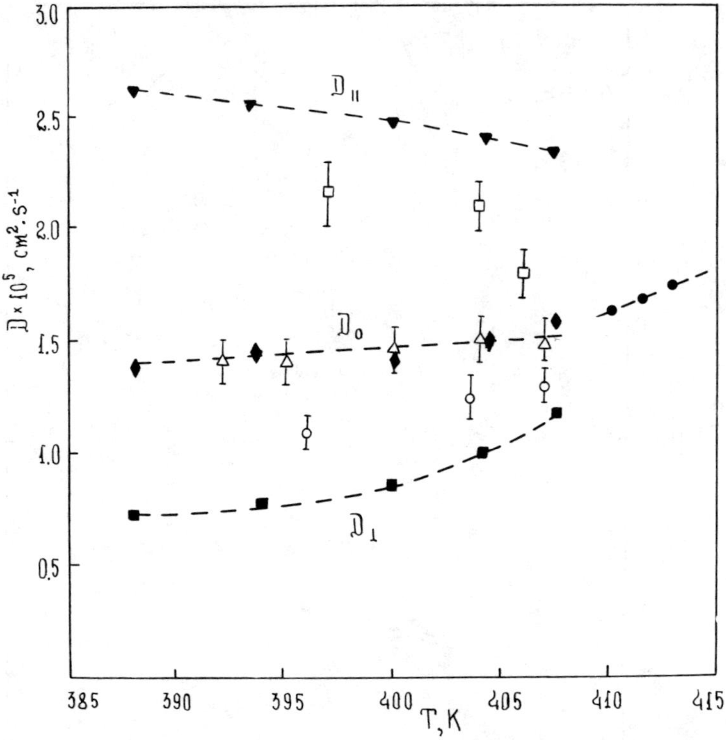

Figure 4.35. Comparison of the experimental values of the self-diffusion tensor components with the results of the molecular dynamics modelling for PAA: open symbols are experimental data (X-ray diffraction,[253] the vertical lines correspond to tolerance incorrections); dark symbols are molecular dynamics results.[30]

molecular-dynamics calculation, however, leads to higher values of $D_{\parallel}$ and lower values of $D_{\perp}$ compared with the experimental values; the ratio $D_{\parallel}/D_{\perp}$ is higher than the real value for PAA. This is explained by the fact that the experiments of Otnes *et al.*[253] were performed with relatively small orienting actions (the magnetic field induction was 1.75 kGs). New, more accurate experimental data on the self-diffusion of PAA will help to refine the form of the pair-interaction potential and its parameters.

It is even more difficult to describe the viscous characteristics of NLC taking into account the possible relative orientations of the director and the velocity gradients. A widely used description is the representation obtained by Ericksen[263] and Leslie[264] for incompressible NLC in which viscous properties can be described by six quantities α_i (Leslie coefficients). Parodi[265] derived a relationship for these coefficients. Thus for an incompressible NLC there are five independent coefficients which describe its viscous properties and have the dimensions of viscosity. Other representations of the viscous characteristics of NLC are also possible.[13]

Miesowicz[266] was one of the first researchers to perform exact measurements. He determined the viscosity coefficients of *p*-azoxyanisole and *p*-azoxyphenetole for the three principal orientations of the director and the velocity gradient (Miesowicz's coefficients): the director and the flow are parallel ($\eta_{\|}$); the director and velocity gradient are parallel ($\eta_{\perp}$); and, the director is perpendicular to the flow and the velocity gradient (η_x). Later, Gohwiller[267] peformed detailed studies for MBBA (Fig. 4.36). The viscosity of PAA was studied carefully in Ref. 268. All the Leslie coefficients are known for PAA and MBBA.

The determined temperature dependences of the viscosity of PAA are presented in Fig. 4.37.[266,269,270] Gahwiller's[271] data, obtained by the method of V. N. Tsvetkov and G. M. Mikhailov,[272,273] who studied the anisotropy of the viscosity of PAA, were used to calculate

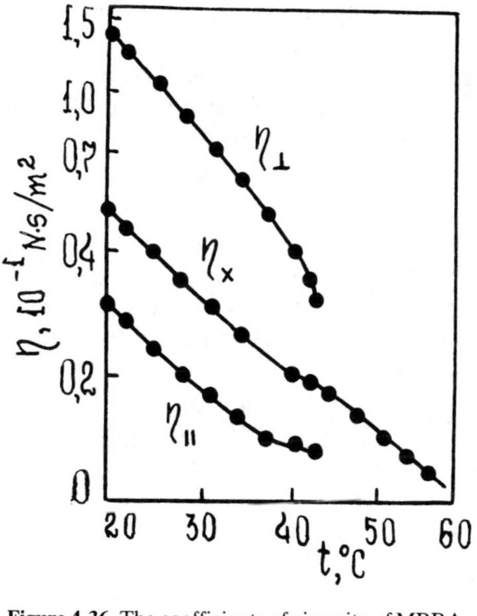

Figure 4.36. The coefficients of viscosity of MBBA as a function of temperature.[267]

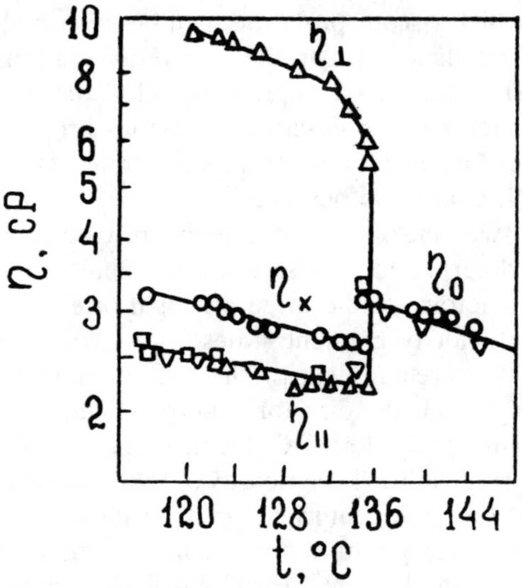

Figure 4.37. Temperature dependences of the viscosity of PAA[266,268,270] (η_x is the viscosity of the unoriented mesophase and η_0 is the viscosity of IL).

the coefficients of viscosity of MBBA. The results of such calculations for PAA, MBBA, and PAP are summarized in Table 4.13.

The rotational viscosity η_ω, which is most distinctly manifested in the "entrainment" effect, is of great interest. The classical method for making use of this effect was used by V. N. Tsvetkov to determine the temperature dependence of PAA in the nematic phase.[278] This phenomenon was studied theoretically by a number of authors.[275,279,280] Not only was it possible to measure experimentally the temperature dependence of η_ω for PAA and MBBA (Fig. 4.38), but it was also possible to obtain a working equation for this quantity.[277] A new method for determining η_ω, based on a study of the

Table 4.13
Coefficients of Viscosity (*cP*) of PAA, MBBA and PAP

Viscosity Coefficient	PAA (122°C)		MBBA (25°C)		PAP (144.4°C)	
	Value	Reference	Value	Reference	Value	Reference
$\eta_\perp(\eta_1)$	9.2 ± 0.4	[274]	103.5 ± 0.05	[271]	8.3 ± 0.4	[287]
$\eta_\parallel(\eta_2)$	2.4 ± 0.05	[274–276]	23.8 ± 0.3	[271]	1.3 ± 0.05	[287]
$\eta_x(\eta_3)$	3.4 ± 0.3	[274]	41.6 ± 0.7	[271]	2.5 ± 0.3	[287]
η_ω	6.8	[277]	9.6	[277]		
α_1					6.5 ± 4	[271]

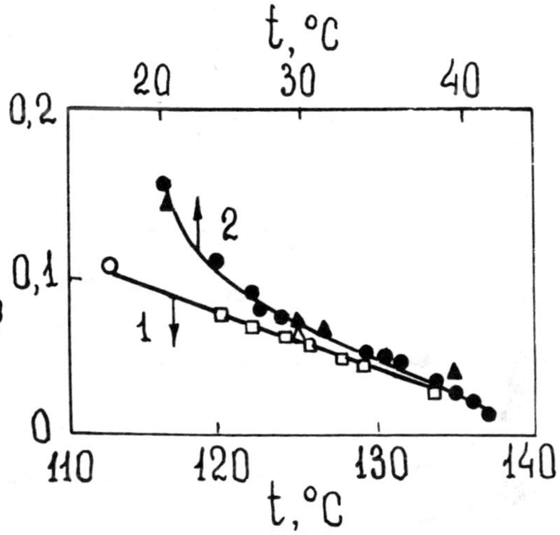

Figure 4.38. Temperature dependences of the rotational viscosity of PAA (1) (open symbols) and of MBBA (2) (dark symbols).[277,281–283]

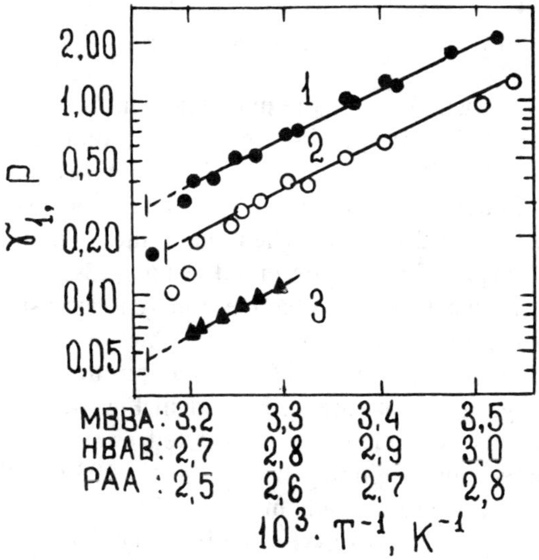

Figure 4.39. Coefficients of viscosity γ_1 of some NLC[288,289] 1—MBBA, 2—HBAB, 3—PAA.

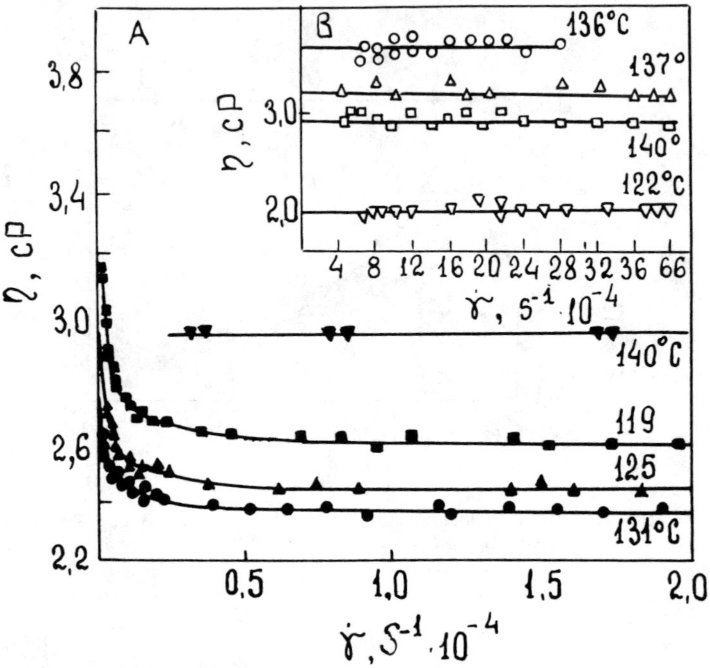

Figure 4.40. The apparent viscosity of PAA versus the gradient of the rate at different temperatures:[270,285,286] A) for low gradients, B) for high gradients.

relaxation of "twisting" after the magnetic field is switched off, was proposed in Ref. 284. It was observed that the rotational viscosity of MBBA, similarly to the viscosity of ordinary liquids, is proportional to exp(1/T); in addition, the energy of activation for MBBA turned out to be close to 48.2 kJ/mole. The temperature dependences of the rotational viscosity of γ_1 are presented in Fig. 4.39.

It is well known that the effective apparent viscosity of a NLC depends on the gradient of the rate $\dot{\gamma}$, which is characteristic for non-Newtonian liquids. The results of measurements for PAA for small and large gradients are presented in Fig. 4.40.[270,285,286] It is evident that this dependence is manifested especially strongly for small gradients, and beginning from $\dot{\gamma} = 5 \cdot 103\ s^{-1}$ the viscosity no longer depends on the gradient.

The viscosity of 4-*n*-octyloxy-4-cyanobiphenyl in the nematic phase and in the region of the NLC–SLC-A transition was studied in detail in Ref. 290. The coefficients of viscosity η_1, η_2, η_3, and γ_1 were

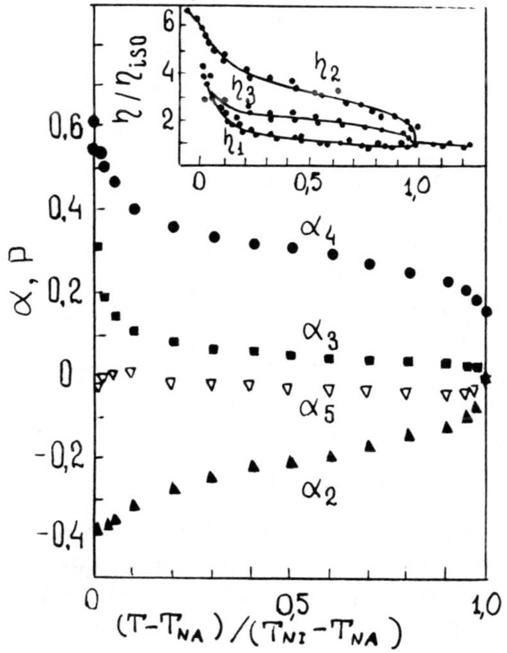

Figure 4.41. Leslie coefficients of viscosity α and the coefficients of viscosity η_i[290] of 4-*n*-octyloxy-4′-cyano-biphenyl (η_{iso} = 78 cP).

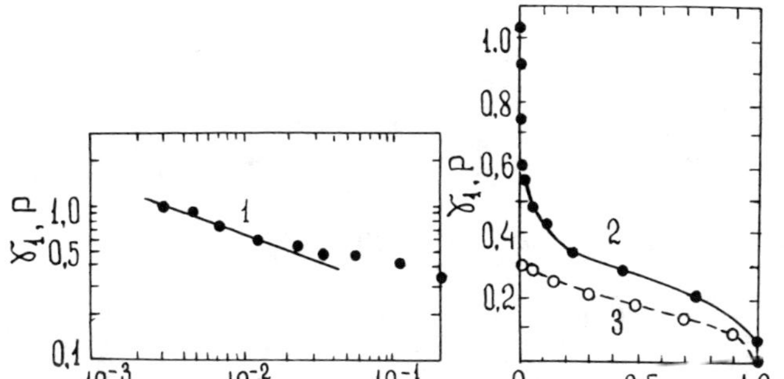

Figure 4.42. Temperature dependences of the coefficient of viscosity γ_1 of 4-*n*-octyloxy-4′-cyanobiphenyl (1 and 2); curve 3 is the "regular" part of γ_1, obtained from the measurements of η_1.[290]

studied. Some results of this study are presented in Figs. 4.41–4.43; in addition, the Leslie coefficients, which, as is well known, are interrelated with one another by the relations

$$\eta_1 = \alpha_3 + (\alpha_2 + \alpha_1 + \alpha_5)/2, \qquad \eta_2 = (-\alpha_2 + \alpha_4 + \alpha_5)/2$$

$$\eta_3 = \alpha_4/2 \qquad \text{and} \qquad \gamma_1 = \alpha_3 - \alpha_2$$

are also presented. It turned out that in the region of the phase transition the behavior of η_2 and η_3 remains regular, while η_1 and ν_1 exhibit distinct divergences.[290]

The twist viscosity γ_1 was measured as a function of the temperature for the series di-alkoxyazoxybenzenes (mOAB), which exhibit a nematic phase, was measured by Diogo and Martins.[291] These data are presented in Fig. 4.44. Diogo and Martins[291] developed a theory, based on which they obtained an expression for calculating γ_1:

$$\gamma_1(T) = g\langle P_2\rangle^2 \exp[E/kT + \theta\langle P_2\rangle^2/(T - T_0)],$$

where E is the height of the Maier–Saupe potential, T_0 is the temperature of "freezing" of the motion of the director of the NLC, and g and θ are virtually constant parameters, which can be determined based on other physical properties. It was shown that this relationship

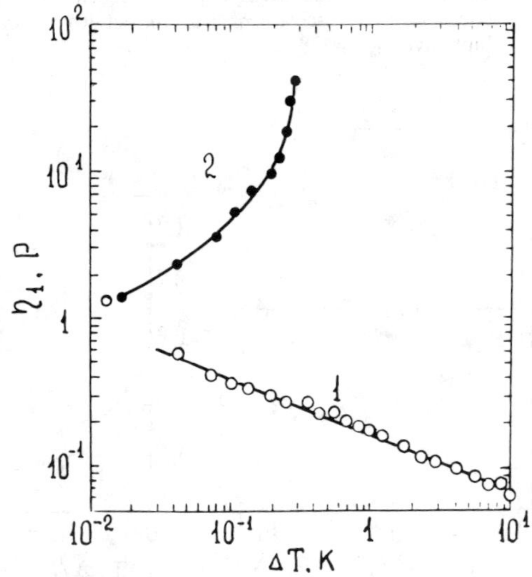

Figure 4.43. Coefficient of viscosity η_1 of 4-*n*-octyloxy-4′-cyanobiphenyl as a function of $\Delta T = T - T_{NA}$:[290] 1—NLC, 2—SLC. The law $\Delta T^{-\nu}$, where $\nu = 0.36$, is valid throughout the nematic phase.

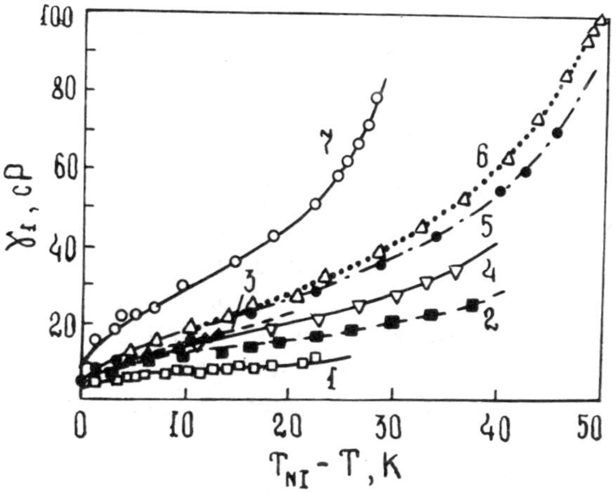

Figure 4.44. Temperature dependences of the viscosity γ_1 for a series of di-alkoxy-azoxybenzenes (mOAB).[291] The numbers next to the curves correspond to the numbers in the terms of the series (m = 1–7).

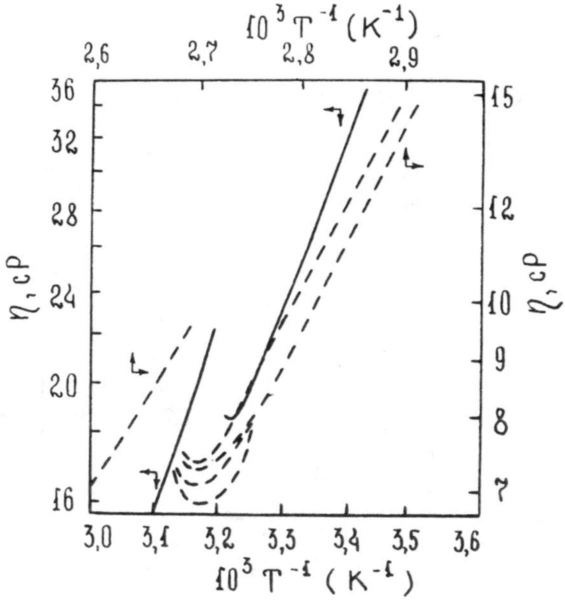

Figure 4.45. Temperature dependences of the apparent viscosity of MBBA (solid line) and HBAB (broken line).[292] The different broken lines correspond to the results obtained with the help of different viscosimeters.

describes very well the experimental data obtained for the region of existence of the nematic phase of the substances mentioned.[291]

The data on the apparent viscosity of liquid crystals is of considerable practical interest. Detailed measurements of this viscosity for methoxybenzylidene-butylaniline (MBBA) and hexyloxybenzylidene-aminobenzonitrile (HBAB) in the nematic and isotropic phases, as well as for 4-cyanobenzylidene-4′-octyloxyaniline (CBOOA) in the smectic A, nematic, and isotropic phases were performed in Ref. 292. The data obtained, which turned out to be sensitive to the choice of the experimental apparatus, are presented in Figs. 4.45 and 4.46. A very detailed and careful study of the apparent viscosity of MBBA and EBBA as a function of temperature and pressure was performed by Kuss.[293] The main results of this study are presented in Figs. 4.47 and 4.48. The existence of sharp peaks, corresponding to the points of the NLC–IL transition, was established, the isotherms of the vis-

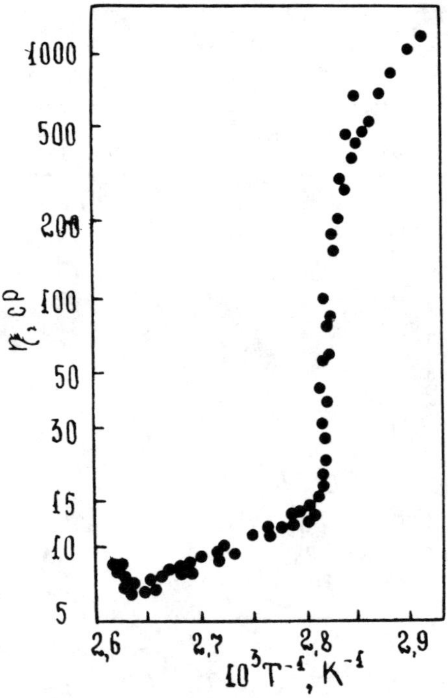

Figure 4.46. Temperature dependence of the apparent viscosity of CBOOA (data obtained with the help of three different viscosimeters).[292]

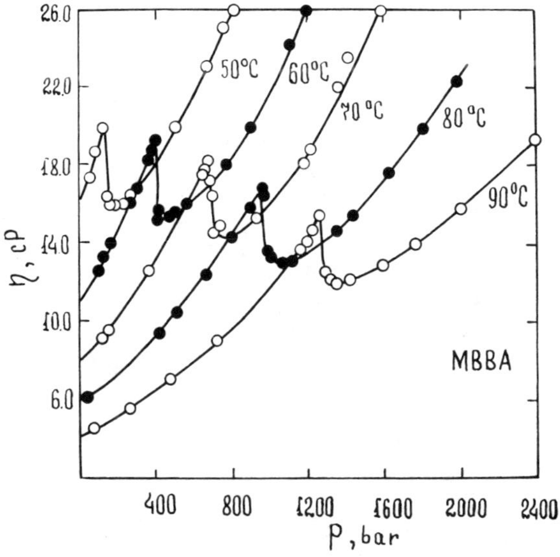

Figure 4.47. Pressure dependence of the coefficient of viscosity of MBBA on the isotherms.[293]

cosity of EBBA are approximately linear in the plane defined by the logarithm of the viscosity and the pressure (Fig. 4.48). It is surprising that the viscosity of the nematic phase can be less than the viscosity of the less dense isotropic phase. This can apparently be explained by the effect of orientational ordering of the molecules of the NLC. The difference in the behavior of the viscosity of MBBA from the viscosity of EBBA is interesting: the points of the maxima of the viscosity of EBBA form an ascending curve, while the points for the viscosity of MBBA form a descending curve.[293]

In most of the NLC studied, the experimentally measured values of the Leslie coefficient α_3 were negative, but at low temperatures, especially in the region adjacent to the transition into the SLC-A, positive values of α_3 can be expected. Indeed, the measurements for octyl-cyanobiphenyl (8CB) showed that this coefficient is positive in a wide range of temperatures of existence of NLC, with the exception of a small region adjacent to the point of the transition into the isotropic phase.[294] The temperature dependence of α_2 was also determined in Ref. 294 (Fig. 4.49).

A new method for determining the coefficient of viscosity γ_1 based on data on the rate of reorientation of molecules in the field of a

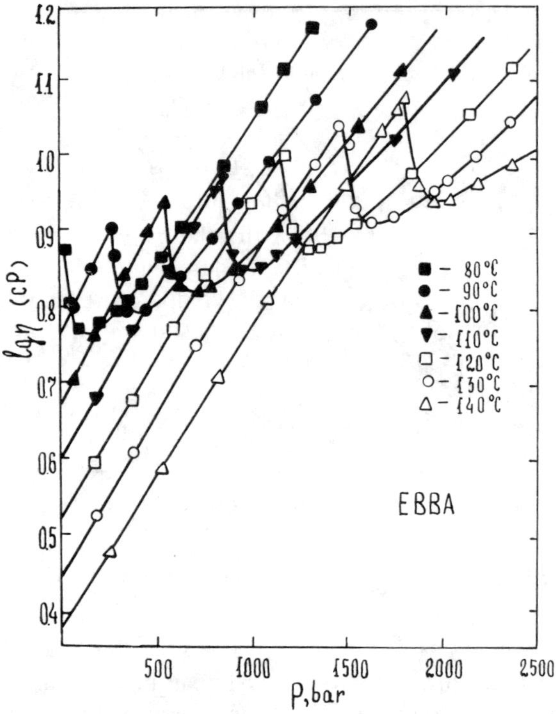

Figure 4.48. Pressure dependence of the coefficient of viscosity of EBBA on isotherms.[293]

light beam was proposed.[295] The value for OCBP was determined by this method (approximately 0.75–106 $cm^2 \cdot s^{-1}$).

In conclusion, we note that the temperature dependence of the coefficients of viscosity of mixtures of NLC is analogous to that of the pure substance (Fig. 4.50).[296]

A new method of the measurement of the rotational viscosity η_ω from the director pattern relaxation in twisted nematic cells is of interest.[297] This method makes it possible to define the temperature dependence η_ω and is useful both for pure NLC and mixtures.[297]

Among theoretical calculations of the viscosity of NLC, one should distinguish the nonequilibrium molecular dynamics studies of the anisotropic viscosity of perfectly ordered NLC.[138] Mesovich's coefficients, Leslie's coefficients and rotational viscosity were found. The agreement of simulation results with experimental data was obtained.

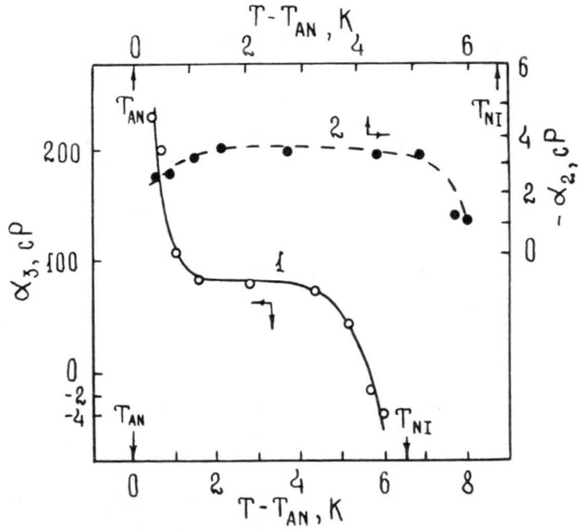

Figure 4.49. Temperature dependences of the Leslie coefficients of viscosity of 4-*n*-octyl-4′-cyanobiphenyl (8CB) between the transition temperatures T_{AN} and T_{NI}:[294] 1) α_3 (negative values are presented in an enlarged scale); 2) α_2.

The calculation of the pressure dependence of the complete set of viscosity coefficients of NLC was made on the basis of a simple model within the free volume theory.[298] The comparison of the results of such a calculation with experimental data on viscosity of MBBA indicates a satisfactory agreement up to the pressure of 1 kbar and

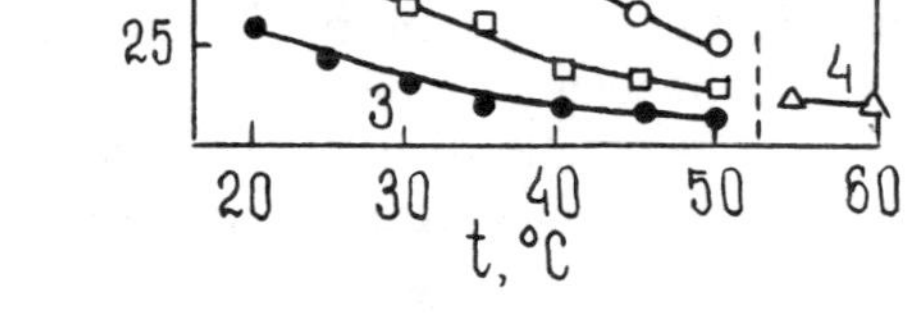

Figure 4.50. Temperature dependences of the coefficients of viscosity of a mixture of MBBA (66,7%) and EBBA (33,3%):[296] 1) η_1, 2) η_3, 3) η_2, 4) IL.

temperatures of 355 K. The approximate linear correlation between the value of viscosity of NLC and parameters of the effective potential of intermolecular interaction established by the Japanese authors[166] are of practical interest.

Viscosity properties of liquid crystals were considered on the basis of thermodynamics of irreversible processes in paper 299. The role of the molecular flexibility in NLC and the direct determination of separate molecular core and alkyl tail rotational relaxation times could be found out from the infra-red spectrum data.[300]

A number of attempts have been made to establish a relationship between the Leslie viscosity coefficients and the order parameters $\langle P_2 \rangle$.[301–303] Many of these results are approximate and require the use of a number of parameters and, sometimes, empirical temperature dependences.[301] The results obtained by statistical methods for studying transport in NLC,[304–306] although of great interest, cannot predict the viscous properties. Molecular-dynamics modeling can in principle yield complete information on the viscosity of NLC, but at the present time this method is beset with the difficulties of inadequate computational speed of existing computers and the length of time required for computer experiments

Very little is yet known about the thermal conductivity of NLC. All that has been established is that the thermal conductivity of the nematic phase of PAA in a magnetic field is anisotropic and depends on the field intensity up to saturation. Some experimental data have been obtained for PAA.[249] It has also been established that a field does not affect the thermal conductivity of the isotropic phase of PAA.[307,308]

Thermal conductivity of the liquid crystal copolyester prepared from 4-hydroxybenzoic acid (HBA) and 2-hydroxy-6-naphthoic acid (HNA) was studied.[309] It appeared that the anisotropy of the heat conductivity of this copolyester was small (the results of the measurements of the heat conductivity in the direction parallel to the director orientation and perpendicular to it did not differ practically). The authors of this investigation proposed the theoretical model of the heat conductivity mechanism in the copolymer HBA/HNA.

The construction of a quantitative molecular theory of the thermal conductivity of NLC is an important but unsolved problem.

References for Part I

CHAPTER I

1. I. G. Chistyakov, *Zhidkie kristally (Liquid Crystals)*, Nauka, Moscow (1966).
2. L. K. Vistin', *Ustroistva na zhidkikh kristallakh v sistemakh upravleniha, svyazi, i informatsii. Obzornaya informatsiya (Liquid-Crystal Devices in Control, Communication and Information Systems. General Information)*, TsNIIPI, Moscow (1977).
3. I. N Kompanets and V. V. Nikitin, *Nematicheskie zhidkie kristally v optoelektricheskikh ustroistvakh (Nematic Liquid Crystals in Optoelectronic Devices)*, Fiz. Inst. Akad. Nauk SSSR, Moscow (1973).
4. L. K. Vistin' in: *Tekhnika indikatsii (Display Technology)*, Naukova Dumka, Kiev (1976), p. 24.
5. L. K. Vistin' and I. G. Chistyakov, *Pribory i sistemy upravleniya (Control Systems and Devices)*, No. 3 (1975).
6. A. S. Sonin in: *Svoistva materialov, primenyaemykh v ustroistvakh optoelektroniki (Properties of Materials Used in Optoelectronics)*, Krasnoyarsk (1975), p. 112.
7. E. L. Williams, *Liquid Crystals for Electronic Devices*, John Wiley and Sons, New York (1975).
8. I. G. Chistyakov and V. A. Usol'tseva, *Zhidkie kristally i ikh rol' v meditsine i biologii (Liquid Crystals and Their Role in Medicine and Biology)*, Ivanovo (1962).
9. R. K. Mishra, *MCLC*, **29**, 201 (1975).
10. I. G. Chistyakov, V. A. Usol'tseva, S. A. Seleznev, and N. M. Maksimova, *Usp. Sovremen. Biol.*, **82**, 89 (1976).
11. B. K. Vainshtein and I. G. Chistyakov in: *Problemy sovremennoi kristallografii (Problems of Modern Crystallography)*, Nauka, Moscow (1975), p. 12.
12. A. P. Kapustin, *Elektroopticheskie i akusticheskie svoistva zhidkikh kristallov (Electrooptical and Acoustical Properties of Liquid Crystals)*, Nauka, Moscow (1973).
13. P. G. de Gennes, *The Physics of Liquid Crystals*, Clarendon Press, Oxford (1974).
14. L. M. Blinov, *Usp. Fiz. Nauk*, **114**, 67 (1974) [*Sov. Phys. Usp.*, **17**, 658 (1975)].
15. L. M. Blinov, *Elektro- i Magnitooptika zhidkikh kristallov (Electro- and Magnetooptics of Liquid Crystals)*, Nauka, Moscow (1978).
16. M. J. Stephen and J. P. Straley, *Rev. Mod. Phys.*, **46**, 617 (1974).
17. S. Chandrasekhar, *Rep. Prog. Phys.*, **30**, 613 (1976).
18. A. I. Larkin and S. A. Pikin, *Zh. Eksp. Teor. Fiz.*, **56**, 1664 (1969) [*Sov. Phys. JETP*, **29**, 891 (1969)].
19. S. A. Pikin, *Zh. Eksp. Teor. Fiz.*, **58**, 1406 (1970) [*Sov. Phys. JETP*, **58**, 753 (1970)].
20. T. C. Lubensky, *J. Phys. Chem. Solids*, **34**, 365 (1973).
21. F. P. Price and C. S. Bak, *MCLC*, **29**, 225 (1975).

22. A. L. Tsykalo and A. D. Bagmet, *Zh. Fiz. Khim.*, **50**, 751 (1976) [*Russ. J. Phys. Chem.*, **50**, 439 (1976)].
23. S. Diele, P. Brand, and H. Sackmann, *MCLC*, **17**, 163 (1972).
24. D. Demus, *et al.*, *MCLC*, **15**, 161 (1971).

CHAPTER II

25. E. B. Priestley, *RCA Rev.*, **35**, 144 (1974).
26. I. G. Chistyakov and V. M. Chaikovskii, *Kristallografiya*, **18**, 293 (1973) [*Sov. Phys. Crystallogr.*, **18**, 181 (1973)].
27. M. Kohli, K. Otness, R. Pynn, and T. Z. Riste, *Z. Physik.*, **24**, 147 (1976).
28. V. N. Tsvetkov, *Zh. Fiz. Khim.*, **16**, 132 (1942).
29. S. Jen, N. A. Clark, P. S. Pershan, and E. B. Priestley, *Phys. Rev. Lett.*, **31**, 1552 (1973).
30. A. L. Tsykalo and A. D. Bagmet, *Fiz. Tverd. Tela.*, **20**, 1326 (1978) [*Sov. Phys. Solid State*, **20,** 762 (1978)].
31. V. V. Andrushkenich in: *Zhidkie kristally (Liquid Crystals)*, Ivanovo (1976), p. 42.
32. Ya. I. Frenkel', *Kineticheskaya Teoriya Zhidkostei*, Nauka, Moscow (1975) [English translation: *Kinetic Theory of Liquids*, Dover Publications, New York (1955)].
33. N. V. Efimov, *Kvadratichnye formy i matritsy (Quadratic Forms and Matrices)*, Nauka, Moscow (1975).
34. A. Kobinata, Y. Nakajima, H. Yoshida, and S. Naeda, *Mol. Cryst. Liq. Cryst.*, **66**, 67 (1981).
35. E. M. Aver'yanov and V. F. Shabanov, *Kristallografiya*, **23**, 320 (1978).
36. E. M. Aver'yanov, K. S. Aleksandrov, and V. F. Shabanov, *Doklady AN SSSR*, **242**, 84 (1978).
37. S. O. Durbin and Y. R. Shen, *Phys. Rev. A.: Gen. Phys.*, **30**, 1419 (1984).
38. A. L. Tsykalo and A. D. Bagmet, *Kristallografiya*, **21**, 1101 (1976) [*Sov. Phys. Crystallogr.*, **21**, 637 (1976)].
39. K. A. Valiev and E. N. Ivanov, *Usp. Fiz. Nauk*, **109**, 31 (1973) [*Sov. Phys. Usp.*, **16**, 1 (1973)].
40. D. N. Zubarev, *Neravnovesnaya statisticheskaya mekhanika* (*Nonequilibrium Statistical Mechanics*), Nauka, Moscow (1971).

CHAPTER III

41. C. A. Croxton, *Introduction to Liquid State Physics*, Wiley-Interscience, N.Y. (1975).
42. A. de Vries, *J. Phys. (Paris)*, **36**, Colloq. C1, C1-1 (1975).
43. M. Born, *Sitz. Preus. Akad.*, **30**, 614 (1916); *Ann. Phys. (Leipzig)*, **55**, 221 (1918).
44. N. V. Madhusudana and S. Chandrasekhar, *Pramana Supplement 1*, 57 (1975).
45. Y. Poggi, *MCLC*, **37**, 1 (1976).
46. F. Grandjean, *C.R. Acad. Sci.*, **164**, 280 (1917).
47. W. Maier and A. Z. Saupe, *Naturforsch. A*, **13**, 564 (1958); *ibid.* **A14**, 882 (1959); *ibid.* **A15**, 287 (1960).

48. I. V. Sushkin and I. G. Chistyakov in: *Struktura i svoistva zhidkikh kristallov i vysokomolekulyarnykh veshchestv* (Structure and Properties of Liquid Crystals and High-Molecular Substances), Ivanovo (1967), p. 22.
49. R. L. Humphries, P. G. James, and G. R. Luckhurst, *J. Chem. Soc. Faraday Trans.*, **68**, 1031 (1972).
50. R. L. Humphries and G. R. Luckhurst, *Chem. Phys. Lett.*, **17**, 514 (1972).
51. P. J. Wojtowicz, *RCA Rev.*, **35**, 118 (1974).
52. G. R. Luckhurst and C. Zannoni, *Nature*, **267**, 412 (1977).
53. B. Deloche, B. Cabane, and D. Jerome, *MCLC*, **15**, 197 (1971).
54. J. R. McColl and C. S. Shih, *Phys. Rev. Lett.*, **29**, 85 (1972).
55. J. R. McColl, *Phys. Lett. A*, **38**, 55 (1972).
56. S. Chandrasekhar and N. V. Madhusudana, *Acta Crystallogr. A*, **27**, 303 (1971).
57. S. Chandrasekhar and N. V. Madhusudana, *MCLC*, **10**, 151 (1970).
58. S. Chandrasekhar and N. V. Madhusudana, *MCLC*, **17**, 37 (1972).
59. R. G. Priest, *Phys. Rev. Lett.*, **26**, 423 (1971).
60. T. D. Schultz, *MCLC*, **14**, 147 (1971).
61. J. C. Raich, R. D. Etters, and L. Flax, *Chem. Phys. Lett.*, **6**, 491 (1970).
62. S. Chandrasekhar and N. V. Madhusudana, *MCLC*, **24**, 179 (1973).
63. K. V. Semenchenko and B. Chotchaev, *Zh. Fiz. Khim.*, **51**, 718 (1977) [*Russ. J. Phys. Chem.*, **51**, 420 (1977)].
64. G. R. Luckhurst, C. Zannoni, P. L. Nordio, and U. Segre, *Mol. Phys.*, **30**, 1345 (1975).
65. G. Vertogen and B. W. Van der Meer, *Physica A*, **99**, 237 (1979).
66. Udai P. Singh and Yashwant Singh, *Phys. Rev. A.: Gen. Phys.*, **33**, 2725 (1986).
67. K. Sokalski, *Acta phys. pol.*, **A70**, 737 (1986).
68. D. K. Remler and A. D. J. Haymet, *J. Phys. Chem.*, **90**, 5426 (1986).
69. G. R. Luckhurst, *Nucl. Magn. Res. Liq. Cryst. Proc.*, NATO Adv. Study Inst. San Miniato, July 26–Aug. 7, 1983. Dordrecht e.a. 53 (1985).
70. G. R. Luckhurst, *Recent Adv. Liq. Cryst. Polym. Proc. Eur. Sci.*, Found. 6 Polym. Lyngby, 12–14 Sept. 1983. London, New York, 105 (1985).
71. Kimura Hatsuo and Nakano Huzio, *J. Phys. Soc. Jap.*, **54**, 1730 (1985).
72. L. Onsager, *Ann. N.Y. Acad. Sci.*, **51**, 627 (1949).
73. P. J. Flory, *Proc. Roy. Soc. A*, **234**, 73 (1956).
74. P. J. Flory and G. Ronca, *Mol. Cryst. Liq. Cryst.*, **54**, 289 (1979).
75. P. J. Flory and G. Ronca, *Mol. Cryst. Liq. Cryst.*, **54**, 311 (1979).
76. E. Di Marzio, *J. Chem. Phys.*, **35**, 658 (1961).
77. M. A. Cotter, *MCLC*, **35**, 33 (1976).
78. E. H. Lieb, *J. Math. Phys.*, **8**, 2339 (1967).
79. P. Shukla and T. J. Sluckin, *J. Phys. A: Math. and Gen.*, **18**, 93 (1985).
80. G. F. Kventsel and T. Katriel, *Phys. Rev. A: Gen. Phys.*, **34**, 518 (1986).
81. G. Foex, *Trans. Faraday Soc.*, **29**, 958 (1933).
82. P. G. de Gennes, *MCLC*, **12**, 193 (1971).
83. L. D. Landau and E. M. Lifshitz, *Statisticheskaya Fizika, Nauka*, Moscow (1976) [English translation: *Statistical Physics*, Pergamon Press, New York (1980)].
84. S. Chandrasekhar, *et al.*, *Proc. Int. Liquid Crystals Conf.*, Bangalore, 1973; Pramana Supplement 1, p. 117.
85. S. Chandrasekhar, *et al.*, *MCLC*, **24**, 201 (1973).
86. S. Chandrasekhar, G. S. Ranganath, and K. A. Suresh, *Proc. Int. Liquid Crystals Conf.*, Bangalore, 1973; Pramana Supplement 1, p. 341.

87. J. G. Ypma and G. Vertogen, *Solid State Commun.*, **18**, 475 (1976).
88. A. Mauger, G. Zribi, D. L. Mills and John Toner, *Phys. Rev. Lett.*, **53**, 2485 (1984).
89. A. Poniewierski and T. J. Sluckin, *Mol. Phys.*, **55**, 1113 (1985).
90. R. M. Hornreich, *Phys. Lett. A*, **109**, 232 (1985).
91. E. M. Aver'yanov, P. V. Adomenas, V. A. Shuikov and V. Ya. Zirjanov. *Zh. Eksp. Teor. Fiz.*, **91**, 551 (1986).
92. H. A. Bethe, *Proc. Roy. Soc.*, **149**, 1 (1935).
93. M. Metropolis, *et al.*, *J. Chem. Phys.*, **21**, 1087 (1953).
94. M. Rosenbluth and A. Rosenbluth, *J. Chem. Phys.*, **22**, 881 (1954).
95. W. Wood, F. Parker, and J. Jacobson, *Nuovo Cimento*, **9**, 133 (1958).
96. I. Z. Fisher, *Statistical Theory of Liquids*. Chicago University Press, Chicago (1964).
97. W. Wood in: *Physics of Simple Liquids*, Ed. H. N. V. Temperley, J. S. Rowlinson, and G. S. Rushbrooke, North-Holland Publishing Co., Amsterdam (1968), p. 115.
98. V. M. Zamalin, G. E. Norman, and V. S. Filinov, *Metod Monte-Karlo v statisticheskoi termodinamike* (Monte-Carlo Method in Statistical Thermodynamics), Nauka, Moscow (1977).
99. G. Lasher, *Phys. Rev. A*, **5**, 1350 (1972).
100. P. A. Lebwohl and G. A. Lasher, *Phys. Rev. A*, **6**, 426 (1972).
101. H. J. Jansen, G. Vergogen, and J. G. Ypma, *MCLC*, **38**, 87 (1977).
102. J. Vieillard-Baron, *J. Chem. Phys.*, **56**, 4729 (1972).
103. J. Vieillard-Baron, *Mol. Phys.*, **28**, 809 (1974).
104. M. A. Cotter and D. E. Martire, *J. Chem. Phys.*, **52**, 1902 (1970); *ibid.* 52, 1909 (1970); *ibid.* 53, 4500 (1970).
105. T. Boublik, J. Nezbeda, and O. Trnka, *Czech. J. Phys. B*, **26**, 1081 (1976).
106. G. A. Few and M. Rigby, *Chem. Phys. Lett.*, **20**, 433 (1973).
107. R. Kubo in: *Thermodynamics of Irreversible Processes* [Russian translation], Inostr. Lit., Moscow (1962).
108. A. M. Evseev in: *Sovremennye problemy fizicheskoi khimii* (Contemporary Problems in Physical Chemistry), Izd-vo MGU, Moscow (1972), Vol. 6, p. 34.
109. Computer Modeling in the Study of Materials [Russian translation], Mir, Moscow (1974).
110. J. Kushick and B. J. Berne, *J. Chem. Phys.*, **64**, 1362 (1976).
111. B. J. Berne and P. Pechukas, *J. Chem. Phys.*, **56**, 4213 (1972).
112. H. Jmura and K. Okano, *J. Chem. Phys.*, **58**, 2763 (1973).
113. D. Beeman, *J. Comput. Phys.*, **20**, 130 (1976).
114. J. Chistyakov, *Adv. Liq. Cryst.*, **1**, 143 (1975).
115. E. Boilini and S. K. Ghosh, *J. Appl. Phys.*, **45**, 78 (1975).
116. L. M. Blinov, *et al.*, *Kristallografiya*, **20**, 1245 (1975) [*Sov. Phys. Crystallogr.*, **20**, 750 (1975)].
117. H. Yoshida, Y. Nakajima, S. Kobinata and S. Naeda. 8-th Int. Liq. Cryst. Conference. Kyoto, Japan, 1980, p. 85.
118. V. M. Chaikovskii and I. G. Chistyakov, *Uch. Zapiski Ivanovskogo Gos. Ped. In-ta*, **99**, 301 (1972).
119. I. G. Chistyakov in: *Sb. dokladov II Vsesoyuznoi konferentsii po zhidkim kristallam* (Contributions to the 2nd All-Union Conference on Liquid Crystals), Ivanovo (1973), p. 3.

120. V. M. Chaikovskii and I. G. Chistyakov, *Uch. Zapiski Ivanovskogo Gos. un-ta* **128**, 5 (1974).
121. G. R. Alms, T. D. Glerke, and W. H. Flygare, *J. Chem. Phys.*, **61**, 4083 (1974).
122. S. M. Stishov, V. A. Ivanov, and V. N. Kachinskii, *Pis'ma Zh. Eksp. Teor. Fiz.*, **24**, 329 (1976) [*JETP Lett.*, **24**, 2997 (1976)].
123. A. L. Tsykalo, *Doklady AN SSSR*, **249**, 1348 (1979).
124. A. L. Tsykalo and A. D. Bagmet, *Inzhenerno-Fisicheskii Zhurnal*, **4**, 685 (1980).
125. A. L. Tsykalo, *MCLC*, **129**, 409 (1985).
126. A. L. Tsykalo and A. D. Bagmet, *Acta physica polonica*, **A55**, 111 (1979).
127. A. L. Tsykalo and A. D. Bagmet, *Zhurnal Technicheskoi Fiziki*, **50**, 1480 (1980).
128. G. R. Luckhurst and S. Romano, *Proc. R. Soc. Lond. A*, **373**, 111 (1980).
129. G. R. Luckhurst, P. Simpson, and C. Zannoni, *Chem. Phys. Lett.*, **78**, 429 (1981).
130. C. Zannoni and M. Guerra, *Molec. Phys.*, **44**, 849 (1981).
131. R. L. Humphries, G. R. Luckhurst and S. Romano, *Mol. Phys.*, **42**, 1205 (1981).
132. G. J. Fuller, G. R. Luckhurst and C. Zannoni, *Chem. Phys.*, **92**, 105 (1985).
133. G. R. Luckhurst, T. J. Sluckin and H. B. Zewdie, *Mol. Phys.*, **59**, 657 (1986).
134. D. Frenkel and R. Eppenga, *Phys. Rev. A.: Gen. Phys.*, **31**, 1776 (1985).
135. V. P. Chasovskikh, M. I. Salakhutdinov, A. Nurullaev and A. A. Adkhamov, *Doklady Akademii Nauk Tadzhikskoi SSR*, **29**, 406 (1986).
136. Chandan Dasgupta, *Phys. Rev. Lett.*, **55**, 1771 (1985).
137. Andrzej Koliński, *Chem. Phys. Lett.*, **116**, 160 (1985).
138. Dieter Baalss and Siegfried Hess, *Phys. Rev. Lett.*, **57**, 86 (1986).
139. A. L. Tsykalo and A. D. Bagmet, *Ukrainskii Fizicheskii Zhurnal*, **31**, 387 (1986).
140. A. L. Tsykalo and A. D. Bagmet, *Doklady AN SSSR*, **287**, 324 (1986).
141. N. P. Tumanyan and E. P. Sokolova, *Zh. Fir. Khim.*, **58**, 2444 (1984).
142. N. P. Tumanyan, E. P. Sokolova, and A. G. Morachevsky, *Ibid.*, **58**, 2448 (1984).
143. E. P. Sokolova and N. P. Tumanyan, *Ibid.*, **60**, 1087 (1986).
144. E. P. Sokolova and N. P. Tumanyan, *Ibid.*, **61**, 1349 (1987).
145. A. Derzhanski, A. G. Petrov and M. D. Mitov, Wiss. Beitr. M.-Luther-Univ. Halle-Wittenberg **17**, 170 (1986).
146. Nitish K. Sanyal, S. N. Tiwari and M. Roychondhury, *J. Phys. Soc. Jap.*, **54**, 4586 (1985).
147. James S. Walker and Jonathan L. Mace, *Phys. Lett.*, **A115**, 281 (1986).
148. Vl. K. Pershin and V. A. Konoplev, *Pisma Zh. Eksp. Teor. Fiz.*, **18**, 146 (1987).
149. E. M. Aver'yanov, *Zh. Fiz. Lhim.*, **59**, 2145 (1985).
150. N. Oranth and F. Strohbusch, *Ber. Bunsenges. phys. chem.*, **91**, 211 (1987).
151. V. M. Samsonov and L. M. Shcherbakov, *Ultrazvuk i termodinamicheskie svoistva veshchestva* (Ultrasound and thermodynamic properties of substance), Kursk, **90** (1986).
152. T. J. Sluckin and A. Poniewierski, *Phys. Rev. Lett.*, **55**, 2907 (1985).
153. L. Georgescu, Popa-Nita Vlad and R. Moldovan, *Rev. roum. phys.*, **29**, 553 (1984).
154. L. Georgescu and Popa-Nita Vlad, *Ann. Univ. Bucuresti. Fiz.*, **34**, 29 (1985).
155. J. Murakami, *J. Phys. Soc. Jap.*, **42**, 210 (1977).
156. F. Sagues and M. Miguel, *Phys. Rev. A: Gen. Phys.*, **33**, 2769 (1986).
157. W. Franklin, *MCLC*, **14**, 227 (1971); *Phys. Lett. A*, **48**, 247 (1974); *Phys. Rev. A*, **11**, 2156 (1975).
158. K. S. Chu, *et al.*, *Liquid Crystals and Ordered Fluids*, New York (1974), Vol. 2.

159. K. S. Chu and D. Moroi, *J. Phys. (Paris) Suppl.*, **36**, 1 (1975).
160. Giorgio Moro, Ulderico Segre and Pier Luigi Nordio, *Nucl. Magn. Resonance Liquid Cryst. Proc.*, NATO Adv. Study Inst., San Miniato, July 26–Aug. 7, 1983. Dordrecht e.a. 207 (1985).
161. Giorgio Moro and Pier Luigi Nordio, *Mol. Phys.*, **56**, 255 (1985).
162. D. Wong, *Phys. Lett.*, **98A**, 259 (1983).
163. Lin Lei, Shangqing Shu and Gang Xu, *Phys. Lett.*, **A109**, 277 (1985).
164. G. Zhy, *Phys. Rev. Lett.*, **49**, 1332 (1982).
165. J. Verhás, *Acta phys. hung.*, **55**, 275 (1984).
166. Hiromochi Muramotsu, Shinya Ohmi and Kunihiko Hara, *Jap. J. Appl. Phys.*, **24**, 1583 (1985).

CHAPTER IV

167. R. Pynn, *J. Chem. Phys.*, **60**, 4579 (1974).
168. S. Chandrasekhar, *et al.*, *MCLC*, **10**, 327 (1970).
169. N. A. Tikhomirova, L. K. Vistin' and V. N. Nosov, *Kristallografiya*, **17**, 1000 (1972) [Sov. Phys. Crystallogr., **17**, 878 (1973)].
170. H. Arnold, *Z. Phys. Chem.*, **226**, 146 (1964).
171. J. Robberecht, *Bull. Soc. Chem. Belges.*, **47**, 597 (1938).
172. B. Delicha, B. Cabane and D. Jerome, *MCLC*, **15**, 197 (1971).
173. P. H. Keyes, *et al.*, *J. Chem. Phys.*, **63**, 5006 (1975).
174. P. H. Keyes, H. T. Weston, W. J. Lin and W. B. Daniels, *J. Chem. Phys.*, **63**, 5006 (1975).
175. Philip J. Bullock, David J. Byron, David J. Harwood, Robert C. Wilson and Adrian M. Woodward, *J. Chem. Soc. Perkin Trans.*, Pt. 2, 2121 (1984).
176. David J. Byron, Robert C. Wilson, Philip A. Baker, Ighinas Danilewicz, Gians S. Hayer, David A. Hewison, James M. Taylor and Julian M. Wyer, *J. Chem. Soc. Perkin Trans.*, Pt. 2, 297 (1985).
177. Akira Takase and Sakumitsu Sakagami, *Jap. J. Appl. Phys.*, **23**, 1532 (1984).
178. Prasad S. Krishna, S. Pfeiffer, G. Heppke and R. Shashidhar, *Z. Naturforsch.*, **40**, 632 (1985).
179. R. Shashidhar, *Physica*, **BC 139**, 609 (1986).
180. Shunsuke Kobinata, Takamichi Kobayashi, Hiroshi Yoshida, A. D. L. Chandani and Shiro Naeda, *J. Mol. Struct.*, **146**, 373 (1986).
181. H. Hsiung, Th. Rasing and Y. R. Shen, *Phys. Rev. Lett.*, **57**, 3065 (1986).
182. E. M. Aver'yanov, P. V. Adomenas, V. A. Zhuilov and V. Ya. Ziryanov, *Pisma Zh. Eksp. Teor. Fiz.*, **43**, 89 (1986).
183. Vl. K. Pershin, *Zh. Fiz. Khim.*, **60**, 2540 (1986).
184. I. E. Tomashevskii, S. V. Shiyanovskii and A. M. D'yachenko, *Ukraincki Fizicheskii Zhurnal.*, **31**, 1222 (1986).
185. F. M. Aliev, I. K. Meshkovskii and V. I. Kuznetzov, *Doklady AN SSSR*, **279**, 848 (1984).
186. H. Yokoyana, S. Kobayashi and H. Kamei, *J. Appl. Phys.*, **56**, 2645 (1984).
187. P. R. Kishore and C. R. Murty, *Acta Phys. Pol.*, **A67**, 667 (1985).
188. A. C. Zawisza and J. Stecki, *Solid State Commun.*, **19**, 1173 (1976).

189. V. Ya. Baskakov, V. K. Semenchenko, and V. M. Byankin, *Zh. Eksp. Teor. Fiz.*, **66**, 792 (1974) [*Sov. Phys. JETP*, **39**, 383 (1974)].
190. R. G. Horn and T. E. Faber, *Proc. Roy. Soc. (London) A*, **368**, 199 (1979).
191. E. Kuss, *Mol. Cryst. Liq. Cryst.*, **47**, 71 (1978).
192. A. C. Zawisza and J. Stecki, *Solid State Commun.*, **19**, 1173 (1976).
193. B. Bahadur, *J. Chem. Phys. Phys.-Chem. Biol.*, **73**, 255 (1976).
194. B. Bahadur, *MCLC*, **35**, 83 (1976).
195. S. K. Ghosh and S. Amadesi, *Phys. Lett. A*, **59**, 282 (1976).
196. A. E. White, P. E. Cladis, and S. Torza, *Mol. Cryst. Liq. Cryst.*, **43**, 13 (1977).
197. C. Cabos and J. Sicard, *C. R. Acad. Scie. (Paris)*, **281**, 109 (1975).
198. B. Bahadur, *Z. Naturforsch. A*, **30**, 1094 (1975).
199. K. W. Sadowska, A. Zywocinski, and J. Stecki, *J. Physique*, **43**, 1673 (1982).
200. V. G. K. Pisipati and N. V. S. Rao, *Z. Naturforsch*, **A39**, 696 (1984).
201. K. R. K. Rao, J. V. Rao, P. Venkatacharyulu and V. Baliah, *Cryst. Res. and Technol.*, **21**, 771 (1986).
202. D. Demus, M. Pohl, S. Schönberg, L. Weber, A. Wiegeleben and W. Weissflog, *Wiss. Beitr. M.-Luther-Univ. Halle-Wittenberg.*, **12**, 18 (1983).
203. N. V. S. Rao and V. G. K. Pisipati, *Z. Naturforsch.*, **A40**, 466 (1985).
204. D. Demus, S. Diele, A. Hauser, I. Latif, Ch. Selbmann and W. Weissflog, *Cryst. Res. and Technol.*, **20**, 1547 (1985).
205. T. Shirakawa, H. Eura, H. Ichimura, T. Ito, K. Toi and T. Seimiya, *Thermochim. acta*, **105**, 251 (1985).
206. Matthew W. Lampe and Peter J. Collings, *Phys. Rev. A: Gen. Phys.*, **34**, 524 (1986).
207. J. G. Ypma and G. Vertogen, *Phys. Lett. A*, **61**, 45 (1977).
208. M. A. Cotter, 6th Int. Liquid Crystals Conf., Kent, August, 1977, p. 18.
209. F. M. Jaeger, *Z. anorg. algg. Chem.*, **101**, 1 (1917).
210. F. K. Gorskii and N. M. Sakevich, *Kristallografiya*, **12**, 586 (1968).
211. A. Fergason and S. Kennedy, *J. Phil. Mag.*, **26**, 41 (1938).
212. V. Naggiar, *Ann. de Phys.*, **18**, 5 (1943).
213. W. M. Schwartz and H. W. Moseley, *J. Phys. Coll. Chem.*, **51**, 826 (1947).
214. D. Langevin, *J. Phys. (Paris)*, **33**, 249 (1972).
215. I. Haller, *Appl. Phys. Lett.*, **24**, 349 (1974).
216. D. Meyerhofer, A. Sussman, and R. Williams, *J. Appl. Phys.*, **43**, 3685 (1972).
217. F. J. Kahn, G. N. Taylor, and H. Schonjorn, *Proc. IEEE*, **61**, 823 (1973).
218. I. Haller, *Progress in Solid-State Chemistry (1975)*, **10**, Part 2, p. 103.
219. G. Foex and L. Royer, *C. R. Acad. Sci.*, **180**, 1912 (1925).
220. V. Tsvetkov and A. Sosnovskii, *Acta Phys. Chem. URSS*, **18**, 358 (1943).
221. V. N. Tsvetkov, *Zh. Eksp. Teor. Fiz.*, **9**, No. 5 (1939).
222. J. Prost and H. Gasparoux, *Phys. Lett. A*, **36**, 245 (1971).
223. L. Pauling, *J. Chem. Phys.*, **4**, 673 (1936).
224. C. H. Massen and J. A. Poulis, *Ordered Fluids and Liquid Crystals*, Washington (1967), p. 72.
225. V. A. Korzhenevsky and M. G. Tomilin, *Voprosy phiziki formoobrazovaniya i fazovych prevrashchenij (Problems of Formation and of Phase Transformations)*. Kalinin, 60 (1986).
226. S. Faetti and V. Palleschi, *J. Chem. Phys.*, **81**, 6254 (1984).
227. S. Faetti and V. Palleschi, *J. Phys. Lett. (Paris)*, **45**, L313 (1984).
228. Barbara Stryla and Wojciech Kuczynski, *Post. fiz.*, **37**, 369 (1986).

229. V. O. Oganesyan and V. B. Pakhalov, *Izvestiya Akademii Nauk Armyanskoi SSR. Phizika.*, **21**, 29 (1986).
230. H. Kimura and H. Nakano, *J. Phys. Soc. Jap.*, **54**, 1730 (1985).
231. L. Georgescu, Popa-Nita Vlad and R. Moldovan, *Rev. roum. phys.*, **29**, 553 (1984).
232. E. M. Barall, R. S. Porter, and J. F. Johnson, *J. Phys. Chem.*, **71**, 895 (1967).
233. N. M. Sakevich, *Izv. Vyssh. Uchebn. Zaved. Fiz.*, **10**, 52 (1967).
234. L. C. Chow and D. E. Martire, *J. Phys. Chem.*, **73**, 1127 (1969).
235. S. Chandrasekhar and N. V. Madhusudana, *Liquid Crystals*, Proc. 3rd Int. Liq. Cryst. Conf., Vol. 3, Part 1, Berlin (1970).
236. B. I. Ostrovskii, *et al.*, *Zh. Eksp. Teor. Fiz.*, **71**, 692 (1976) [*Sov. Phys. JETP*, **44**, 363 (1976)].
237. V. N. Kachinskii, V. A. Ivanov, A. N. Zisman, and S. M. Stishov, *Zh. Eksp. Teor. Fiz.*, **75**, 545 (1978).
238. S. M. Stishov, *Usp. Fiz. Nauk*, **114**, 3 (1974).
239. J. O. Hirschfelder, Ch. F. Curtiss and R. B. Bird, *Molecular Theory of Gases and Liquids*, J. Wiley and Sons, Inc., New York, Chapman and Hall, Lim., London, 1954.
240. K. Hirakawa and S. Kai, *J. Phys. Soc. Jpn.*, **37**, 1472 (1974).
241. B. Bahadur, *Z. Naturforsch. A*, **30**, 1094 (1975).
242. J. Gabrielle and L. Verdini, *Nuovo Cim.*, **11**, 526 (1955).
243. W. A. Hoyer and A. W. Nolle, *J. Chem. Phys.*, **24**, 803 (1956).
244. A. P. Kapustin and N. T. Bykova, *Kristallografiya*, **11**, 297 (1966), [*Phys. Crystallogr.*, **11**, 300 (1966)].
245. E. L. Aero and A. N. Vulygin, *Tidromekhanika*, Itogi nauki i tekhniki, VINITI AN SSSR, Moscow, Vol. 7, 106 (1973).
246. O. Ya. Shmelev and V. A. Balandin. *Sbornic nauchnykh trudov Vsesouznogo zaochnogo mashinostroitelnogo instituta (Collection of Scientific Works of All-Union Correspondence Institute of Machine-Building)*, **36**, 53 (1984).
247. J. R. Opia and A. R. K. L. Padmini, *Indian J. Pure Appl. Phys.*, **11**, 1973 (1973).
248. A. A. Sitnov, Candidate's Dissertation. Institute of Heat and Mass Transfer, Academy of Sciences of the Belorussian SSR, Minsk (1975).
249. A. P. Kapustin, *Eksperimental'nye issledovaniya zhidkikh kristallov (Experimental Investigations of Liquid Crystals*, Nauka, Moscow (1978).
250. V. M. Samsonov, L. M. Shcherbakov and O. B. Kuz'michev, *Zh. Fiz. Khim.*, **59**, 2582 (1985).
251. J. Topler, B. Alefeld, and T. Springer, *MCLC*, **26**, 297 (1974).
252. C. K. Yun and A. G. Fredrickson, *MCLC*, **12**, 73 (1970).
253. K. Otnes, R. Pynn, J. A. Janik, and J. M. Janik, *Phys. Lett. A*, **38**, 335 (1972).
254. G. J. Kruger and R. Weiss, *J. Phys. (Paris)*, **38**, 353 (1977).
255. C. K. Yun and A. G. Fredrickson, *Mol. Cryst. Liq. Cryst.*, **12**, 73 (1970).
256. H. Hakemi, *J. Chem. Phys.*, **63**, 3708 (1975).
257. H. Hervet, W. Urbach, and F. Rondelez, *J. Chem. Phys.*, **68**, 2725 (1978).
258. R. Blinc, V. Dimic, J. Pirs, M. Vilfan, and I. Zupancic, *Liquid Crystals, Proc. 3rd Internat. Cryst. Conf.*, Berlin (1970), Vol. 3, Part 1, p. 365.
259. J. A. Janik, *Wiss. Beitr. M. Luther-Univ. Halle-Wittenberg*, **4**, 43 (1976).
260. G. Rollman, *Diss. Dokt. Naturwiss. Fak. Phys. Univ. Stuttgart*, 142 s. (1984).
261. M. Hara, S. Ichikawa, H. Takezoe and A. Fukuda. *Jap. J. Appl. Phys.*, **23**, 1420 (1984).

262. A. L. Tsykalo and A. D. Bagmet, *MCLC*, **50**, 217 (1978).
263. J. L. Ericksen, *Phys. Fluids*, **9**, 1205 (1966).
264. F. M. Leslie, *Archs. Ration. Mech. Analysis*, **28**, 265 (1968).
265. O. Parodi, *J. Phys. (Paris)*, **31**, 581 (1970).
266. M. Miesowicz, *Nature*, **17**, 261 (1935); **158**, 27 (1946).
267. C. Gohwiller, *MCLC*, **20**, 301 (1973).
268. J. Johnson and R. Porter, Orsay Liquid Crystal Group, *Liquid Crystals and Ordered Fluids*, Plenum Press, New York (1970), p. 195.
269. G. Becherer and W. Kast, *Ann. Physik*, **41**, 355 (1942).
270. R. S. Porter and J. F. Johnson, *J. Phys. Chem.*, **66**, 1826 (1962).
271. C. Gahwiller, *Phys. Lett. A*, **36**, 311 (1971).
272. V. N. Tsvetkov and G. M. Mikhailov, *Zh. Eksp. Teor. Fiz.*, **7**, 1399 (1937).
273. G. M. Mikhailov and V. N. Tsvetkov, *Zh. Eksp. Teor. Fiz.*, **9**, 597 (1939).
274. M. Miesowicz, *Bull. Acad. Pol. Sic. Math Nat. Ser. A*, **229** (1936).
275. H. C. Tseng, D. L. Silver, and B. A. Finlayson, *Phys. Fluids*, **15**, 1213 (1972).
276. V. N. Tsvetkov and V. M. Marinin, *Zh. Eksp. Teor. Fiz.*, **10**, 929 (1940).
277. H. Gasparoux and J. Prost, *J. Phys. (France)*, **32**, 953 (1971).
278. V. N. Tsvetkov, *Zh. Eksp. Teor. Fiz.*, **9**, 603 (1939).
279. E. L. Aero and A. N. Bulygin, *Prikladnaya matematika i mekhanika*, **35**, 879 (1971).
280. J. Prost and H. Gasparoux, *Phys. Lett. A*, **36**, 245 (1971).
281. G. Heppke and F. Schneider, *Z. Naturforsch. A*, **27**, 976 (1972).
282. C. M. Dafermos, *J. Appl. Math.*, **16**, 1305 (1968).
283. V. N. Tsvetkov, *Zh. Eksp. Teor. Fiz.*, **9**, 947 (1939).
284. P. E. Cladis, *Phys. Rev. Lett.*, **28**, 1629 (1972).
285. S. Peter and H. Peters, *Z. Phys. Chem.*, **3**, 103 (1955).
286. E. L. Aero and A. N. Bulygin, *Gidromekhanika, Itogi nauki i tekhniki*, VINITI AN SSSR, Moscow (1973), Vol. 7, p. 106.
287. M. Miesowicz, *Nature*, **158**, 27 (1946).
288. M. Meiboom and R. Hewitt, *Phys. Rev. Lett.*, **30**, 261 (1973).
289. A. Saupe, *Ann. Rev. Phys. Chem.*, **24**, 441 (1973).
290. L. Leger and A. Martinet, *J. Phys. (France)*, **37**, C3 (1976).
291. A. C. Diogo and A. F. Martins, *Mol. Cryst. Liq. Cryst.*, **66**, 133 (1981).
292. A. E. White, P. E. Cladis, and S. Torza, *Mol. Cryst. Liq. Cryst.*, **43**, 13 (1977).
293. E. Kuss, *Mol. Cryst. Liq. Cryst.*, **47**, 71 (1978).
294. T. Skarp, T. Carlsson, S. T. Lagerwall, and B. Stebler, *Mol. Cryst. Liq. Cryst.*, **66**, 199 (1981).
295. A. S. Zolot'ko in: *Dinamika pereorientatsii direktora NZhK v pole svetovogo lucha*, Fizicheskii institut AN SSSR, Preprint, Moscow (1982), No. 139, p. 1.
296. V. A. Tsvetkov and G. A. Beresner, *Pribory i tekhnika eksperimenta*, No. 5, 223 (1977).
297. E. Leenhouts, *J. Appl. Phys.*, **58**, 2180 (1985).
298. A. C. Diogo, *Solid State Commun.*, **50**, 895 (1984).
299. S. Hess. *J. Non-Equil. Thermodyn.*, **11**, 175 (1986).
300. M. P. Fontana and N. Kirov, *J. Phys. Lett. (Fr.)*, **46**, L341 (1985).
301. T. Lubensky, *Phys. Rev. A*, **2**, 2497 (1970); *ibid. A*, **6**, 452 (1972).
302. H. Ymura and K. Okano, *Jpn. J. Appl. Phys.*, **11**, 1442 (1972).
303. A. F. Martins, *Portugal. Phys.*, **9**, 1 (1974).
304. D. Forster, *et al.*, *Phys. Rev. Lett.*, **26**, 1016 (1971).

305. P. C. Martin, O. Parodi, and P. J. Pershan, *Phys. Rev. A*, **6**, 401 (1972).
306. V. B. Nemtsov, *Teor. Matem. Fiz.*, **25**, 118 (1975) [*Theor. Math. Phys.*, **25**, 1019 (1975)].
307. J. Fischer and A. G. Fredrickson, *MCLC*, **6**, 255 (1969).
308. C. K. Yun, J. J. C. Picot, and A. G. Fredrickson, *J. Appl. Phys.*, **42**, 4767 (1971).
309. A. J. Crispin and D. Greig, *Polym. Commun.*, **27**, 264 (1986).

PART II.

Smectic and Cholesteric Liquid Crystals

Why the unbroken spiral virtuoso,
Like a trick sonnet in one long, versatile sentence?
Robert Francis, "*Apple Peeler*"

In recent years, engineers and doctors have taken a great interest in smectic and cholesteric liquid crystals. These crystals have unique thermophysical and physicochemical properties, they play an important role in biological systems, and they have extensive practical applications in technology and medicine. At the same time the large number of different modifications of such crystals (SLC A, B, C, G, H, etc., levo- and dextrorotary SLC and "blue" phases) and the corresponding phase transitions between different mesophases make it very difficult to study them. The relatively complicated nature of the intermolecular interaction is also of great significance.

At the same time SLC and CLC encompass interesting and instructive objects, whose study enriches and deepens our understanding of the nature of the condensed state.[1-8]

CHAPTER V.

Structural Characteristics of Molecules and the Structure and Classification of Smectic and Cholesteric Liquid Crystals

First, there's the room you can see through the glass— that's just the same as our drawing-room, only the things go the other way. . . . Well then, the books are something like our books, only the words go the wrong way; I know that, *because I've held up one of our books to the glass, and then they hold up one in the other room.*

Lewis Carroll, *Through the Looking Glass*

5.1. STRUCTURE AND PROPERTIES OF MOLECULES

Though the relationship of many details of molecular structure to the ability to make a transition into a liquid-crystalline phase is not yet clear, it is known that liquid crystals are made up of elongated (rod- or plate-like) molecules. In addition, various features of fragments of molecule or its shape are characteristic of different types of liquid crystals. For example in many cases, the presence of a rigid molecular center and soft (flexible) end chains are characteristic of SLC.[7] Examples are ethyl ester of *p*-azoxybenzoic acid and *p*-*p*′-dialkyloxyazoxybenzene, whose molecular shape is shown in Fig. 5.1.

Many liquid crystals are formed by aromatic compounds, primarily, *para*-substituted benzene rings. As the number of connected benzene rings increases, the ability to form liquid crystals usually increases also. Compounds containing naphthalene rings exhibit a lower ability to form mesophases. Liquid-crystalline substances are also observed among aliphatic compounds in some cases.

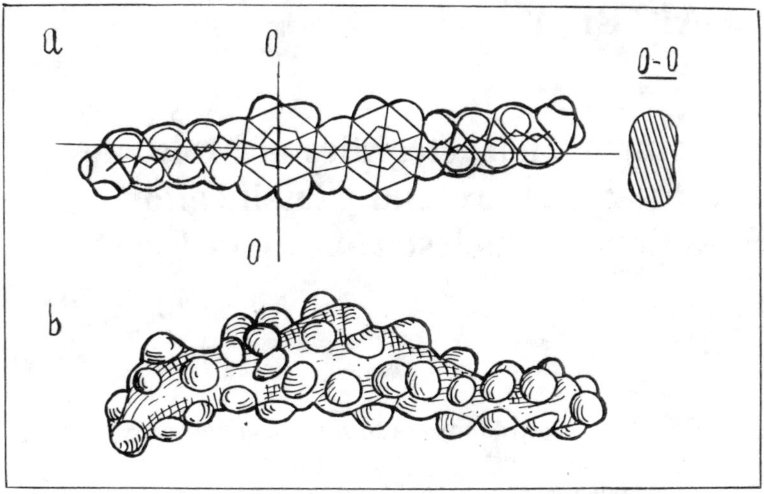

Figure 5.1. Models of molecules of a smectic liquid crystal (a: *n,n'*-dialkoxybenzene) and of a cholesteric liquid crystal (b: cholesteryl nonanoate).[9]

Cholesteryl esters, substances containing the cyclopentanone-perhydrophenanthrene group and a long side chain, are among the important cholesteric liquid crystals. The system of rings can be viewed as a rigid formation to which relatively flexible tails, namely a saturated chain and radicals, are connected.[7] In the first approximation the molecule is similar to a slightly twisted rod and differs from its mirror image (Ref. 1 and Fig. 5.1).

Molecular models developed to explain observed properties of liquid crystals or to predict these properties are usually based on the characteristics of the molecular structure. However, there is often no clear correlation between the molecular characteristics and the properties of the smectic and cholesteric liquid crystals (for example, between the direction of rotation of the CLC helix and the shape of individual molecules[10]). In addition, in many cases a relationship has been found experimentally between the transition temperatures and the number of carbon atoms (or molecular mass) in the homologs (see Sec. 8.1).

The dipole moments of liquid crystals usually have values up to $(1.0\text{–}1.2) \cdot 10^{-29}$ C·m.[11] Though no direct correlation has been observed between the molecular dipole moment and the ability of the substance to form liquid crystals,[12] some current theories explain the appearance of smectic mesophases and a cholesteric structure in terms of the characteristics of electron density distribution and the corresponding nature of the intermolecular interaction.[4–6]

It has been established that liquid-crystalline mesophases are formed by molecules that exhibit an anisotropy of polarizability.[10] Here, transition temperatures increase with polarizability. Longer terminal substituents distinguishes SLC from NLC. For this reason in many homologous series the lowest terms usually exhibit the properties of NLC, the middle terms exhibit the properties of SLC and NLC, and the higher terms exhibit the properties of SLC. The constant dipole moments and variations in the shape of a molecule along its length play a definite role in the formation of the smectic ordering. A detailed discussion of the relation between the structures of molecules with mesogenous properties is given in Ref. 10.

Table 5.1 presents structural formulas, temperatures, and, in some cases, the heats of the phase transitions for sufficiently studied SLC and CLC. The thermophysical properties of many of these liquid crystals are examined in the present review.* In addition to those presented in Table 5.1, the following SLC and CLC are of interest and have been investigated:

ethyl-*p*-anisalaminocinnamate (EAAC)[15]:

$$\mathrm{SC} \xrightarrow{380} \mathrm{SLCA} \xleftrightarrow{390} \mathrm{NLC} \xleftrightarrow{411} \mathrm{IL}$$

$$\mathrm{SLCB} \xleftarrow{365} \!\!\rfloor$$

4-cyanobenzylidene-4′-*n*-octyloxyaniline (CBOOA)[16]:

$$\mathrm{SC} \underset{38.1}{\overset{346}{\longleftrightarrow}} \mathrm{SLCA} \overset{356}{\longleftrightarrow} \mathrm{NLC} \underset{1.1}{\overset{380}{\longleftrightarrow}} \mathrm{IL}$$

4,4′-di-*n*-heptyloxyazoxybenzene (HEPTOAB)[16]:

$$\mathrm{SC} \underset{40.9}{\overset{348}{\longleftrightarrow}} \mathrm{SLCC} \underset{1.59}{\overset{369}{\longleftrightarrow}} \mathrm{NLC} \underset{1.02}{\overset{397}{\longleftrightarrow}} \mathrm{IL}$$

---

*In Table 5.1 and in the following text, the numbers above the arrows indicate the transition temperatures in *K*; the numbers below the arrows indicate the specific heats of the transitions in *J*/*g*.

Table 5.1

Structure Formulae and Heats of Phase Transitions of Studied Smectic and Cholesteric Liquid Crystals

Substance	Structure Formulae	Temperatures (above arrows) and Heates (below arrows) of the Phase Transitions	Reference
Ethyl ester of *n*-azoxy-benzoic acid	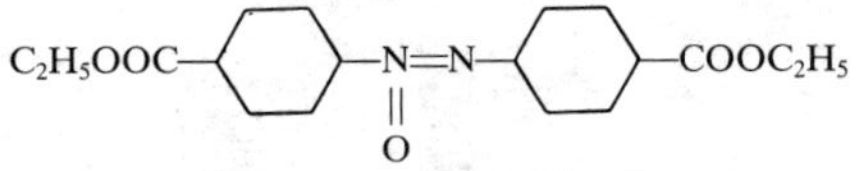	$SC \xleftrightarrow{387} SLC \xleftrightarrow{393} IL$	11
4-*n*-nitrophenyl ester of 4-*n*-oxybenzoic acid	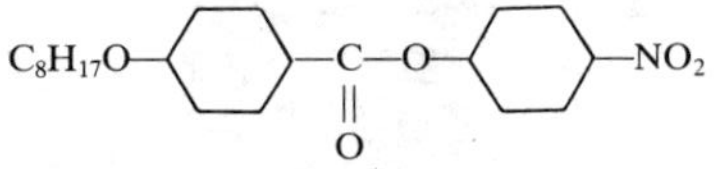	$SC \xleftrightarrow{321} SLC \xleftrightarrow{333} NLC \xleftrightarrow{340} IL$	13
n-heptyloxybenzoic acid (HOBA)	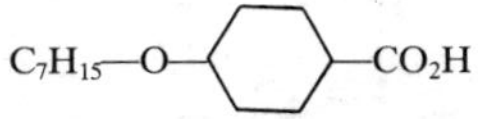	$SC \xleftrightarrow{365} SLC \xleftrightarrow{371} NLC \xleftrightarrow{419} IL$	14
Amyl ester of dodecycloxy-benzal-*n*-amino-cinnamatic acid	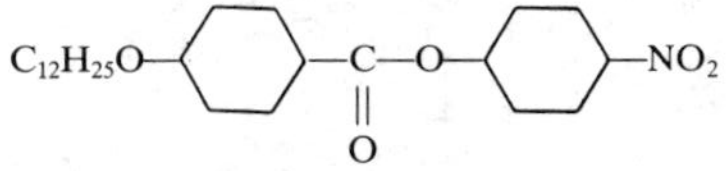	$SC \xleftrightarrow{348} SLCB \xleftrightarrow{370} SLCC \xleftrightarrow{380} SLCA \xleftrightarrow{406} IL$	11
Terephthal-bis-*n*-butylaniline (TBBA)	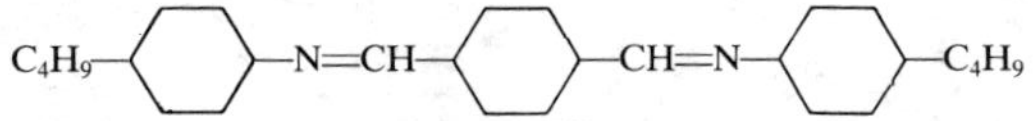	$SC \xleftrightarrow[18.17]{386} SLCB \xleftrightarrow[3.76]{417} SLCC \xleftrightarrow{445} SLCA \xleftrightarrow[0.29]{473} NLC \xleftrightarrow[0.75]{509} IL$	7, 11
Amyl ester of *n*-phenyl-benzylidene-*n*′-amino-cinnamatic acid	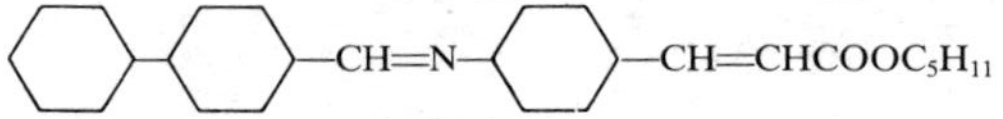	$SC \xleftrightarrow{365} SLCE \xleftrightarrow{375} SLCB \xleftrightarrow{441} SLCA \xleftrightarrow{477} IL$	10
2,5-di(*n*-butylphenyl)-pirazine	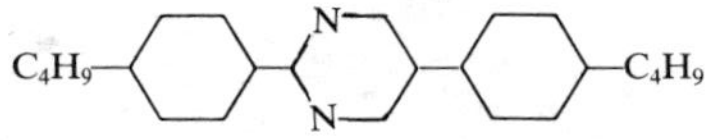	$SC \xleftrightarrow{435} SLCC \xleftrightarrow{440} NLC \xleftrightarrow{455} IL$	10

D-(or *L*-)2-methylbutyl ester of *n*-cetyloxy-benzylidene-*n'*-amino-cinnamatic acid	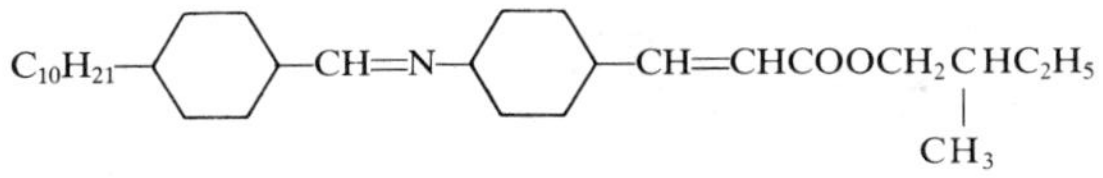	$SC \xleftrightarrow{349} SLCC(X) \xleftrightarrow{368} SLCA \xleftrightarrow{390} IL$ $NLC(X) \xleftarrow{336}$	10
3-nitro-4'-hexadecyloxy-4-biphenylcarbonic acid	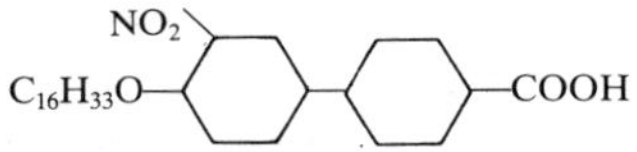	$SC \xleftrightarrow{400} SLCC \xleftrightarrow{444} SLCD \xleftrightarrow{470} SLCA \xleftrightarrow{475} IL$	10
Dipropyl ester of *n*-terphenyl-4,4'-dicarbonic acid	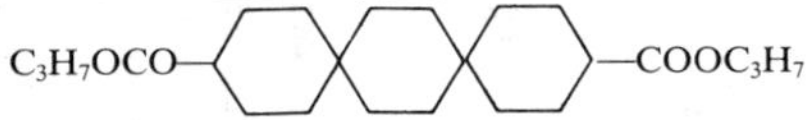	$SC \xleftrightarrow{395} SLCE \xleftrightarrow{410} SLC \xleftrightarrow{512} IL$	10
2-*n*-amylphenyl-5-*n*-amyloxyphenylpyridine	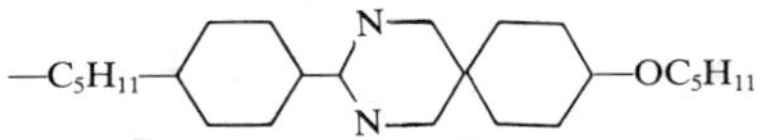	$SC \xleftrightarrow{352} SLCG \xleftrightarrow{376} SLCF \xleftrightarrow{387} SLCC \xleftrightarrow{417} SLCA \xleftrightarrow{483} IL$	10
Amyl-*n*-(4-cyanobenzylidene-amino)-cinnamate	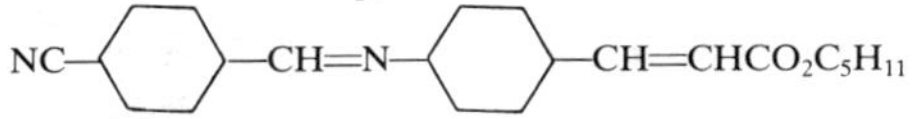	$SC \xleftrightarrow{368} CLC \xleftrightarrow{378} IL$	11
Cholesteryl erucate (CE)	$CH_3(CH_2)_7CH{=}CH(CH_2)_{11}OCO—$	$SC \xleftrightarrow{299} CLC \xleftrightarrow{314} IL$	10
D-(or *L*-)methoxy-benzyliden-*n'*-(2-methyl-butyl)-aniline	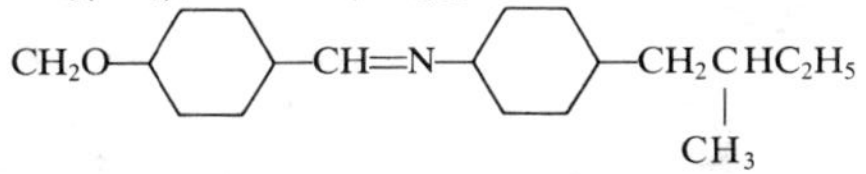	$SC \xleftrightarrow{294} NLC \xleftrightarrow{297} IL$	10

Cholesteryl caprylate[11]:

$$\text{SC} \xleftrightarrow{355} \text{CLC} \longleftrightarrow \text{IL:}$$
$$\text{CLC} \xrightarrow{351} \text{SLC} \xrightarrow{326} \text{SC}$$

Cholesteryl nonanoate (CN)[16]:

$$\text{SC} \xrightarrow{352} \text{CLC} \underset{0.66}{\overset{364}{\longleftrightarrow}} \text{IL}$$
$$\text{CLC} \underset{0.25}{\overset{349}{\longrightarrow}} \text{SLC} \longrightarrow \text{SC}$$

The problems of the chemistry of promising but inadequately studied liquid crystals (including smectic and cholesteric) are examined by G. G. Maidachenko *et al.*[17] Synthesis and results of studies of cholesteryl cinnamates are discussed by V. G. Tishchenko *et al.*[18] The molecular structure and polymorphism of steroid derivatives that represent most of the cholesteric liquid crystals of practical importance are examined by V. G. Tishchenko and R. M. Cherkashina[19]; the molecular structures and properties of a large group of SLC are studied by Gray and Goodby.[20] Detailed information is presented by Gray and McDonnell in Ref. 21 on the synthesis and regions of existence of the liquid-crystalline state of chiral alkylcyanobiphenyls, *p*-terphenyls, and some other chiral compounds, namely, biphenyl derivatives.

5.2. CLASSIFICATION AND STRUCTURAL FEATURES OF SMECTIC LIQUID CRYSTALS

As is well known, NLC exhibit long-range orientational order (existence of a preferred molecular orientation characterized by the axis of nematic order and by a unit vector, the director) in the absence of long-range translational order. In contrast, SLC exhibit a combination of long-range orientational and translational order in the positioning of the centers of mass of the particles. Smectic liquid crystals characteristically exhibit the presence of molecular layers (the centers of molecules in the layers lie in a single plane, while their long axes are perpendicular or tilted relative to this plane). The distance between these layers corresponds to or is shorter than the length of a molecule (Fig. 5.2).

We shall examine the structural characteristics of the basic types of SLC (Table 5.2). In smectic-A (SLC-A), the preferred direction of the long molecules axes (orientation of the director) is perpendic-

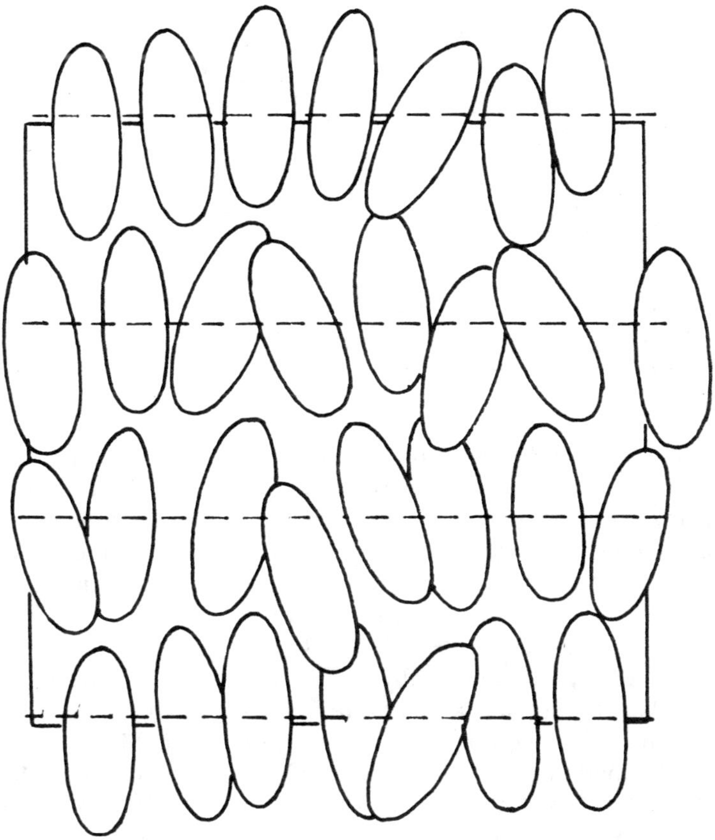

Figure 5.2. Characteristic instantaneous configuration of smectic-A molecules according to the results of a molecular-dynamics experiment.[22]

ular to the planes in which the centers of mass of the molecules are principally situated. There is no long-range translational order in the positioning of the centers of mass of the molecules in each smectic layer and only a short-range order which is typical for liquids occurs. Thus the arrangement of the centers of the molecules within each smectic layer corresponds to a two-dimensional liquid. Long-range translational order is observed in the positioning of the smectic layers. Usually the thickness of these layers is close to the length of the molecules.

Smectic-A liquid crystals are optically uniaxial systems in which the optic axis is perpendicular to the plane of the layers. The $+z$ and

Table 5.2
Molecular Order of Liquid-Crystalline Smectic Phases According to McMillan[3–5]

Characteristic Property	Type of Liquid-Crystalline Phase or Isotropic Liquid (IL)					
	IL	NLC	SLC-A	SLC-B	SLC-C	SLC-H
Parallel long molecular axes	−	+	+	+	+	+
Presence of smectic molecular layers	−	−	+	+	+	+
Orientationally ordered dipoles	−	−	−	−	+	+
Presence of two dimensional lattice	−	−	−	+	−	+

$-z$ directions (the z axis is perpendicular to the plane of the layers) are equivalent. Smectic-A forms are possible in which the long molecular axes are tilted to the plane of the layer (for example the axes rotate around a cone). There is, however, no correlation between the orientations of the long axes of molecules situated in different smectic planes, and this leads to an average uniaxial structure (in contrast to the biaxial tilted smectic-C liquid crystals, see below). Smectic-A molecules can usually rotate around the long axis and move in the plane of the smectic layer. Motion between layers is inhibited.

The smectic-A mesophase is the least ordered and is observed at higher temperatures than other types of SLC. For this reason, SLC-A–NLC, SLC-A–CLC, or SLC-A–IL transitions can occur with heating. Smectic-A liquid crystals are a common type of SLC (see Table 5.1).

Smectic-B liquid crystals exhibit not only long-range orientational order of the long molecular axes and long-range translational order in the positioning of the smectic layers, but also long-range translational order in the positioning of the centers of mass of molecules within a smectic layer. Thus the positioning of the centers of molecules in such a layer corresponds to a two-dimensional solid. Two-dimensional hexagonal order (the centers of molecules are situated at the sites of a hexagonal lattice) is observed in each smectic layer of TBBA (see Table 5.1).

The preferred orientation of the long axes of smectic-A and -B molecules is perpendicular to the planes of the smectic layers. The thickness of each smectic layer corresponds approximately to the length of a molecule.

The difference on a microscopic level between a smectic-B liquid crystal and a solid can be explained using specific models. For example according to the early Meyer-McMillan theory,[5] the difference between a smectic-B liquid crystal and a solid lies in the disordered orientation of the molecular dipoles while the molecules retain the freedom to rotate around their long axes. Differing interpretations of the structure of smectic-B liquid crystals have been proposed. For example, the possibility of the combined existence of long-range orientational and short-range positional order is discussed in Ref. 23. On the basing of x-ray scattering experiments it is shown that smectic-B liquid crystals have a three-dimensional long-range order.[24] The interlayer interaction energy is small in this case. Examples of smectic-B liquid crystals are EEDOC and TBBA (see Table 5.1).

The arrangement of the centers of the molecules in each molecular layer of a smectic-C liquid crystal corresponds to a two-dimensional liquid (presence of short-range translational order in the absence of long-range order). The preferred orientation of the long axes of the molecules is tilted by an angle β to the normal layers. For this reason, the thickness $l\cos\beta$ of each smectic layer is less than the length of a molecule l. A smectic-C liquid crystal is an optically biaxial system. According to some theoretical representations,[4] the inability of molecules to rotate around their long axes is an important feature of smectic-C liquid crystals. Examples of substances that form smectic-C liquid crystals are presented in Table 5.1.

In some cases a special modification of a smectic-C liquid crystal is observed, namely, the chiral smectic-C mesophase [SLC-C(Ch)]. The long molecular axes are tilted to the normal to the plane of the smectic layer while the direction of the preferred orientation changes from layer to layer and a helical structure is formed with some pitch. These compounds usually exhibit optical activity. An example is D-(or L-) 2-methyl butyl ester of *n*-cetyloxybenzylidene-*n*′-aminocinnamic acid (see Table 5.1).

Smectic-D liquid crystals are characterized by a quasicubic order and there is no layered structure. This type of SLC is optically isotropic or has a mosaic structure and has been observed for only a few compounds, which have not been adequately studied. An example is 3′-nitro-4′-hexadecyloxy-4-biphenyl carboxylic acid (see Table 5.1).

Three-dimensional order is observed in the smectic-E mesophase where the centers of the molecules form an orthorhombic lattice.[10] Free rotation of molecules around their long axes is absent. Examples are amyl ester of *n*-phenylbenzylidene-*n*′-aminocinnamic acid and dipropyl ester of *n*-terphenyl-4,4′-dicarboxylic acid (see Table 5.1).

Smectic-F and -G liquid crystals are characterized by the presence of a layered structure. Another characteristic feature of smectic-G liquid crystals is the ordered positioning of the centers of molecules in the smectic layers. Many structural characteristics, particle dynamics, and physicochemical properties of these yet unstudied systems are not understood. An example is 2-*n*-amylphenyl-5-*n*-amyloxyphenylpiridine (see Table 5.1). A series of data related to the structural characteristics of these mesophases and a discussion (sometimes of a controversial nature) are presented in the Proceedings of the International Conference on Liquid Crystals (Bordeaux, 1978).*

A layered structure is observed in smectic-H liquid crystals. The centers of the molecules form a hexagonal lattice in the plane of the layer (analogous to smectic-B liquid crystals). The director is inclined relative to the normal to the plane of the layer. According to some theoretical representations,[5] the dipole moments are oriented in this mesophase. Smectic-H liquid crystals are optically biaxial. In this case if the mesophase is formed by optically active molecules, then a chiral smectic-H mesophase [SLC-H(Ch)] is observed.

The most common SLC are usually classified by the properties indicated in Table 5.3. Some modifications of SLC are known (for example, modifications I and II[36] that differ by type of texture and dimensions of regions of molecular ordering), as well as double-layer smectic systems.[37]

While discussing the structural features of the SLC one can of course be interested in defects of some extent. For example, G. S. Ranganath studied defects in smectic liquid crystals: disclinations (defects in director field), dislocations (defects in the field of displacements of smectic layers) and dispirations (mixed defects).[38] In this work the energy of interactions between these defects was also calculated.

5.3. STRUCTURE OF CHOLESTERIC LIQUID CRYSTALS

The long axes of molecules in CLC are approximately oriented parallel to some plane in space. If the cholesteric specimen is imagined as separated in layers parallel to this plane (whose thickness corre-

*See J. de Physique Coll., *40*, N C3 (1979).

Table 5.3
Classification of Smectic Liquid Crystals[25–35]

N	Name	Recommended Mark[27]	Brief Description	Reference
1.	Smectic-A (SLC-A)	S_A	Long molecular axes are perpendicular to layer planes. In each layer particle arrangement is disordered (correctly only short-range "liquid" order in absence of long-range order is observed).	28, 29
2.	Smectic-B (SLC-B)	S_B	Long axes of molecules are perpendicular to layer planes. Ordered arrangements of molecules in layer (two-dimensional hexagonal lattice)* is observed.	28, 29
3.	Smectic-C (SLC-C)	S_C	Long molecular axes are inclined to layer planes. Disordered arrangement of molecules in each layer occurs.**	28, 29
4.	Smectic-D (SLC-D)	S_D	Molecular order within each layer is not determined (apparently three-dimensional cubic lattice micelle arrangement occurs).	29–31
5.	Smectic-E (SLC-E)	S_E	Long molecular axes are perpendicular (they may be inclined) to layer planes. Within each layer ordered arrangement of molecules is observed (three-dimensional orthorhombic lattice).***	29, 31–33
6.	Smectic-F (SLC-F)	S_F	Molecular orientation is not determined. Order of arrangement of molecules in each layer is absent.	34
			Other data indicate, that molecules are inclined to layer planes and form quasi-hexagonal lattice.	27
7.	Smectic-G (SLC-G)	S_G	Molecular arrangement in each layer is ordered. Three-dimensional structure is not determined.	34
			Other data indicate, that molecules are inclined to layer planes and form orthorhombic lattice.	27
8.	Smectic-H (SLC-H)	S_H	Long molecular axes are inclined to layer planes. Ordered arrangement of molecules in each layer is observed (three-dimensional monoclinic lattice, according to other data-hexagonal packing).	35

*Optically uniaxial system
**Optically biaxial system
***System can be optically uniaxial or biaxial

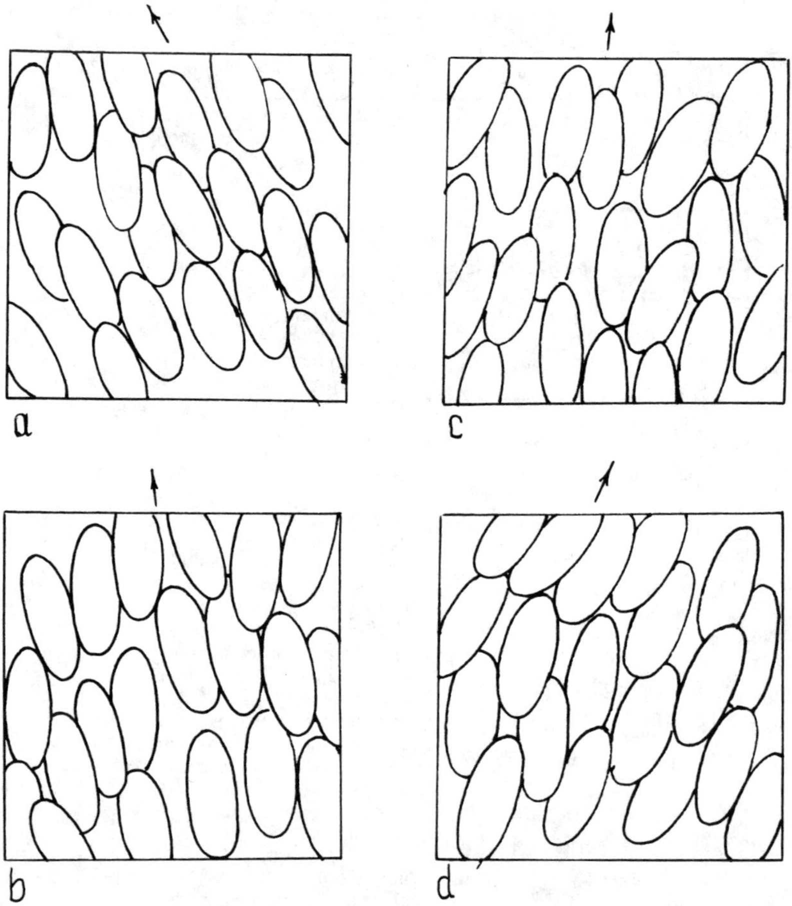

Figure 5.3. The typical instantaneous configuration of the molecules of a CLC in layers, into which the main specimen is arbitrarily separated (according to the results of molecular-dynamic modeling): a) first layer; b) third layer; c) fifth layer; d) seventh layer. The arrows indicate the orientation of the director of the layer.

sponds approximately to the thickness of a molecule), then the arrangement of the molecules within each layer is similar to the arrangement in NLC. As in NLC, there is no long-range translational order in the positioning of the centers of mass of the molecules. There is, however, a long-range orientational order (preferred orientation of the long axes of molecules, situated in an arbitrary layer). The orientation of **r** changes from layer to layer and the director rotates so that a helix typical of CLC is formed (Fig. 5.3).

If the orientation of the axis of the helix corresponds to the orientation of the z axis, then the components of the vector **n** will be as follows:

$$n_x = \cos(q_0 z + \phi);\ n_y = \sin(q_0 z + \phi);\ n_z = 0,$$

where the orientation of the x axis and the value of q_0 are arbitrary. Such a cholesteric structure is periodic with respect to z, while the directions **n** and **-n** are equivalent. A typical value of the spatial period is 3000–4000 Å, i.e., of the order of the wavelength of light in the visible range and much greater than the characteristic size of a molecule. Bragg scattering of light is usually observed in this connection.

The pitch of the helix of a CLC depends on the nature of the molecules and external parameters (temperature, pressure, etc.). For this reason, the wavelength of the reflected light depends, in particular, on temperature. As a result, the color of the substance changes considerably with a small change in temperature. This phenomenon is used in medical thermography and in technology. Cholesteric liquid crystals also have an enormous ability to rotate the plane of polarization (up to 10^5 deg/mm) compared with the rotation characteristic of organic liquids ($3 \cdot 10^2$ deg/mm).

Cholesteric liquid crystals are usually formed by optically active, chiral molecules. Both dextro- and levorotary helices are observed. Some characteristic CLC are presented in Table 5.1.

Some optically active molecules with structures analogous to molecules in NLC form so-called chiral NLC [NLC(Ch)]. An example is D- (or L-) n-methoxybenzylidene-n'-(2-methylbutyl)-aniline[10] (see Table 5.1).

CHAPTER VI.

Orientational and Translational Order in Smectic and Cholesteric Liquid Crystals and Dynamics of Molecules

Sing not, O wind, your fearful song
Of chaos, for the hungry spirit,
Into night's world of shadows flung
Exults in it and strains to hear it.
The bounds of mortal flesh 'twould fly
And merge with boundless ocean sweeping.
Take heed! Let slumbering tempests lie:
Beneath them chaos stirs unsleeping.

F. Tyutchev. Translated by I. Zheleznova

6.1. DISTRIBUTION OF THE CENTERS OF MASS OF MOLECULES

The centers of mass of the molecules in NLC and CLC are distributed in space in a manner similar to their distribution in an isotropic liquid. Molecular layers are observed in SLC. The usual unary and binary functions, often used for homogeneous liquids and systems located near an interface, can be used for the distribution of the centers of mass of the molecules.

The unary distribution function of a NLC is nearly a constant independent of the direction along which the local number density of particles is calculated. This is demonstrated in Fig. 6.1 (curve 4). In a smectic-A liquid crystal, distinct peaks are observed along the director that indicate a local concentration of centers of mass and, in turn, the presence of molecular smectic layers (Fig. 6.1, curve 3).

The determination of the binary distribution function in an ima-

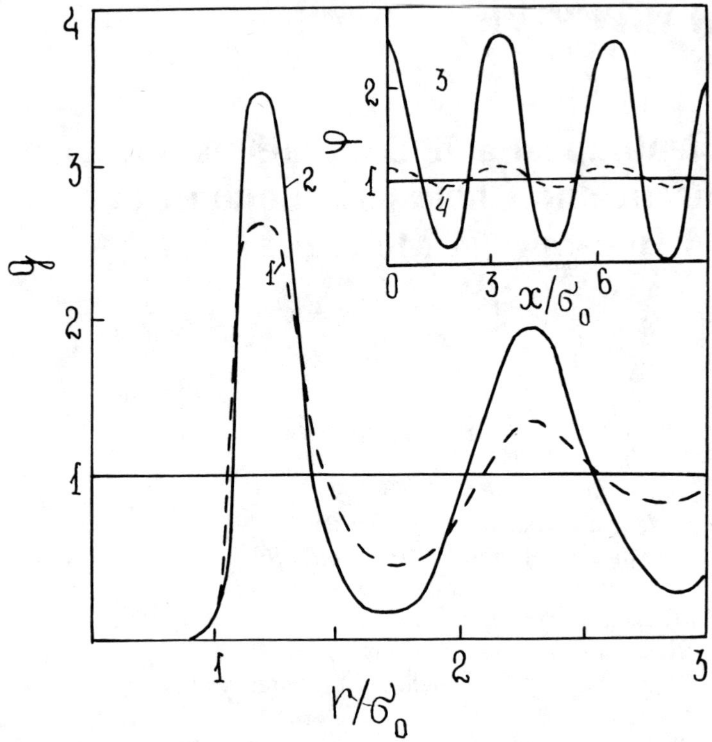

Figure 6.1. Binary distribution functions of the projections of the centers of mass of molecules in a smectic layer on a plane parallel to the layer (1 is for SLC–A and 2 is for SLC–B) and unary density distribution functions of the centers of mass of molecules in SLC–A (3) and NLC (4) along the director (x axis), obtained by the method of molecular dynamics.[22,39]

ginary layer of a NLC perpendicular to the director leads to a typical result for an isotropic liquid (Ref. 1, Fig. 5.4): a distinct first maximum indicating the presence of short-range translational order and rapidly decreasing, small subsequent maxima indicating the absence of long-range translational ordering. A similar situation is also observed in SLC-A when determining the binary distribution function in a smectic molecular layer (see Fig. 6.1, curve 1).

The characteristic form of the binary distribution function for a smectic layer in SLC-B (see Fig. 6.1, curve 2) contrasts with that of SLC-A: the first peak is higher and subsequent peaks are more distinct and damp out more slowly. This indicates much greater translational order in the arrangement of the centers of mass of molecules in the smectic layers of SLC-B and the presence of long-range translational order.

It may often be useful to introduce ellipsoidal binary distribution functions, which are obtained by counting the centers of particles located in the volumes of elementary layers of ellipsoids. The ratio between the lengths of the long and short axes corresponds to the shape of the molecule in the liquid crystal.* This approach makes it possible to represent graphically the spatial distribution of the centers of molecules and to give a natural definition of the important statistical characteristics, for example the average coordination number.

6.2. ORIENTATIONAL AND TRANSLATIONAL ORDERING IN SMECTIC LIQUID CRYSTALS

A detailed discussion of the method using the orientational distribution functions and expressions for parameters of orientational order in NLC to describe orientational ordering in NLC is given in Ref. 1. We shall examine an extension of this approach applied to SLC, restricting our attention to the relatively simple case of SLC-A.

According to the above presentation, a smectic liquid crystal is a system that exhibits both long-range orientational order (the presence of a preferred orientation of the long axes of molecules) and long-range translational order (ordered arrangement of smectic molecular layers). The molecular distribution function must describe the tendencies of the molecules in the smectic liquid crystal to be oriented along the director **n** and to form smectic layers perpendicular (in the case of SLC-A) to **n**. Using an idea of Kirkwood,[40,41] who examined ordering in ordinary melting crystals, Wojtowicz[42] proposed a distribution function in the form of a double series depending on cos θ and z (θ is the angle between the long axis of a molecule and the director, while the z axis is parallel to the director **n**):

$$f(\cos\theta, z) = \sum_{L=0}\sum_{n=0} A_{Ln}P_L(\cos\theta)\cos(2\pi nz/d), \qquad (6.1)$$

*It is assumed that in this case the molecules rotate freely around their long axis and their shape can be represented to a good approximation by an ellipsoid of revolution.

where d is the distance between the smectic layers. The normalization condition has the form:

$$\int_{-1}^{1} \int_{0}^{d} f(\cos \theta, z)\, dz\, d(\cos \theta) = 1.$$

Due to the periodicity in the positioning of the smectic layers, the inner integration is carried out over a single period.

The coefficients A_{Ln} can be determined by multiplying the left and right sides of Eq. (6.1) by

$$P_k(\cos \theta) \cos(2\pi mz/d),$$

and integrating, taking into account the following definition of an average value:

$$\langle X \rangle \equiv \int_{-1}^{1} \int_{0}^{d} X f(\cos \theta, z)\, dz d(\cos \theta).$$

As a result, we obtain

$$A_{Ln} = (2L + 1) \langle P_L (\cos \theta) \cos (2\pi nz/d)\rangle/2d.$$

It is evident that in addition to the usual orientational order parameters (function of $\cos \theta$) and translational order parameters (functions of $2\pi nz/d$), there arise mixed order parameters that depend on both arguments. These mixed parameters describe the correlation between degrees of orientational and translational order.

It is shown in Ref. 1 that to satisfactorily describe the orientational distribution function of NLC it may be necessary to take into account a number of order parameters. An analogous situation occurs for SLC. A good approximation of the distribution function $f(\cos \theta, z)$ must consider a large number of terms on the right side of Eq. (6.1) and a correspondingly large number of order parameters.[42] Only the low-order order parameters, determined using existing molecular theories, have found practical application:

orientational order parameter

$$\eta \equiv \langle P_2 (\cos \theta)\rangle;$$

translational order parameter

$$\tau \equiv \langle \cos (2 \pi z/d)\rangle;$$

mixed order parameter

$$\sigma \equiv \langle P_2 (\cos \theta) \cos (2\pi z/d)\rangle.$$

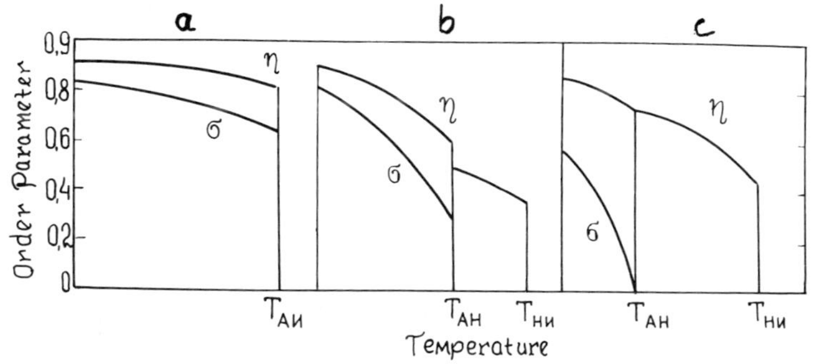

Figure 6.2. Orientational and mixed order parameters[28] according to McMillan's theory.[3] The following parameters of the theory are used: $\delta = 0$ and $\alpha = 1.1$ (a), 0.85 (b), and 0.6 (c). The transition temperatures are as follows: T_{AI} is for SLC–A-isotropic liquid; T_{AN} is for SLC–A–NLC; and, T_{NI} is for NLC-isotropic liquid.

An important task of the molecular theory of SLC is to correctly predict the values and temperature dependences of the order parameters. Figure 6.2 shows the temperature dependences of the orientational order η and mixed order σ parameters obtained using McMillan's theory.[3] The change in these parameters at the SLC-A–isotropic liquid, SLC-A–NLC, and NLC–isotropic-liquid phase transitions is shown for different values of the parameters of the theory.

Meyer and McMillan[5] introduced a different definition of translational and orientational order parameters in order to characterize the order in SLC-B, SLC-C, and SLC-H (see Fig. 7.6).

The tilt angle of the preferred orientation of the long molecular axes (director) is an important characteristic of SLC-C. This quantity was experimentally determined by an optical method[43] (Fig. 6.3) and by x-ray diffraction analysis.[44]

A special analysis of the characteristics of smectic layers of 4-cyanobenzylidene-4′-*n*-octyloxyaniline (CBOOA), based on the results obtained from studies of the scattering of x-rays and the volume behavior, was performed by Guillon *et al.*[45] Figure 6.4 shows the changes in the thicknesses of the smectic layers found in this study. Guillon *et al.* note that upon cooling the SLC-A can be observed in a supercooled state up to 58°C (although as the temperature is increased, this phase appears only at 73°C). Jump-like changes in the thicknesses of the layers at temperatures of 58°C and 40°C (Fig. 6.4) are characteristic. To explain this behavior and also the change in

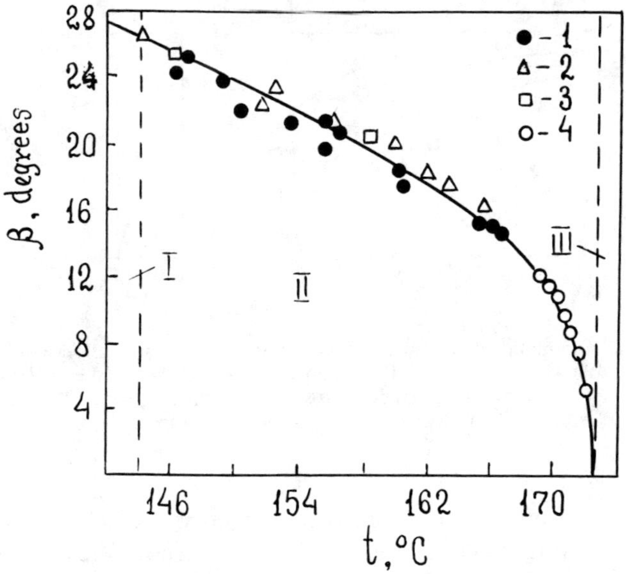

Figure 6.3. The tilt angle of the long axes of molecules in TBBA as a function of temperature according to data in Ref. 43: 1) experimental data at a wavelength of 450 nm; 2) 525 nm; 3) 625 nm; 4) interferences pattern. I, II, and III correspond to SLC–B, SLC–C, and SLC–A, respectively.

the thickness of the SLC layer at 70°C (Fig. 6.4), Guillon *et al.* proposed a double-layer model, which takes into account the inclination of the SBOOA molecules with respect to the plane of the layers.[45]

The structure of smectic phases induced in the NLC by pressure is of great interest. Thus a new interpretation of x-ray data for these compounds was proposed.[46] It was concluded that the SLC-A layers in a number of nitroesters are formed by two types of molecules: short and long (for nitroesters, these two lengths correspond to the length of the antiparallel associated pairs and the length of a single molecule). This leads to the formation of layers with an average interplanar distance. An increase in the pressure leads to the appearance and increase in the number of associates.

Ordering, orientation and specific features of the molecular motion in smectic F phases of PBHA were studied by the spin label method.[47]

The increase of the order parameter from 0.795 to 0.833 at the transition from low temperature F-phase to B-phase is explained by the authors by the change of molecular orientation (the tilt of the long molecular axis in F-phase is nearly 26° and the tilt in B-phase is absent). Order parameters and molecular orientations in PBHA (smectic G,F,B,C,A-phases and nematic phase) are discussed. The separate study of ordering effects of the aromatic skeleton and long aliphatic end chains in smectic F and J phases of TBDA made it possible to estimate the influence of the intramolecular structure on smectic ordering.[48]

Precise measurements of polarization (P) and tilt-angle (θ) near the smectic A–chiral smectic C transition in DOBAMBC made it possible to find that the ratio of P/θ remained nearly constant at $T_c - T > 2$K and fell abruptly near the temperature of the transition T_c.[49] Molecular ordering and molecular reorientation process were studied in the chiral smectic phase of ferroelectric liquid crystals.[50]

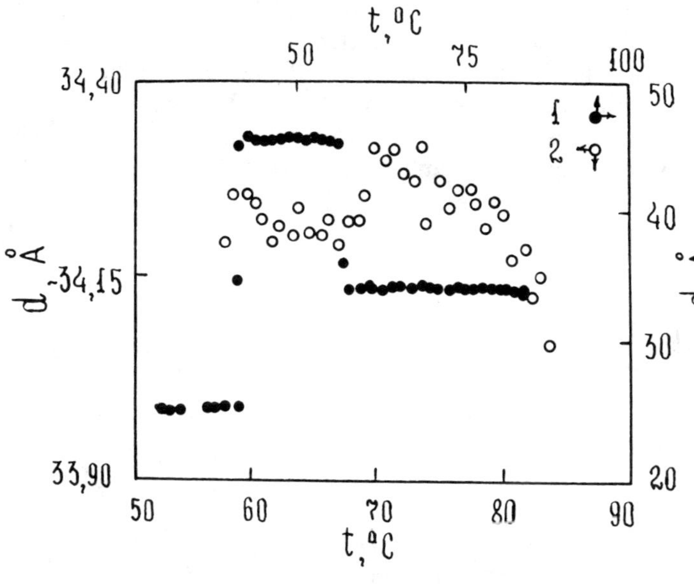

Figure 6.4. The change in the thickness of the layers accompanying a decrease of the temperature in 4-cyanobenzylidene-4′-n-octyloxyaniline (CBOOA)[45]: 1) for a wide temperature range, 2) for the smectic-A phase.

6.3. PITCH OF THE CHOLESTERIC HELIX AND ORDERING OF MOLECULES IN CHOLESTERIC LIQUID CRYSTALS

One of the most important characteristics of the structure of CLC is the pitch of the cholesteric helix. This quantity can significantly depend on temperature, pressure, external fields, and impurities contained in the substance.[7] The temperature dependence of the pitch of the helix is important for this dependence is what permits the use of CLC for medical and technical thermography as well as to convert infrared to visible images.

The pitch h of the helix decreases with increasing temperature for most CLC (derivatives of cholesterol). The characteristic dependences are shown in Figs. 6.5 and 6.6. Here, the quantity

$$\frac{1}{h}\left|\frac{dh}{dT}\right|$$

can vary over a wide range from 10^{-3} to 10^{2} K^{-1}. Using appropriate CLC, this permits determining temperature with high accuracy (to hundredths and thousandths of a degree) from the frequency of selectively reflected light (the wavelength of visible light for Bragg reflection from a helical structure equals the pitch of the helix multiplied by the index of refraction). Although the nature of this behavior of $h(T)$ is not completely clear, an increase in pitch with a decrease in temperature could be related to establishment of short-

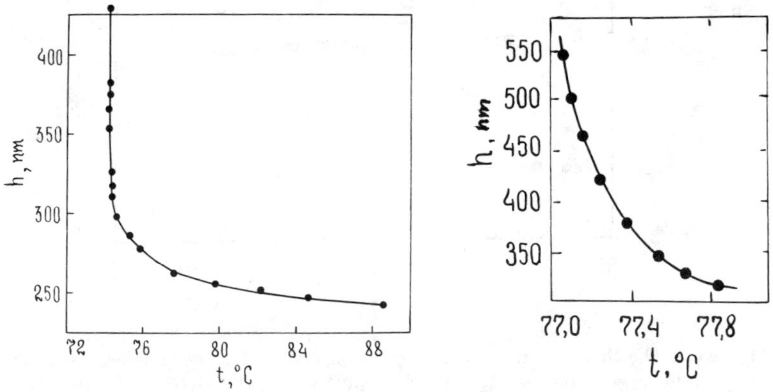

Figure 6.5. The temperature dependences of the pitch of the helix of cholesteryl nonanoate[51] (a) and cholesteryl pelargonate[52] (b).

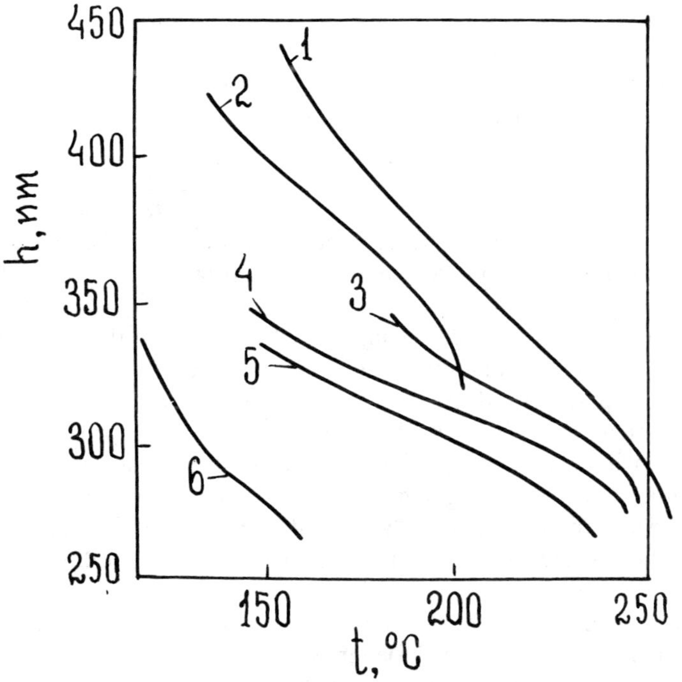

Figure 6.6. The temperature dependences of the pitch of the cholesteric helix of CLC (Ref. 53): 1) *n*-nitrobenzoate; 2) cinnamate; 3) *n*-bromobenzoate; 4) chlorobenzoate; 5) *n*-methylbenzoate; 6) cholesterol *m*-nitrobenzoate.

range order of the smectic type.[7] We note that application of molecular-dynamic modeling based on a realistic molecular interaction potential also leads to a decrease in h with increasing T (Ref. 54) (see Sec. 7.5).

The temperature of dependence of the pitch is usually insignificant for CLC consisting of optically active molecules that do not belong to the class of cholesterol derivatives. Such substances are of greatest interest for electrooptical applications.[10]

Theoretical explanations of observed behavior of $h(T)$ encounter certain difficulties. According to Wulf's model,[55] pitch should be independent of temperature, while Goossens' widely used theory[6] predicts only a weak increase in pitch near the CLC–isotropic-liquid transition. Keating[56,57] first proposed an acceptable description of the observed behavior of $h(T)$ of many CLC. It is assumed in Keating's theory that molecules in the CLC execute a continuous vibrational motion that is anharmonic due to asymmetry (chirality) of the molecules. The solution of the equation of motion of a molecule leads to a nonvanishing angle relative to the twisting of the neighboring quasinematic layers. In the first approximation

$$h = \gamma[1 + \beta/(T - T_c)],$$

where γ, β, and T_c are constants, the appropriate choice of which usually yields acceptable agreement with experiment. Bottcher[58] developed this theory by considering the approximate effect of long-range order.

De Gennes[59] proposed, using an analogy between the CLC–SLC-A and normal-metal–superconductor phase transition, that the increase in pitch with decreasing temperature is due to pretransitional phenomena near the transition temperature. This approach yielded the dependence $h \sim (T - T_c)^{-0,66}$, that well describes many experimental data. P. V. Vigman and V. M. Filev,[60] analyzing the fluctuations in the experimentally studied temperature range, proposed a temperature dependence of the pitch that does not follow a power law.

A molecular–statistical theory of the temperature dependence of the pitch of CLC was developed in Refs. 61 and 62 based on a model proposed for the interaction of chiral molecules. In the molecular field approximation it was established that the pitch for molecules with cylindrical symmetry does not depend on temperature. It is necessary to consider higher order terms that account for the asymmetry of the potential energy in order to describe the temperature dependence of the pitch.[62] Direct calculations can be performed and compared with the experimental dependences in some cases. The temperature dependence of the pitch is calculated explicitly for cholesteryl decanoate in Ref. 63.

The pitch of the helix can increase with increasing temperature (for example, in cholesteryl-2-(ethoxyethoxy)-ethyl-carbonate). Reasons considered for the increase in pitch with temperature included rotation of molecules around the long axis,[55] anharmonic vibrations of molecules that give rise to an effect competing with twisting,[64] and increase in the spread of molecules relative to the axis of preferred orientation.[65,66] Thus, it is concluded in Ref. 66 that under certain conditions pitch can increase with temperature 0.4% per degree, taking into account the spread of molecules relative to the axis of preferred orientation in the mean molecular-field approximation. This and the rotation of molecules around the long axis cannot, however, explain the increase in pitch with temperature observed in cholesteryl-2-(ethoxyethoxy)-ethyl-carbonate ($\simeq$4% per degree[67]). Probable reasons for this (compensation of right-handed twisting by anharmonic vibrations, related to the asymmetric component of the dispersion interaction) are indicated in Ref. 61.

The temperature dependence of the pitch of the cholesteric helix is studied in Ref. 68 using the general form of intermolecular potential containing chiral components. It was shown that the temperature dependence of the pitch can be expressed in terms of the order parameters. The dependences $h(T)$ including increasing and/or decreasing dependences derived from the theory are discussed. Satisfactory agreement with experiment was obtained in the case of h increasing with temperature (the cholesteric liquid crystal CEEC, Fig. 6.7).

The pressure dependence of the pitch of the helix in cholesteryl oleyl-carbonate was theoretically calculated in Ref. 69 and satisfactory agreement was achieved with previously obtained experimental data.[70] Accordingly, when pressure is varied from 0.1 to 200 MPa, the wavelength of the maximum selective reflection by the CLC can vary over the entire visible range of the spectrum.

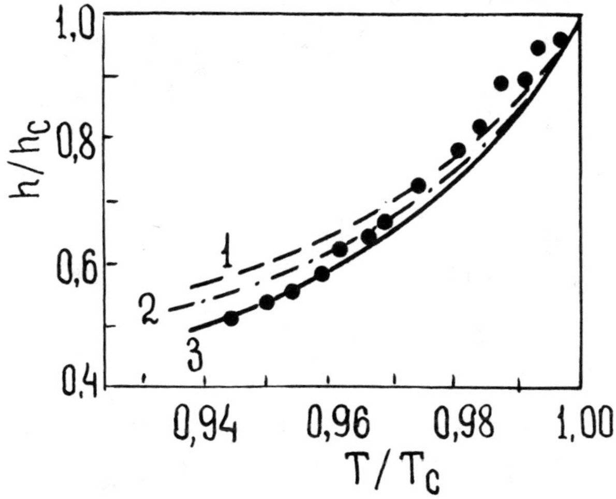

Figure 6.7. The temperature dependence of the pitch of the cholesteric helix of CEEC[68]: the dark circles indicate the experimental data[67]; curves 1, 2, and 3 show the results of a calculation[68] for different parameters of the theory.

The effect of impurities on the pitch of the cholesteric helix was discussed in Refs. 68 and 71–74. Interesting effects were discovered that have not yet been explained theoretically: formation of cholesteric structures by adding impurities to NLC; obtaining nematocholesteric mixtures with the same sign of rotation by adding cholesteric components with different signs of rotation to the NLC; and obtaining levorotary mixtures from two dextrorotary components. However, a discussion of these problems falls outside the scope of this review.

It was assumed above that the ideal cholesteric helix was formed by ordered quasinematic layers. In reality, however, there is a certain degree of molecular order in these layers that depends on temperature, pressure, etc. The helical structure of CLC can also break down to some extent.

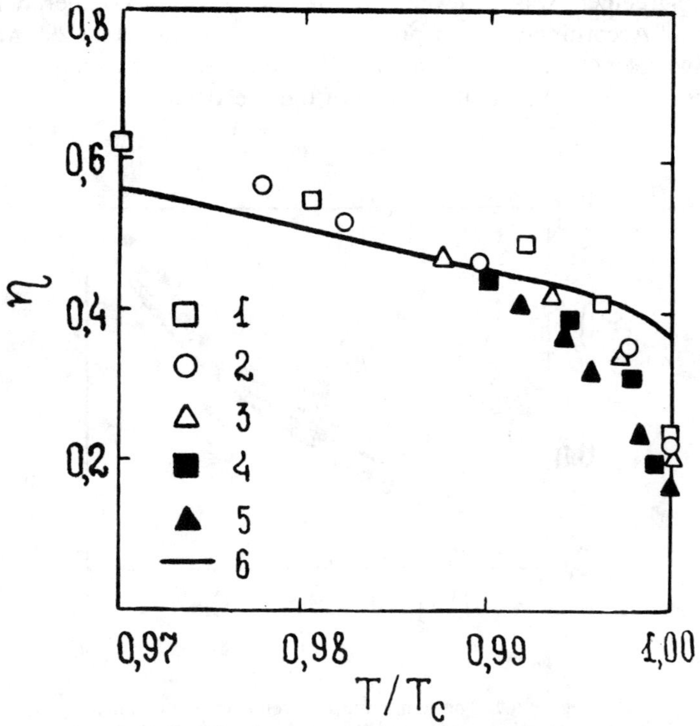

Figure 6.8. The temperature dependence of the orientational order parameter of a quasinematic layer $\eta \equiv \langle P_2 \rangle$ in a CLC (Ref. 78): 1) cholesteryl pelargonate; 2) caprinate; 3) laurate; 4) myristate; 5) palmitate; 6) Maier-Saupe theory.

The degree of orientational order of CLC can be described in the first approximation by using an orientational order parameter defined within a quasinematic layer. Unfortunately, most experimental methods used for determining $\eta \equiv \langle P_2 \rangle$ in NLC are not suitable for CLC, and there are few experimental data on this parameter for CLC. Some values were obtained from an analysis of double refraction measurements,[75] as well as from NMR measurements.[76,77] A. V. Tolmachev, V. G. Tishchenko, and L. N. Lisetskii[78] proposed a method for determining $\langle P_2 \rangle$ in CLC layers by analyzing data from an investigation of the circular dichroism (selective transmission) of a planar texture.[79]

The results for $\langle P_2 \rangle$ determined within a layer of some CLC are presented in Fig. 6.8. These results, on the whole, agree satisfactorily with results of the Maier–Saupe theory. The somewhat lower values of $\langle P_2 \rangle$, compared with the theoretical values near the temperature of transition into an isotropic liquid, are explained by imperfections in the planar texture that increase as the phase transition point is approached.[79]

It is noteworthy that as the length of the chains of cholesteryl alkanoate increases, the decrease in $\langle P_2 \rangle$ becomes still more distinct. The minimum in the angular dependence of the interaction energy becomes diffuse as a result of the conformation of the alkyl chain and leads to additional disordering.[78,80–82] As is well known, at temperatures below the temperature of bleaching the so-called "blue phase" (BP), whose name originates from the specific bluish-violet coloring, can be observed in cholesteric liquid crystals.[83] It has been established with the help of thermodynamic and optical studies that this phase is stable in a narrow temperature interval (of the order of 0.5°.[84–86] The most accurate results for cholesteryl myristate and cholesteryl nonanoate were obtained in Refs. 87–89. It was demonstrated that polymorphism (two blue phases BP1 and BP2) is observed. For example, the sequence of transitions accompanying a decrease of the temperature has the following form for cholesteryl nonanoate[87,88]:

$$CLC \xrightarrow[17]{364.25} BP1 \xrightarrow{364.55} BP2 \xrightarrow[0.53]{364.70} IL,$$

where the transition temperatures (K) are shown in the numerators and the corresponding enthalpy changes (kJ/mole) are shown in the denominators. The enthalpy changes accompany the CLC–BP and BP–IL transitions for cholesteryl myristate are equal to 34 and 1.1

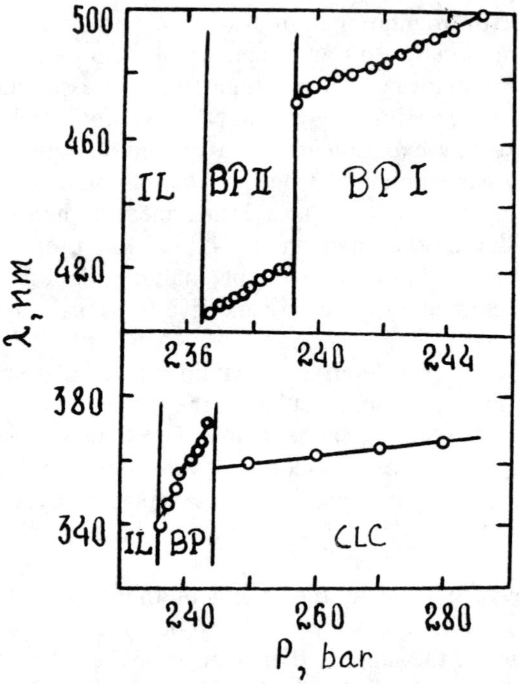

Figure 6.9. Dependence of the wavelength of maximum intensity of reflected light λ at constant temperature (102°C) for cholesteryl nonanoate[89]: IL is the isotropic liquid; BP is the blue phase (BPI and BPII); and, CLC is the cholesteric phase.

kJ/mole, respectively. The selective reflection of light was also studied in Ref. 89 and an interesting dependence of the wavelength on the pressure with constant temperature was observed (Fig. 6.9).

Structural features of the blue phases as well as the optical activity, transmission and reflection spectrums were thoroughly studied during recent years.[90–95] The effect of an electric field on the selective reflection bands of liquid crystalline blue phases was also studied.[96]

CHAPTER VII.

Computational-Theoretical Investigations of the Thermophysical Properties, Structure, and Molecular Dynamics of Smectic and Cholesteric Liquid Crystals

'The Gods of the earth and sea
Sought thro' Nature to find this tree;
But their search was all in vain:
There grows one in the Human brain.

William Blake, "The Human Abstract"

7.1. THE KOBAYASHI–McMILLAN THEORY OF THE PROPERTIES OF SMECTIC-A LIQUID CRYSTALS

Structural features of smectic-A liquid crystals are related to the nature of the intermolecular interactions that lead to parallel arrangement of molecules and the formation of parallel layers. A simple form for the pair interaction potential was proposed by Kobayashi[2]:

$$V_{12} = u(r) + w(r)P_2(\cos\theta_{12}), \qquad (7.1)$$

where r is the distance between the centers of mass of the molecules; θ is the angle between their long axes; $u(r)$ describes the short range central forces; and $w(r)$ describes the orientational interaction (in particular anisotropic dispersion interaction).

Since an exact solution of the statistical problem using the potential (7.1) is too complicated, the mean field approximation was used. For this, it is necessary to average V_{12} with respect to the positions and orientations of the second molecule according to the CLC-A structure. Thus, we use the distribution function $f(\cos\theta, z)$ examined in Chap. 6. McMillan[3] used a particular form of the model interaction potential

$$V_{12} = \delta w(r) + w(r)\, P_2(\cos\theta_{12}), \tag{7.2}$$

where

$$w(r) = -\frac{v}{r_o^3\pi^{3/2}} \exp\left[-(r/r_o)^2\right];$$

and v and δ are constants that characterize the values of the components of the interaction energy, and r_o is a constant that characterizes the range of the central forces (which usually corresponds to the length of the molecules).* Then, the following expression is valid for the energy of molecules in the mean molecular field:

$$V(\cos\theta, z) = -v\{\delta\alpha\tau\cos(2\pi z/d) + [\eta + \alpha\sigma\cos(2\pi z/d)]P_2(\cos\theta)\}, \tag{7.3}$$

η, τ, and σ are the orientational, translational, and mixed order parameters examined in Chap. 6. The parameter α depends on the ratio r_0/d:

$$\alpha = 2\exp\left[-(\pi r_o/d)^2\right].$$

For the given homologous series, $r_0 \simeq$ const and the quantity α characterizes length of a chain.[42,82]

According to the definitions of the order parameters, we can write down the following system of equations of consistency:

$$\eta = \int_0^1\int_0^d P_2(\cos\theta)f(\cos\theta, z)\, dz\, d(\cos\theta);$$

$$\tau = \int_o^1\int_o^d \cos(2\pi z/d)f(\cos\theta, z)\, dz\, d(\cos\theta); \tag{7.4}$$

$$\alpha = \int_o^1\int_o^d P_2(\cos\theta)\cos(2\pi z/d)f(\cos\theta, z)\, dz\, d(\cos\theta),$$

*In addition to the very simple form of the potential presented above, a more complicated variant of the potential is also examined in Ref. 3.

where

$$f(\cos\theta,z) = \frac{\exp[-V(\cos\theta, z)/kT]}{\int_0^1\int_0^d \exp[-V(\cos\theta, z)/kT]\, dz d(\cos\theta)}, \quad (7.5)$$

It is possible to obtain the temperature dependences of the order parameters $\eta(T)$, $\tau(T)$, and $\sigma(T)$ by solving system (7.4). Solutions are then obtained corresponding to a minimum of the free energy per molecule

$$F = \frac{1}{2}v(\eta^2 + \alpha\delta\tau^2 + \alpha\sigma^2) - kT\ln\int_0^1\int_0^d \exp[-V \times (\cos\theta, z)/kT]\, dz d(\cos\theta) \quad (7.6)$$

The temperature dependences $\eta(T)$ and $\sigma(T)$ are shown in Fig. 6.2, and the computed and experimental dependences of the Bragg scattering intensities (qualitatively corresponding to the parameter τ) are compared in Fig. 7.1.

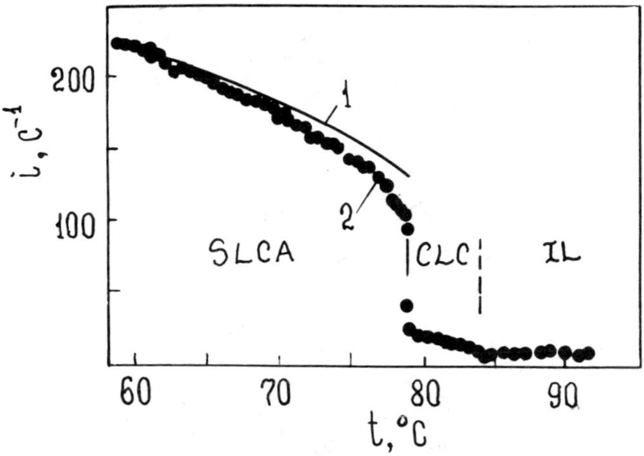

Figure 7.1. The temperature dependence of the Bragg scattering intensity of x-rays[16]: the dark circles indicate the experimental data for cholesteryl myristate; 1) calculation using the simple potential in Ref. 3; 2) calculation using the refined potential in Ref. 3.

It is possible to determine the phase diagram of the system knowing the nature of the variation of the order parameters. For SLC-A–NLC or SLC–CLC transitions the parameters τ and σ vanish, and for NLC–IL, CLC–IL, and SLC-A–IL transitions the parameter η vanishes as well. A variant of procedure for constructing phase diagrams is proposed in Ref. 97. Figure 7.2 compares the phase diagrams

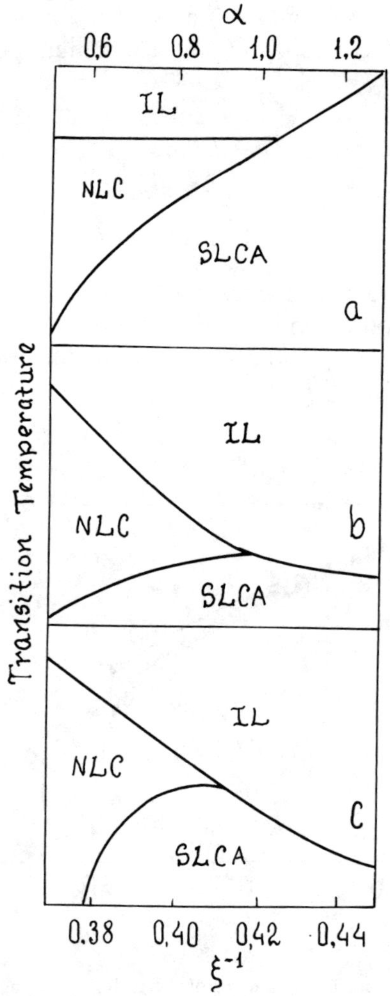

Figure 7.2. Comparison of phase diagrams[42]: a) McMillan's theory[3]; b) theory of Lee et al.[97]; c) typical experimental diagram; ζ is a parameter that is proportional to the length of the chain.

obtained from the methods in Refs. 3 and 97 with a typical experimental diagram. The values of the heat capacity and entropy computed with the theories in Refs. 3 and 97 are compared with the experimental values in Fig. 7.3. Qualitative agreement is evident.

The Kobayashi–McMillan theory has serious defects. Use of an approximate self-consistent mean molecular field neglects the effects of short range order and fluctuations of the order parameters.[3] The theory must be further complicated to account for these, and additional adjustable parameters are required that decrease the value of such an approach, as happens in the case of NLC.[1] Thus, this theory suffers from the same basic shortcomings as the Maier-Saupe theory.[1] In this approximation, the molecular potential in the form (7.3) does not exhibit rotational and translational invariance, which makes the theory difficult to substantiate.[82] Pretransitional phenomena are also not taken into account in the molecular-field approximation although an attempt to account for them was undertaken for the orientational phase transition.[99] Another disadvantage is the very simplified form of the intermolecular potential, and the relationship of this to mo-

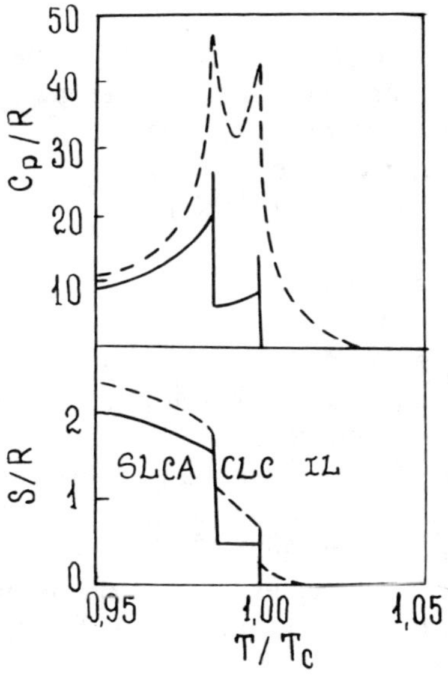

Figure 7.3. Comparison of theoretical and experimental values of the heat capacity and entropy[3]; the continuous line shows the theoretical values (α = 0.95)[3]; the dashed line shows the experimental data for cholesteryl myristate.[98]

lecular structure and properties remains unclear. It is assumed in this approximation that the molecules are rigid so that the effect of internal degrees of freedom on phase transitions is not taken into account.

The Kobayashi–McMillan method was further developed in Ref. 100 where phase diagrams of the same type as in the theory[3] were obtained.

7.2. THE McMILLAN, WULF, AND de GENNES THEORIES OF THE PROPERTIES OF SMECTIC-C LIQUID CRYSTALS AND THE MEYER-McMILLAN THEORY OF THE PROPERTIES OF SMECTIC-C, -B, AND -H LIQUID CRYSTALS

A molecular theory of SLC-C using a simple molecular model and the mean molecular-field approximation was proposed by McMillan.[4] The rotation of molecules around the long axis in SLC-C was assumed to be absent in contrast to SLC-A where this rotation occurs. As McMillan showed, this leads to tilting of the director to the plane of the smectic layers.

A two-dimensional model (a single smectic molecular layer) is studied in this theory. Each molecule is rod-shaped with two or three electric dipole moments rigidly coupled to the molecule along its central axis. Usually molecules in liquid crystals have two or three dipole moments associated with the oxygen and nitrogen atoms in their structure.

The configuration shown in Fig. 7.4 is observed in the smectic-C mesophase when rotation of molecules around the long axes is restrained. The long molecular axes are tilted relative to the direction of polarization by an angle β, that according to the results of this theory is found to be proportional to $(T_c - T)^{1/2}$. Experimental data[43] are not sufficiently reliable to determine the exponent exactly, though some indicate an exponent of about 1/3. Such a model predicts a SLC-A–SLC-C phase transition to be a second order transition (if its temperature does not coincide with the temperature of other mesomorphic transitions), which agrees with the measurements indicated in Ref. 43.

De Gennes[7,101] proposed an approach to smectic-C liquid crystals and SLC-C–SLC-A phase transitions based on the analogy to helium (i.e., the analogy to the Bose condensation in liquid helium) and using tilt angle as the order parameter as in Landau's theory. This

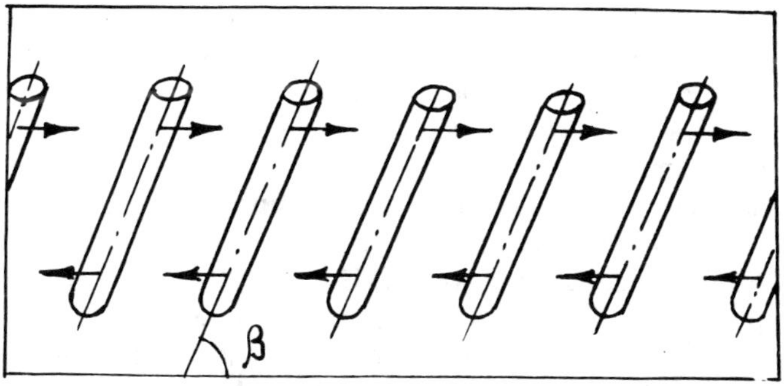

Figure 7.4. The molecular model of a smectic-C layer used in McMillan's theory[4] (the arrows show the orientation of the dipole moments).

approach led to the conclusion that the SLC-C–SLC-A transition must be continuous, while the specific heat capacity has a logarithmic singularity at the point T_{CA}. A temperature dependence of the tilt angle is predicted of the form

$$\beta \sim |T - T_{CA}|^{\alpha},$$

where $\alpha = 0.35$ (the experimental value of α is 0.40 ± 0.05 (Ref. 102).

Wulf[103,104] proposed a new model of the SLC-A and SLC-C in which molecules are assumed to be shaped like ellipsoids of revolution whose centers of mass move in the plane of the layer. Infinite repulsion (due to the rigid shape of the molecules) and dispersive attraction are active between molecules. A variational expression is used for the orientational distribution function of the molecules and the Onsager approximation is used for the configurational entropy. It is established that the transition out of the ordered state (in respect to the tilt angles of the molecules) into a disordered state is a second-order transition.

McMillan's theory of SLC-A was developed and extended to SLC-C, SLC-B, and SLC-H by Meyer and McMillan.[5] A single smectic molecular layer is examined in this method, and the molecules have an elongated rod-like shape with two dipoles placed on the molecular axis at some distance from the center of mass (see Fig. 7.4). The following form of the two-particle interaction is used:

$$u_{12} = \left[\frac{\boldsymbol{\mu}_1\boldsymbol{\mu}_2}{r_{12}^3} - \frac{(\boldsymbol{\mu}_1\mathbf{r}_{12})(\boldsymbol{\mu}_2\mathbf{r}_{12})}{r_{12}^5}\right] S(r_{12}) + T(r_{12}).$$

Here, μ is the dipole moment; $S(r_{12})$ is the two-particle correlation function (its special forms were examined in Ref. 5); $T(r_{12})$ is the repulsion potential, which plays a fundamental role in this theory. The statistical problem is solved within the self-consistent molecular-field approximation.

According to this theory, the phase transitions SLC-A–SLC-C and SLC-B–SLC-H are second-order transitions; the remaining transitions (SLC-C–SLC-H, SLC-A–SLC-H, and SLC-A–SLC-B) are first-order phase transitions (Fig. 7.5). Figure 7.6 presents the temperature dependences of the heat capacity, entropy, and order parameters of smectic mesophases according to the Meyer-McMillan theory.[5]

The computed values of the entropy changes for transitions SLC-A–SLC-B and SLC-C–SLC-H are lower than the experimental values. According to experiment, for ethyl ethoxybenzylidene-amino-cinnamate (EEBAC) $\Delta S_{AB} = 0.65R$, where R is the universal gas constant.[105] The experimental value of ΔS_{AB} for p,n-octyloxy-benzylidene-p'-toluidine (OBT) is 1.08 R. The computed value of ΔS_{AB} does not depend on the parameters of the theory and equals 0.56 R. The experimental value of ΔS_{CH} for TBBA equals to 1.4 R, while the computed value is 0.75–0.90 R, depending on the parameters of the model.

The Meyer-McMillan theory has serious shortcomings. The mean molecular-field approximation neglects order at small distances and

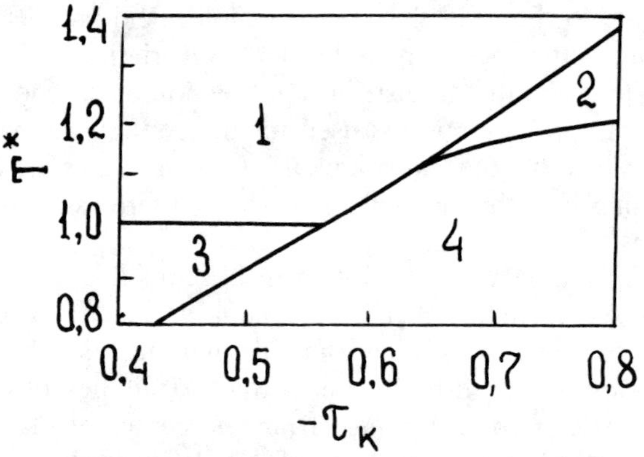

Figure 7.5. Phase diagram of the smectic mesophases according to the Meyer-McMillan model[5] (T^* is the reduced temperature and τ_k is a parameter in the theory): 1) SLC–A; 2) SLC–B; 3) SLC–C; and, 4) SLC–H.

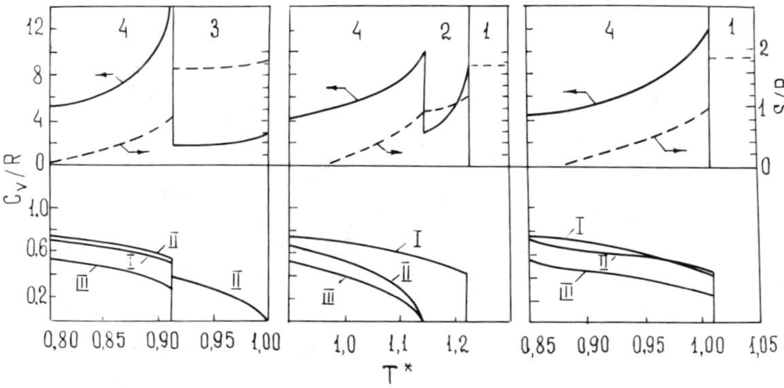

Figure 7.6. The heat capacity C_V, entropy S, and the order parameters as a function of the reduced temperature for smectic mesophases according to the Meyer-McMillan calculations[5]: 1) SLC–A; 2) SLC–B; 3) SLC–C; 4) SLC–H; I) two-dimensional translational order parameter α; II) orientational order parameter β; III) mixed order parameter γ.

ignores fluctuations of order parameters. Approximate and largely arbitrary models are used for the two-particle correlation function $S(r_{12})$. A hard core in the repulsion potential $T(r_{12})$ is neglected since it is difficult to incorporate it in this theory. The effect of internal molecular motions is also neglected, which could account for disagreement between the computed and experimental values of ΔS noted above.

In addition, the Meyer-McMillan theory is hypothetical regarding structural models. Thus, it has been shown that tilting of molecules in SLC-C does not freeze molecular rotation relative to the long axes of the molecules. Quasielastic neutron scattering in SLC-H for TBBA[106] has shown that the phenyl rings in this mesophase do not have a preferred orientation, and therefore the model in Ref. 5, which assumes that the dipoles in this mesophase are oriented, is not applicable to TBBA.

Of course, the model used by McMillan is not the only possible model for constructing a theory of SLC-C. For example, it is possible to imagine the simple molecular model of a smectic layer shown in Fig. 7.7. Increasing the width of the center of the molecule leads to a shorter distance between the long axes of neighboring molecules in a tilted structure. Such a structure is energetically more favorable (the energy of mutual attraction of molecules is higher in this case).

A molecular theory of pretransitional behaviour in smectogenic liquid crystal was developed.[107] In this work, a mean field theory with

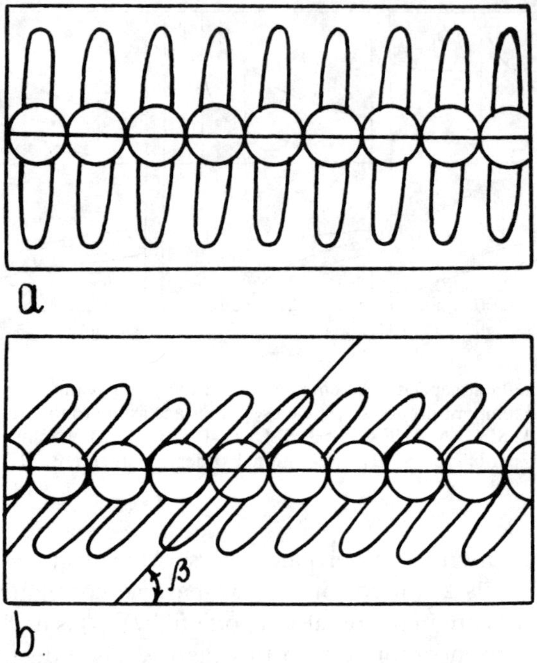

Figure 7.7. Simple model of the smectic layer in a SLC–C mesophase.

orientational, one-dimensional transitional and mixed order parameters was used. Results of this investigation made it possible to explain a large difference in the Kerr's constants of 6OCB (nematic-isotropic transition) and of 12OCB (smectic A-isotropic transition).

7.3. OTHER THEORIES OF SMECTIC LIQUID CRYSTALS

Among other trends in the theoretical study of SLC, it is necessary to note a simple lattice model of the system of rigid rods with one-dimensional translational periodicity and long-range orientational order.[108] This method was of great help in considering the role of the attractive and steric repulsive forces in stabilization of model analog of SLCA as well as in calculating thermodynamic characteristics of transitions between model analogs of smectic, nematic and isotropic phases.

Another simple model of SLCA was proposed on the basis of the notion of smectic as a system of rigid spherocylinders and a statistical problem was solved in the approximation of the self-consistent field.[109] Expression for calculating the equilibrium state between the smectic layers and order translational, orientational and mixed parameters were obtained. The role of the attraction was discussed and the results for the case when the ratio of the lengths of spherocylinders to their diameters equals two were considered. A steric molecular theory for multipole smectic A phases was proposed.[110] Each molecule was assumed to consist of **m** cubic fragments, one part of which belongs to the rigid core of the molecule, and another to self-flexible end chains. Two types of the layer packing were considered: A-phase, when the layer thickness L equals molecular length d, and phase A^1d, when $d < L <$ 2d. The configurational statistical sum for such a lattice system was calculated in the mean field approximation and a phase diagram for the system at different model parameters was drawn.

The Hubbard-Stratanovich transform was used for expressing the statistical sum of the system of uniaxial interacting molecules.[111] The calculation of this sum was performed in the mean field approximation. It enabled to obtain order parameters for the nematic, smectic A and smectic C phases and to consider possible types of phase diagrams. The free energy of the system was found by the expansion method of Landau-de Gennes.

A simple model of smectic A and smectic C liquid crystals[112] was based on the self-consistent field approximation and the account of the pair quadrupole-quadrupole interactions.

A simple alternative to the Mac-Millan theory is the molecular statistical theory of smectic A liquid crystals stated in papers.[113,114] By using the expansion of the internal energy by the orientational and translational order parameters the complete phase diagram was drawn which included the range of stability of the isotropic liquid, nematic, smectic A phases and plastic crystal. It was found that the reentrant behaviour could be described if the interaction parameter depended on the temperature. It was shown that the results of this theory agreed qualitatively with the results of the more complex Mac-Millan theory.

7.4. MOLECULAR THEORIES OF THE CHOLESTERIC MESOPHASE

The theory developed by Goossens is presently most widely used. This theory is based on representation of a CLC as a system of quasinematic layers, and the fundamental point of this theory is to examine the dispersion interaction energy of chiral (mirror- antisymmetric) molecules. This energy is determined using the quantum mechanical perturbation theory. Terms that correspond to the dipole–dipole (pp) and dipole–quadrupole (pq) interactions are taken into account in the multipole expansion:

$$V_{ij}^{pp} = \sum_{\nu,\mu} (p_\alpha^i | p_{\alpha'}^i)_\nu (p_\beta^j | p_{\beta'}^j)_\mu C_{\alpha\beta}^{ij} C_{\alpha'\beta'}^{ij} / (r^{ij})^6 E_{00\nu,\mu};$$

$$V^{pq} = \sum_{\nu,\mu} 3(p_\alpha^i | p_{\beta'}^i)_\nu (p_\beta^i | q_{\alpha'\gamma'}^j)_\mu C_{\alpha\beta}^{ij} D_{\alpha'\beta'\gamma'}^{ij} / 2(r^{ij})^7 E_{00\nu,\mu}{}^+$$

$$+ \sum_{\nu,\mu} 3(p_\alpha^i | p_{\alpha'}^i)_\nu (p_\beta^j | q_{\gamma\gamma}^j)_\mu C_{\alpha\beta}^{ij} (r_{\alpha'}^{ij} | r^{ij}) / 2(r^{ij})^7 E_{00\nu,\mu}{}^+ \ldots,$$

where

$$(p_\alpha^i p_\beta^j)_\nu = \langle \alpha | p_\alpha^i | \nu \rangle \langle \nu | p_\beta^j | 0 \rangle; \; p_\alpha \; -$$

is the electric dipole moment operator; $q_{\alpha\beta}$ is the electric quadrupole moment operator; and r^{ij} is the distance between molecules i and j.

These terms are found to be sufficient to describe the cholesteric twisting of the quasinematic layers. The relative angle of rotation and the pitch of the helix can be determined by minimizing energy. Intermolecular interaction energy is first averaged over all positions of the molecules in neighboring quasinematic layers.

Where molecules in neighboring layers are positioned along a single normal to these surfaces and for a small angle of rotation

$$V_{12} = -vP_2 [\cos(\theta - \theta_0)],$$

where P_2 is a Legendre polynomial. Thus, all the equations of the Maier-Saupe and Kabayashi—McMillan theories can be extended to cholesteric mesophases. Goossens' theory can be successfully applied not only to pure CLC but also to mixtures.[115–117]

The appearance of a cholesteric structure in a bulk specimen of a cholesteric mesophase can be explained following Böttcher.[58] A boundary layer of oriented molecules (a quasinematic layer) arises at the surface as a result of the force exerted by the bounding wall. The molecules in the next layer are located in the force field of the molecules in the boundary layer, which inhibits random thermal rotational fluctuations. As a result of the action of intermolecular interaction forces,[61,62] a finite twisting angle appears for the molecules in the second layer. The effect of the thickness of the CLC layer on the stability of the helical structure is examined in Refs.[22,118,119]

It is interesting to determine the effect of the rotation of molecules in a CLC around the long axis. In contrast to NLC, where this rotation is free, some positions of the short axis of a molecule in CLC are energetically more favorable. Wulf, who examined some discrete positions of molecules located at the sites of a cubic lattice, took this characteristic into account.[55]

Further development of the Goossens' method was performed in Refs. 61, 62, 64, and 68. Possible types of temperature dependences of the pitch of the cholesteric helix are discussed in detail and calculations are performed for specific cases of practical interest based on these models.

In addition to Goossens' theory and its modifications, other explanations also exist for the appearance of a cholesteric structure. Thus, E. I. Dzyaloshinskii, S. G. Dmitriev, and E. I. Kats[120] demonstrated the possibility of the formation of a CLC structure as the result of long range van der Waals forces independent of the chirality of the molecules. A. G. Khachaturyan[121] established the possibility of the existence of cholesteric structures formed by twisted quasi-nematic layers in which parallel positioning of the dipoles with a head-to-tail orientation of the molecules is observed.

Molecular theory of the cholesteric phase became the basis for the development of the theory of tilt and temperature dependence of the pitch in the chiral smectic C phase.[122] This phase has two specific features: the director is tilted with respect to the layer normal, and the phase is optically biaxial.[122,123] When the smectic C phase is composed of chiral molecules, the tilted director rotates around the layer normal, creating a helical structure like that in cholesterics.[124] Theoretical results obtained by W. J. A. Goossens[122] explain the experimental observations.

The features of the cholesteric phase were considered from theoretical point of view in papers 125,126 in which Landau's theory and Ginzburg-Landau's expression were used. In these papers great attention was paid to "blue phases." Detailed reviews devoted to these interesting varieties of cholesteric liquid crystals are given in papers 127, 128. Many features of the structure and properties of the blue phases were also discussed in papers 129–131.

7.5. MOLECULAR-DYNAMICS INVESTIGATION OF SMECTIC AND CHOLESTERIC LIQUID CRYSTALS

The molecular dynamics method was successfully applied to the investigation of NLC.[132–134] This method (discussed in Ref. 1) permits obtaining a rigorous solution of the statistical problem by integrating the system of equations of motion of molecules with a given intermolecular interaction law. It is possible to obtain detailed information about the structure of the substance, nature of the molecular motion, thermodynamic properties, and transport coefficients as a result of such molecular dynamics experiments.

We shall examine the application of this method to the study of smectic liquid crystals carried out by A. L. Tsykalo[135] and continued by A. D. Bagmet and A. L. Tsykalo.[136–138] The basic specimen was a rectangular parallelipiped and the orientation of its long edges coincided with the orientation of the director. The usual periodic boundary conditions were used. The basic specimen contained 168 particles that were shaped like elongated ellipsoids of revolution. As in the case of nematics, the intermolecular interaction potential included anisotropic repulsion and dispersion attraction. In contrast to nematic liquid crystals, however, the possibility (noted by a number of authors) was explored of representing the molecules of smectic liquid crystals as consisting of central parts (benzene rings), which strongly interact with one another (attractive interaction at large distances and repulsive interaction at short distances) and tails (chains), which inhibit the molecules from approaching one another by means of a comparatively soft repulsion.[7]

The form of the pair potential corresponded to the form of the generalized (12-6) Lennard—Jones potential with parameters dependent on the relative orientation of the interacting molecules. The nature of this dependence was proposed by Berne and Pechukas[139]:

$$\Phi_1 = \frac{4\epsilon_0\epsilon_1(\mathbf{u}_i,\mathbf{u}_j)}{\sigma_1^n(\mathbf{u}_i,\mathbf{u}_j,\hat{\mathbf{r}})}\left[\frac{\sigma_\perp\,\sigma_1(\mathbf{u}_i,\mathbf{u}_j,\hat{\mathbf{r}})}{r}\right]^{12} - \frac{4\epsilon_0\epsilon_2(\mathbf{u}_i,\mathbf{u}_j)}{\sigma_2^n(\mathbf{u}_i,\mathbf{u}_j,\hat{\mathbf{r}})}\left[\frac{\sigma_\perp\,\sigma_2(\mathbf{u}_i,\mathbf{u}_j,\hat{\mathbf{r}})}{r}\right]^{6}, \tag{7.7}$$

where

$$\epsilon_\kappa(\mathbf{u}_i,\mathbf{u}_j) = [1 - \chi_\kappa^2(\mathbf{u}_i\mathbf{u}_j)^2]^{-1/2};$$

$$\sigma_\kappa(\mathbf{u}_i,\mathbf{u}_j,\hat{\mathbf{r}}) = \left\{1 - \chi_\kappa\left[\frac{(\hat{\mathbf{r}}\,\mathbf{u}_i + \hat{\mathbf{r}}\,\mathbf{u}_j)^2}{1 + \chi_\kappa(\mathbf{u}_i\mathbf{u}_j)} + \frac{(\hat{\mathbf{r}}\,\mathbf{u}_i - \hat{\mathbf{r}}\,\mathbf{u}_j)^2}{1 - \chi_\kappa(\mathbf{u}_i\mathbf{u}_j)}\right]\right\}^{-1/2};$$

$$\chi_\kappa = \frac{\sigma_{\|\kappa}^2 - \sigma_{\perp\kappa}^2}{\sigma_{\|\kappa}^2 + \sigma_{\perp\kappa}^2};$$

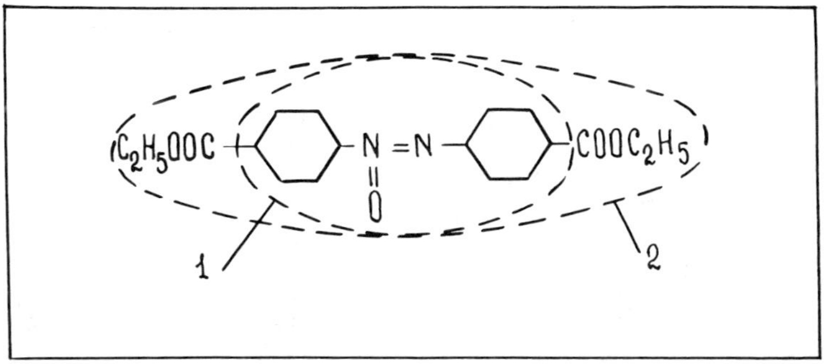

Figure 7.8. The schematic representation of the molecular model of a SLC used in molecular-dynamics modeling (ethyl esther *p*-azoxybenzoic acid): 1) inner ellipsoid; 2) outer ellipsoid.

$\kappa = 1$ or 2; $\sigma_\parallel$ and $\sigma_\perp$ are the lengths of the axes of the ellipsoid; $\hat{\mathbf{r}}$ is the distance between the centers of the particles; $\mathbf{u}_i$ and $\mathbf{u}_j$ are unit vectors oriented along the long axes of the ellipsoids; $\hat{\mathbf{r}}$ is a unit vector oriented along the line connecting the centers of the ellipsoids; χ is the anisotropy parameters; and ϵ_0 is the force parameter of the potential.

In the particular case $\sigma_1 = \sigma_2$ and $\epsilon_1 = \epsilon_2$, the potential (7.7) corresponds to one used earlier to study nematic liquid crystals.[133–134] For the smectic liquid crystals of interest here, $\sigma_1 \neq \sigma_2$. In this case, each of the two interacting molecules is a "molecule within a molecule." Large ellipsoids interact according to a power-law repulsion.* Each of the large ellipsoids contains a smaller ellipsoid and the smaller ellipsoids attract one another (Fig. 7.8). The first term in the potential (7.7), which includes the orientational dependence, accounts for the comparatively soft repulsion of the end chains of molecules and the more rigid repulsion of the molecular centers. Based on data for typical smectic structures, we used the following values of the degree of anisotropy: $\gamma = \sigma_\parallel/\sigma_\perp$; $\gamma_1 = 3.0$; and $\gamma_2 = 2.5$ (the lengths of the short axes $\sigma_\perp$ of the large and small ellipsoids are identical).

Figure 7.9 shows the dependences of the energy on the distance between the centers of molecules with different relative orientations.

*In general, the attraction of large ellipsoids can also be included. However, if the dimensions of the largest and smallest ellipsoids do not differ very much, then it is not necessary to do so.

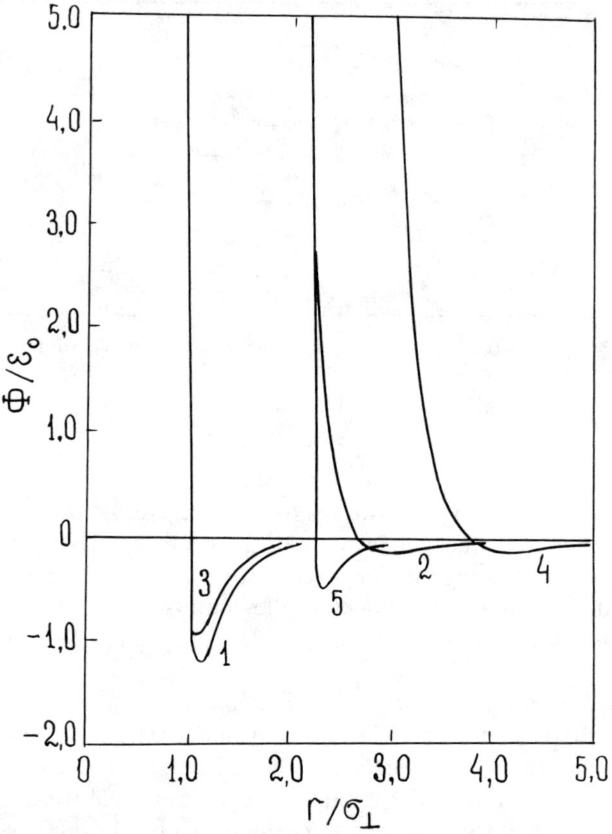

Figure 7.9. The energies of interaction of molecules, interacting according to the potential (7.7), as a function of the distance between their centers with different relative orientations: 1) parallel ||; 2) perpendicular |–; 3) crossed |·; 4) coaxial ––; 5) parallel displaced ||.

Analysis shows that an arrangement of molecules in layers corresponding to smectic order is energetically favored.

All calculations were performed in reduced form; the parameters of the pair potential as well as the mass of a molecule were used as scaling units. Two series of molecular-dynamics experiments were performed: one series was performed for the states corresponding to the smectic phase; a second series was performed for the states of the nematic phase. In the first experiments of each series, the completely ordered configuration was used as the starting point; for subsequent calculations, the initial conditions were taken from the data of preceding experiments. In determining basic properties, the function of dynamic variables were averaged after transition of the system into the equilibrium state was completed over an interval of one thousand time steps with the time step given by $\Delta t^* = 0.0075$.

In performing the molecular-dynamics experiment, pressure, energy, and basic structural characteristics (i.e., translational and orientational distribution functions, and order parameters examined in Sec. 7.1) were determined. The order parameters were as follows:

orientational

$$\eta_1 \equiv \langle P_2 (\cos \theta)\rangle = (1/2)(3 \langle \cos^2 \theta\rangle - 1);$$

$$\eta_2 \equiv \langle P_4 (\cos \theta)\rangle = (1/8)(35 \langle\cos^4 \theta\rangle - 30 \langle\cos^2 \theta\rangle + 3);$$

translational

$$\tau_1 \equiv \langle\cos (2\pi z/d)\rangle;\ \tau_2 \equiv \langle\cos (4 \pi z/d)\rangle$$

and mixed

$$\sigma \equiv \langle P_2 (\cos \theta) \cos (2\pi z/d)\rangle.$$

In addition, temporal correlation functions of linear and angular velocities and reorientation functions characterizing the particle dynamics were also determined.

Figure 7.10 shows the change in the order parameters η_1 and τ_1 with a transition from the initial nonequilibrium state into the equilibrium state for two cases. In the first, the density and temperature corresponded to the smectic mesophase ($\rho^* = 0.245$ and $T^* = 0.42$); in the second, they corresponded to the nematic phase (the temperature was higher ($T^* = 0.57$) while the density remained the same). It is evident that in the first case both order parameters change insignificantly, but in the second case the translational order parameter τ_1 decreases rapidly with a small change in the orientational order parameter η_1. This indicates that the system rapidly loses its layered character and transforms from the smectic (nonequilibrium under the given conditions) into the nematic mesophase. The small residual smectic order ($\tau_1 \simeq 0.12$, Table 7.1) in this case is apparently related to insufficient duration of the molecular-dynamics experiment. Thus the system in the equilibrium state belongs to the nematic or smectic mesophase and does not depend on the initial conditions (coordinates, orientations, and velocities of particles).

Table 7.1 presents a list of the states of the system that we studied.

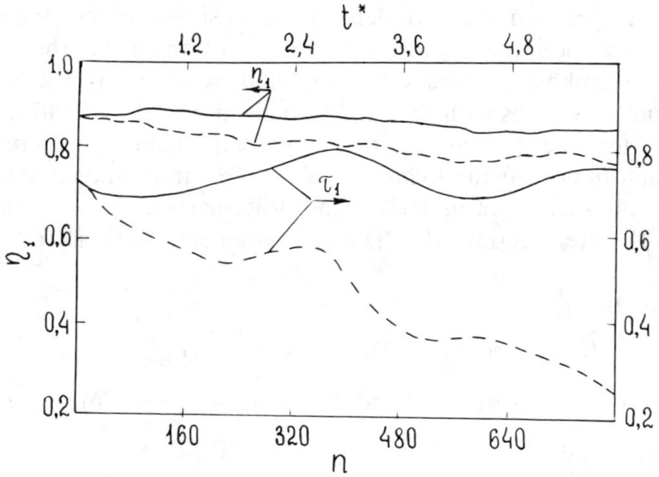

Figure 7.10. The change in the order parameters η_1 and τ_1 accompanying the transition of the system into an equilibrium state: the continuous line corresponds to the smectic phase and the dashed line corresponds to the nematic phase (t^* is the reduced time and n is the number of steps in the experiment).

It was established that the smectic mesophase is observed in the range of reduced densities $\rho^* = 0.236–0.255$ and temperatures $T^* = 0.38–0.44$. This is indicated both by the structural and transport characteristics, which agree reasonably well with known experimental values and with existing representations of the nature of the smectic state. The orientational order of the system was high, and the orientational order parameters exceeded values characteristic of the nematic phase. With an increase in temperature, transformation of the smectic system into the nematic mesophase was observed.

Table 7.1
Some Results of the Determination of the Order Parameters and Self-Diffusion Coefficients of Liquid Crystals Using Molecular Dynamics

Mesophase	T^*	ρ^*	η_1	η_2	τ_1
SLC	0.38	0.236	0.81	0.75	0.55
SLC	0.42	0.245	0.87	0.63	0.73
SLC	0.44	0.255	0.91	0.72	0.76
NLC	0.57	0.245	0.75	0.45	0.12
	τ_2	σ	$D^*_\parallel$	$D^*_\perp$	$D^*_\parallel/D^*_\perp$
SLC	0.12	0.50	0.029	0.037	0.78
SLC	0.29	0.64	0.006	0.011	0.54
SLC	0.33	0.69	0.012	0.009	1.30
NLC	0.00	0.10	0.110	0.025	4.4

The typical instantaneous particle configuration in the equilibrium state corresponding to the smectic liquid crystal is shown in Fig. 7.2 (a layer whose surface is parallel to the director). The tendency toward formation of molecular layers, typical for SLC-A and SLC-B, is clearly evident. This layering is also reflected in the distribution function of the molecular centers of mass ϕ represented by curve 3 in Fig. 6.1: the presence of sharp maxima and minima; the period corresponds approximately to the length of a molecule.

The states of the studied smectic mesophase included states whose structure corresponded to SLC-A and SLC-B. The difference is demonstrated in Fig. 6.1, which shows the binary distribution functions for projections of the centers of mass of the molecules in the smectic layer on the surface parallel to the layer. There is no long-range order in the positioning of molecules in the layers in the case of SLC-A (curve 1); in the case of SLC-B (curve 2), long-range order remains and is manifested in the much smaller damping of oscillations of the binary distribution function at large distances. This separation of smectic phases A and B is now quite common.[7]

Let us compare our modeling results with the experimental data for real liquid-crystal systems. Starting from the well-known temperature range for the smectic phase of ethyl ester of *n*-azoxybenzoic acid (387–394 K) and comparising it to the reduced temperatures for the existence of SLC in the molecular-dynamics experiment, it is possible to find the energy parameter of the potential $\epsilon/k = 960 \pm 60$ K (in the first approximation, we assume $\epsilon_1 \simeq \epsilon_2$). The distance parameter and the values of γ can be estimated from the known geometric characteristics of the molecule ($\sigma_\perp \simeq 5.4$ ang; $\gamma_1 = 3.0$; and $\gamma_2 = 2.5$). The energy parameter for TBBA ($\epsilon/k = 1045 \pm 30$ K) and some other SLC can be estimated in a similar manner. The accuracy of such an estimate is completely acceptable. This is indicated by the correct description by our model of the region of existence of the smectic mesophase.

The molecular-dynamics method gives broad possibilities for studying particle dynamics in models of liquid crystals. We investigated the translational particle motion with the help of time correlation functions of the linear velocities in directions parallel ($\psi_v^{\parallel}$) and perpendicular ($\psi_v^{\perp}$) to the director $\mathbf{n}$, where

$$\psi_v\,(t) = \langle v(0)v(t)\rangle/\langle v(0)v(0)\rangle,$$

and v indicates the projections of the linear velocities on an axis parallel or perpendicular to the director (Fig. 7.11). The form of the

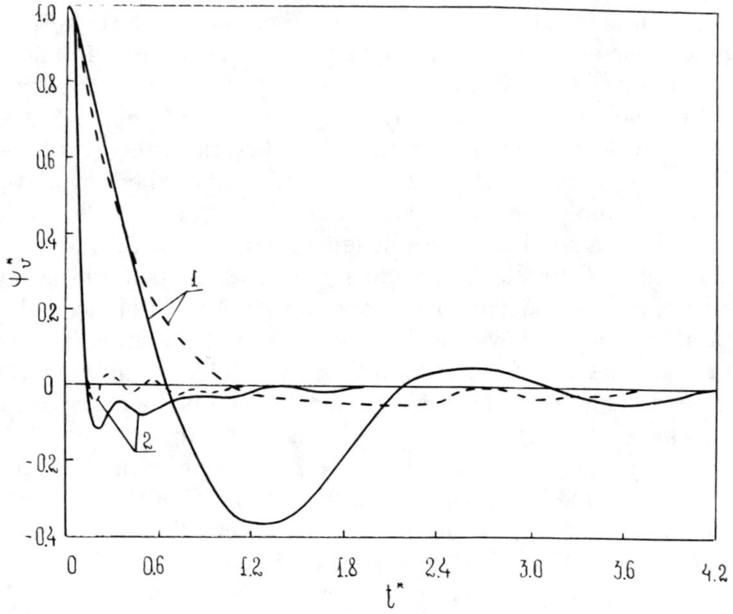

Figure 7.11. The temporal correlation functions of the linear velocities in directions parallel (1) and perpendicular (2) to the director: the continuous line is for SLC–A and the dashed line is for NLC.

functions $\psi_v^{\parallel}$ and $\psi_v^{\perp}$ turned out to be very different from that of the nematic mesophase. Thus, in the case of a nematic crystal, the depth of the negative region of the function $\psi_v^{\perp}$ is much greater than that of $\psi_v^{\parallel}$,[134] but in the case of SLC, the opposite is observed: for $\rho^* = 0.245$ and $T^* = 0.42$ the depth of the negative region of $\psi_v^{\parallel}$ is approximately three times greater than the analogous quantity for the function $\psi_v^{\perp}$. This form of the time correlation function $\psi_v^{\parallel}$ indicates the presence of highly ordered oscillatory motions in a direction perpendicular to the smectic layers. Such a significant difference between the characteristics of particle motion in SLC and NLC, which have the same density and comparatively close temperatures, can be explained by the higher average particle density in smectic layers than in NLC.

The time correlation functions $\psi_v^{\parallel}(t)$ and $\psi_v^{\perp}(t)$ permitted calculation of the corresponding self-diffusion coefficients by integration:

$$D = \langle v^2 \rangle \int_0^\infty \psi_v(t)\, dt.$$

The results of experimental investigations and of molecular-dynamics experiments indicate for NLC the ratio $D_{\parallel}/D_{\perp} > 1$. The opposite situation usually occurs in the case of the smectic mesophase: $D_{\parallel}/D_{\perp} < 1$, except for the region close to the transition state (see Table 7.1). This result agrees qualitatively with experimental data which were obtained for TBBA.[140] However, the measurements of the thermal diffusion coefficients $D^t_{\parallel}$ and $D^t_{\perp}$ that were found experimentally for 8CB indicate that both in smectic A and nematic phases $D^t_{\parallel}$ is larger than $D^t_{\perp}$, although the anisotropy increases as the temperature decreases.[141]

The rotational particle motion in the systems studied was investigated with the help of time correlation functions of the angular velocities:

$$\psi_{\omega}(t) = \langle \boldsymbol{\omega}(0)\cdot\boldsymbol{\omega}(t)\rangle / \langle \boldsymbol{\omega}(0)\cdot\boldsymbol{\omega}(0)\rangle,$$

where ω is the angular velocity, and with the help of reorientational time correlation functions:

$$\psi_l(t) = \langle P_l[\mathbf{n}(0)\cdot\mathbf{u}(t)]\rangle \; (l = 1,2).$$

The typical dependences are presented in Fig. 7.12. The overall behavior of the time correlation functions $\psi_{\omega}(t)$ did not have singularities compared with the corresponding characteristic of the nematic mesophase.[133] The reorientational functions, however, differ from those characteristic of NLC and are more like the corresponding characteristics of solids that consist of nonspherical molecules.[142]

A sensitive criterion of the adequacy of the molecular-dynamics model is the degree of correspondence between the ratios of components of the self-diffusion tensor $D_{\parallel}/D_{\perp}$ (see Table 7.1). In smectic mesophases, in contrast to nematic mesophases, particles characteristically move more freely within the smectic layers than in the direction of the director. The values that we found for $D_{\parallel}/D_{\perp} = 0.52$–$0.78$ for $T^* = 0.38$–0.42 correspond well with the experimental values obtained for TBBA[143] where $D_{\parallel}/D_{\perp} = 0.2$–$1.0$. For other SLC, this ratio can constitute 0.2–0.6 (Ref. 144) and 0.81 (Ref. 145). In the range of temperatures adjacent to the temperature of the transition into the nematic mesophase, this ratio characteristically increases and can greatly exceed unity (according to data in Ref. 143 for TBBA, in this temperature range $D_{\parallel}/D_{\perp}$ attains 1.8). The feature is reflected by our molecular-dynamics model (see Table 7.1, $T^* = 0.44$). On the other hand, at low temperatures some increase in $D_{\parallel}/D_{\perp}$ can also be observed (see Table 7.1, $T^* = 0.38$) that is apparently related to

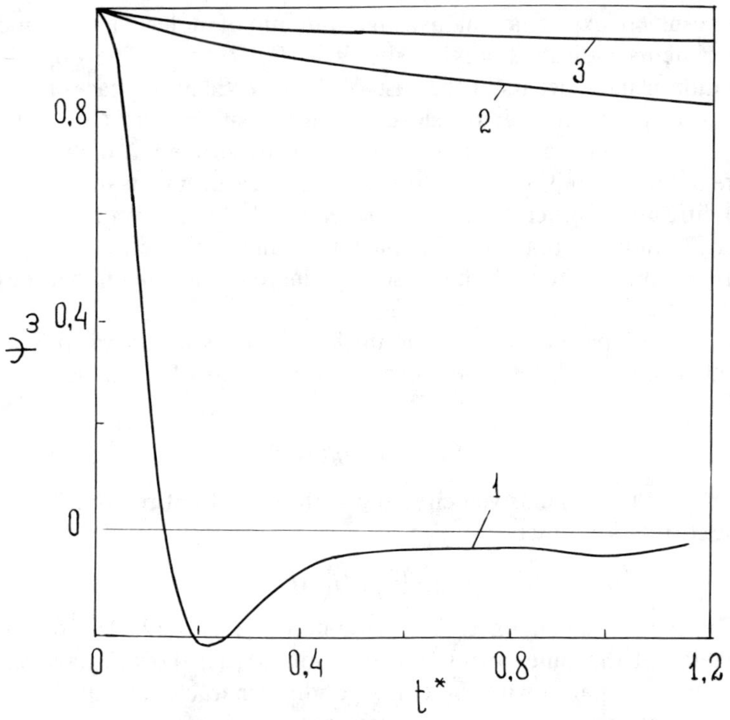

Figure 7.12. The temporal correlation functions of the angular velocities of molecules in SLC–A ψ_ω (curve 1) and the reorientational temporal correlation functions ψ_l for $l = 1$ (curve 3) and $l = 2$ (curve 2).

the translational order within the smectic layers. For the nematic mesophase, the model used agrees completely with experiment and leads to values of $D_\parallel/D_\perp$ that are several times greater than unity (in our calculation $D_\parallel/D_\perp = 4.4$; the experimental value for TBBA is $D_\parallel/D_\perp = 3.0$ (Refs. 143 and 146).

Analysis of results of the molecular-dynamics study of rotational motion of molecules led to the conclusion that these results best correspond with results obtained from the generalized theory of rotation in gases.[147] This theory takes into account the presence of negative values of the function $\psi_\omega(t)$ and replaces the Gaussian form of the function $\psi_l(t)$ by an exponential function for times of the order of $2.5 \cdot 10^{-13}$ s.

During the computer simulation of the system of rigid spherocylinders it appeared that at certain ratios of their lengths to the diameters they managed to find the range of stability of the smectic phase between solid and nematic phases.[148] However, the structural features of this simulated phase do not agree with the accepted theoretical ideas.

Let us now examine the results of investigations of cholesterics using the molecular-dynamics method.[54,149] In this case, we used the following intermolecular pair interaction potential

$$\phi = \phi_1 - 4\epsilon_0 m(\mathbf{u}_i \times \mathbf{u}_j \cdot \check{r})(\mathbf{u}_i \cdot \mathbf{u}_j)\left(\frac{\sigma_\perp}{r}\right)^7, \qquad (7.8)$$

where ϕ_1 is the potential (7.7) and the second term on the right side describes the dipole–quadrupole interaction, which, according to Goossens' theory,[6] primarily determines the features of CLC. The orientational dependence of the energy of this interaction is taken in the form presented in Ref. 62. The parameter m describes the magnitude of the dipole–quadrupole interaction energy and is related to the electronic density distribution of the molecules (in our calculations, $m = 2$). The dependence of the interaction energy on the azimuthal angle is shown in Fig. 7.13.

The statistical problem was solved by numerically integrating the system of equations of motion of 168 ellipsoidal particles placed in the basic sample. This sample was bounded above and below by energetically homogeneous walls while the usual periodic boundary conditions were imposed on the lateral boundaries of the sample, i.e., an infinite film with translational periodicity along two axes was modeled. Each molecule interacted with the walls according to an anisotropic (9-3) potential.[146] The direction of easy orientation was not distinguished in this case. This means that the minima of the interaction energy corresponded to orientations of the long axis of the molecules parallel to the wall; a change in the orientation of molecules lying in the horizontal plane did not change the molecule–wall interaction energy.

The initial conditions (particle coordinates and velocities) corresponded to those taken from molecular-dynamics investigations of a thin nematic film. Thus nematic orientational order was observed in the system, but long-range translational order was absent. During the computer experiment, a transition of the system into the equilibrium state corresponding to a cholesteric mesophase occurred; in addition, spontaneous twisting was observed of the arbitrarily separated molecular layers perpendicular to the vertical z axis into which the basic

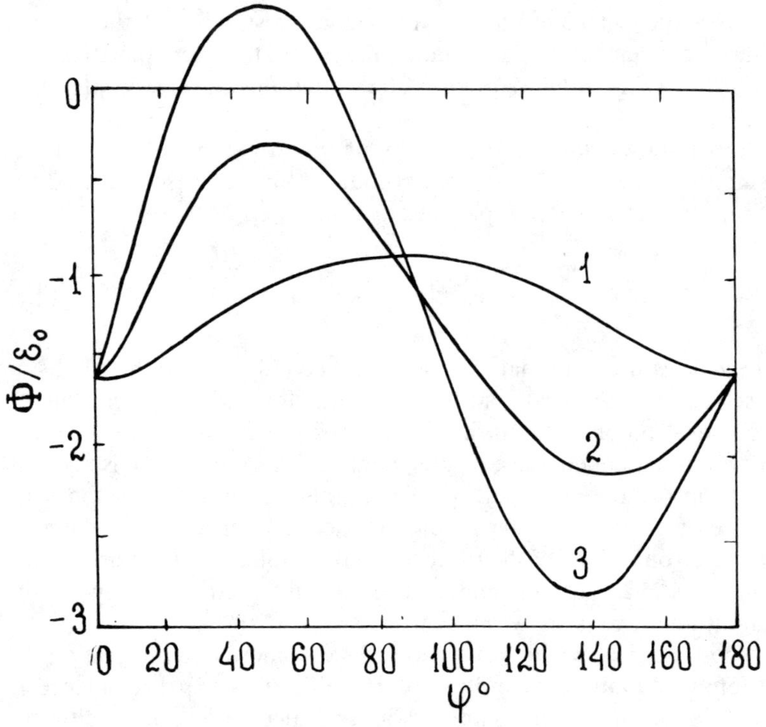

Figure 7.13. The interaction energy for two ellipsoidal molecules according to the potential (7.8) as a function of the azimuthal angle[22,54] (the long molecular axes are perpendicular to the line connecting the molecular centers of mass; the distance between the centers corresponds to the minimum of the potential): 1) first part of the potential ($m = 0$); 2) complete potential ($m = 1$); 3) complete potential ($m = 2$).

specimen was arbitrarily divided (see Fig. 5.3). For each of eight layers (approximate thickness 1.1 $\sigma_\perp$) containing about 20 molecules, the orientation of the director **n** in space, the value of the orientational order parameter, and the orientational distribution functions of the long molecular axes were determined. The computer experiments were performed for the same particle number density $\rho^* = N/v^* = 0.333$ and a series of values of the reduced temperature $T^* = 0.39$, 0.47, 0.61, 0.75, and 1.20.

The data in Fig. 7.14 illustrate the dynamics of formation of the cholesteric helix at $T^* = 0.75$. The initial preferred orientation of the long molecular axes along the x axis was gradually destroyed: an approximately symmetrical (with respect to the middle plane) twisting

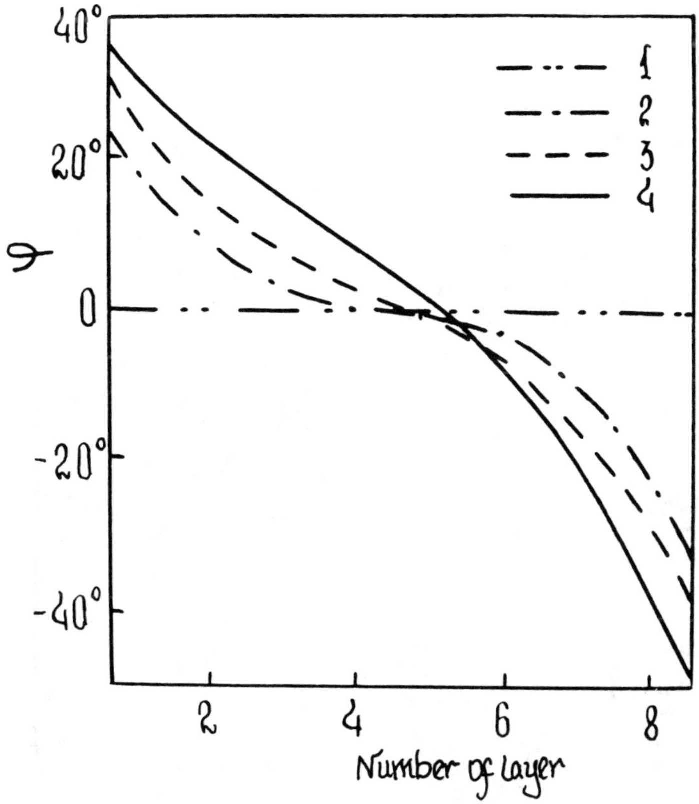

Figure 7.14. The change in the angle of rotation of the projection of the director of a CLC layer on the plane of the layer according to data from a molecular-dynamics experiment[54] at $T^* = 0.75$: 1) initial angle (step number $n = 0$); 2) $n = 675$; 3) $n = 1400$; 4) $n = 1650$.

appeared on different sides of conditionally chosen molecular layers. The directors of different layers were tilted insignificantly from the x-y plane and the tilting was random in nature. This pattern of formation of the cholesteric helix was stable and was observed at different temperatures and for different initial conditions.

Figure 7.15 shows the dependences obtained for the angle of rotation of the projections of the directors of the layers on the x-y plane as a function of the layer number for two values of the temperature. As the temperature increased, the pitch of the helix decreased as in most real cholesteric liquid crystals, which are derivatives of choles-

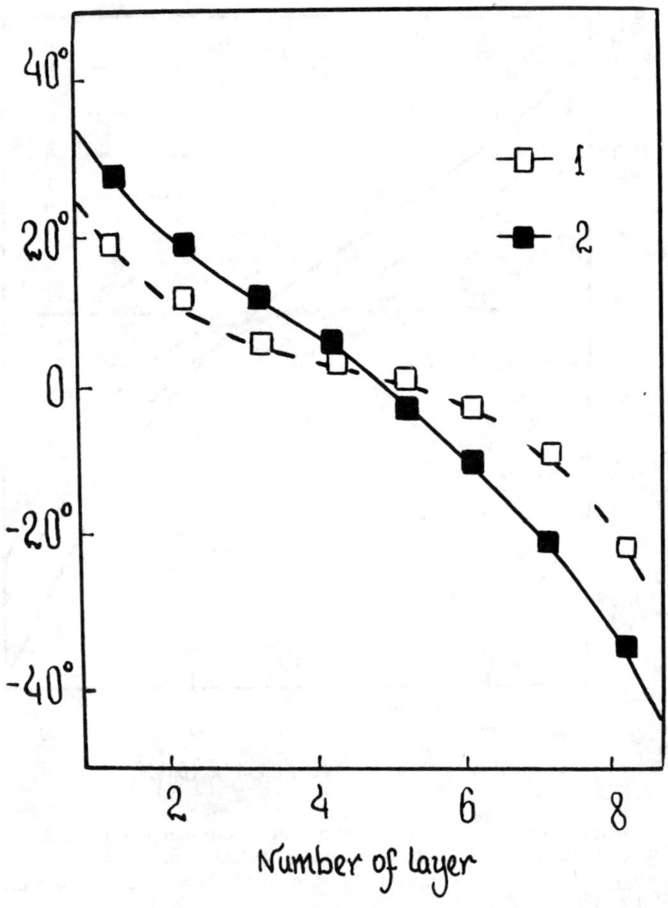

Figure 7.15. The angle of rotation of the projections of directors of CLC layers on the plane of the layers as a function of the layer number according to molecular-dynamics modeling data at $T^* = 0.39$ (1) and $T^* = 0.75$ (2) (Ref. 22,54).

terol. If $\epsilon_1/k \simeq 300$ K and $\sigma_\perp \simeq 5$ Å are chosen as characteristic values, then the quantity

$$\frac{1}{h}\left|\frac{dh}{dT}\right| \simeq 0.5 \cdot 10^{-2}\ \mathrm{K}^{-1}$$

(here h is the pitch of the helix), corresponding in order of magnitude to the experimental values.[7,51–53] The absolute value of the pitch is less than that observed for bulk specimens. In order to obtain better correspondence between the computed and experimental values, it is necessary to choose other values for the parameters of the potential (in particular a smaller m), but then the study of the qualitative characteristics of CLC will require use of more powerful computers and more computing time. The S-shaped dependences of the rotation angles of the projections of the directors of layers on the layer number are apparently related to the influence of the walls. The twisting is greater near the walls, which corresponds to ideas developed in Ref. 117.

The orientational distribution functions $f(\cos\theta)$ that characterize the probability for molecules in a quasinematic layer to be oriented at an angle θ to the director are presented in Fig. 7.16 for three temperatures. These functions change insignificantly from one arbitrarily chosen quasinematic layer to another and correspond qualitatively to those found for NLC. The heights of the maxima of $f(\cos\theta)$ and the orientational order parameters are, however, greater than in the molecular-dynamics models of bulk NLC[134] and real NLC and CLC. This is apparently related to the very thin CLC films being studied, whose structural characteristics are greatly affected by the bounding surfaces. A much higher degree of orientational order is also observed in thin NLC films[146] than in bulk specimens.

When the temperature increased to $T^* = 1.20$, the system transformed into an isotropic liquid state. Figure 7.17 shows the change in the orientational order parameters during the computer experiment with a transition of the system from the cholesteric into the isotropic liquid state.

7.6. THEORETICAL INVESTIGATIONS OF PHASE TRANSITIONS

Phase transitions between SLC-A, CLC (or NLC), and an isotropic liquid can be studied using the Kobayashi–McMillan theory of SLC-A as a starting point. The interesting result here is that the SLC-A–NLC transition can be a first- or second-order transition depending on the value of the order parameter α of the theory. However, it is difficult to confirm this fact from the experimental data for some homologous series (the parameter α varies with molecular dimensions) due to the stepped nature of the change in the length of the alkyl chain.[150] In some substances this phase transition was established to be close to a second-order transition. Thus measurement of the

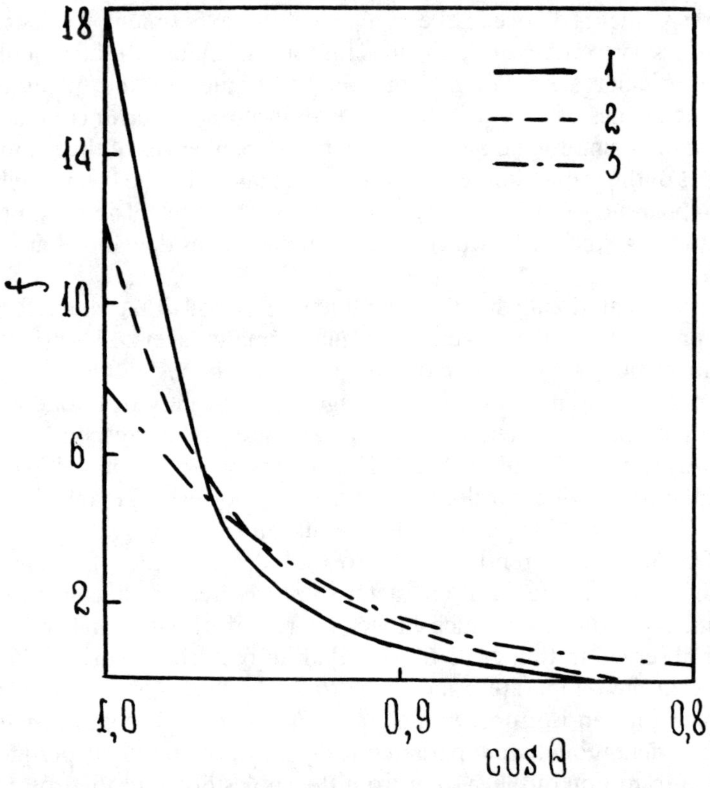

Figure 7.16. The orientational distribution functions in a CLC layer for different temperatures according to data from a molecular dynamics experiment[54] at $T^* = 0.39$ (1), $T^* = 0.47$ (2), and $T^* = 0.75$ (3).

thermal properties of cyanobenzylidenoxyloxyaniline (CBOOA)[151,152] provided the foundation for assuming the presence of a very weak first-order transition. On the other hand, a second-order phase transition was discovered in light scattering experiments. According to the Meyer–McMillan theory developed for SLC-B, SLC-C, and SLC-H mesophases,[5] SLC-A–SLC-C and SLC-B–SLC-H phase transitions are second-order transitions while SLC-C–SLC-H, SLC-A–SLC-H, and SLC-A–SLC-B phase transitions are first-order transitions.

The position of the tricritical point for the phase transition of the second kind nematics-smectics A was found on the basis of numerical and analytical solutions of the self-consistent field McMillan's equations for the three order parameters.[153]

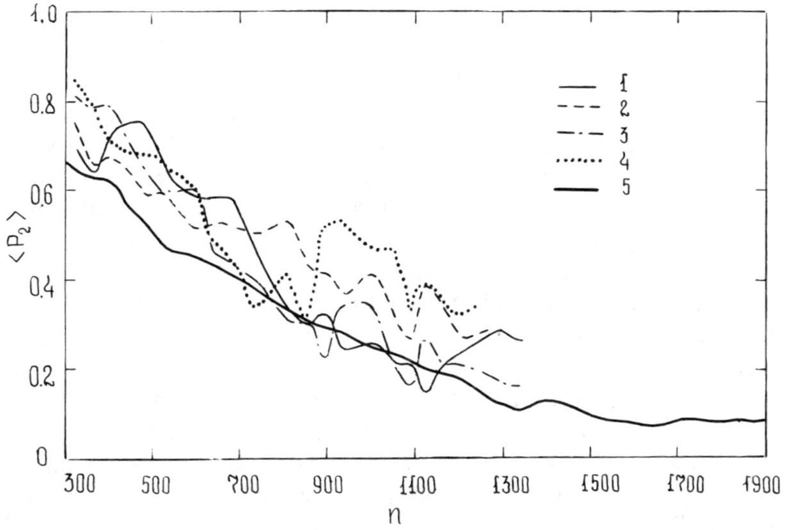

Figure 7.17. The order parameters $\langle P_2 \rangle$ as a function of the step number during the transition of the system out of the CLC state ($T^* = 0.38$) into the IL state ($T^* = 1.2$): 1) second layer; 2) fourth layer; 3) sixth layer; 4) eighth layer; 5) entire specimen (Ref. 22,54).

De Gennes[7] found a general approach to the problem of phase transitions using the analogy between smectic-A liquid crystals and a superconductor, and between the SLC-A–SLC-C transition, the λ transition in liquid ^{4}He (Bose condensation), and L. D. Landau's theory of phase transitions. De Gennes arrived at the following important conclusions based on this method:

1) the SLC-C–SLC-A transition can be a second order transition;
2) if the SLC-C–SLC-A transition is continuous, then the specific heat capacity must have a logarithmic singularity at the point T_{CA}, and a series of interesting analogies between T_{CA} and the λ point in helium is possible;
3) the tilt angle of molecules in SLC-C varies according to the law $\beta \sim |\Delta T|^{\beta}$ (see Sec. 6.2).

The papers[154,155] were devoted to the theoretical discussion of the phase transition SLCA-SLCC. The generalized mean-field model for the smectic A-chiral smectic C phase transition was developed.[156] By the mean-field theory and by using the results of the measurements for the tilt angle of the director and of the polarization of chiral smectic C phase, the authors not only described the above-mentioned phase transition but also found the values of all coefficients of phenomenological expansion of the free energy.

The SLC-C–NLC transition is interesting.[7] For according to L. D. Landau's rules[157] this can be a second-order transition. Very small values of the heat of the transition were indeed discovered for some substances that provided justification for expecting interesting pretransitional phenomena.[7] An obvious analog among the superfluid liquids has not been found here.

It was proposed in Ref. 158 that correlations between the orientations of molecules be considered in constructing a theory of phase transitions in thermotropic liquid crystals. This approach is a variant of thermodynamic perturbation theory. The method of describing a NLC–SLC-A phase transition based on the mean-field theory was proposed by Mayer and Lubensky.[159] Lee and Tan examined the smectic-A–crystalline-phase phase transition in a similar approximation.[160]

S. A. Pikin[161] gave a description of phase transitions between an isotropic liquid, NLC, and SLC-A that was free from the difficulties of the microscopic approach and that generalized results obtained by McMillan[5] and de Gennes.[7] Types of instabilities of the layered smectic structure relative to thermal fluctuations, external actions, and modulations of the structure were also discussed in Ref. 161. The important problem of the change in the nature of the phase transformation (classified according to symmetry considerations as a second-order transition) due to interaction of the order parameter and other degrees of freedom of the system (homogeneous and inhomogeneous elastic deformations, impurities, etc.) was also examined.

A generalized form of Landau's theory for tricritical point was used for the description of the nematic-smectic A phase transition.[162] It was noted that the possibility of the appearance of the reentrant nematic phase was due to the interaction of the transitional and the orientational degrees of freedom. Universality of the reentrant nematic-smectic C-smectic A and nematic-smectic A-smectic C multicritical points was established because the values of critical indexes are equal in both cases.[163] The calculations of reentrant transition enthalpies were made on the basis of the lattice frustrated spin-gas model and the agreement of a number of octyl-oxybenzoiloxy-cyanostilbenes with experimental data was obtained.[164] The interesting correlation between the values of the entropy changes ΔS at the smectic A-nematic, smectic C-nematic, and smectic B-nematic phase transitions, as well as between smectics B-A, B-C, F-B, G-F, F-E and the ratio of the times of the dielectric relaxation was found.[164]

7.7. THEORETICAL STUDY OF TRANSPORT PROCESSES AND THE PERCOLATION EFFECT

The examined theories of SLC and CLC do not enable the study of molecular dynamics and the determination of transport properties. An exception is the molecular-dynamics method, which, however, is a computer-experiment method and not a theoretical approach in the usual sense.

Approximate theoretical methods for estimating the transport coefficients of NLC (primarily useful for calculating self-diffusion coefficients) were examined in Ref. 1. There are practically no molecular-transport theories for SLC and CLC at present. The method of molecular dynamics in principle permits solving this problem for quite realistic interaction potentials. These calculations can be carried out for the self-diffusion coefficients. Information on the procedure of such calculations and the basic results are presented in Sec. 7.5. Calculation of the viscosity and thermal conductivity is, however, a more complicated problem due to greater demands on computer time and the need for high computational speeds that exceed existing capabilities.

We shall briefly examine the problem of the viscosity of SLC and CLC, an interesting solution of which was proposed by Helfrich.[165] It is known from capillary experiments that the measured viscosity coefficients of CLC and NLC are anomalously large compared with the viscosity of ordinary liquids and NLC. Thus the ratio of viscosity of CLC and of an isotropic liquid can attain 10^6. At the same time, the Leslie coefficients and the friction coefficients do not differ significantly from those of NLC. The viscous properties of SLC and CLC in general depend on the shear velocity and do not satisfy Poiseuille's and Newton's laws.

The features of the viscosity of NLC and SLC, in particular the unusually high values of the viscosity coefficients of these substances, were explained by Helfrich[165] who proposed the percolation (or penetration) theory. Assuming that the texture of the substance is "blocked" (spatially fixed) due to coupling with bounding walls while the velocity of the flow (parallel to the axis of the CLC helix or parallel to the CLC-A director) is uniform, Helfrich concluded that the viscosity in this case arises from rubbing of separate molecules of the homogeneous flow and the "blocked" texture. The situation is analogous in the case of CLC to the extrusion of a well-lubricated screw with a large pitch of threads out of a threaded opening; friction arises with rotational motion of the screw while the translational velocity is low. In the case of SLC, the substance "permeates" through smectic layers fixed as a result of interaction with the capillary walls during the effusion process.

De Gennes[7,166] examined the interesting flow features of smectic liquid crystals related to boundary layers, which are completely different from the usual boundary layers formed during the flow of isotropic liquids. These unusual layers arise even in very slow SLC flows and have a characteristic thickness of the order of 10^3 Å.

The theoretical model of molecular motion describing the experimental data by the temperature dependence of the spin-lattice relaxation of smectic A and smectic C phases was developed.[167]

The large-scale properties of the cholesteric and chiral smectic phases were discussed on the theoretical basis.[168] Static and dynamic characteristics were expressed by the small-scale characteristics (elastic moduli, kinetic coefficients, helix pitch) by excluding the rapid degrees of freedom.

CHAPTER VIII.

Thermophysical Properties of Smectic and Cholesteric Liquid Crystals

A sphynx is nature. What that mask serene
Conceals is to man's mind a darker trouble
Because, it may well be, there is no riddle
For him to solve—nor has there ever been.
F. Tyutchev, A Sphynx. Translated by A. Pyman

8.1 REGIONS OF EXISTENCE AND THERMAL PROPERTIES OF SMECTIC AND CHOLESTERIC LIQUID CRYSTALS

The phase diagrams of smectic and cholesteric liquid crystals were studied in Refs. 16, 37, 169–172 and are shown in Figs. 8.1–8.7. The presence in some substances of tricritical points is significant. The results of proton splitting (proportional to the order parameter in NLC and SLC-A) measurements for CBNA showed that as the pressure increases, the differences between the properties of the coexisting SLC-A–NLC mesophases decrease to values that cannot be measured.[14] The tricritical point in this case is characterized by the parameters 140°C and 2.89 MPa (see Fig. 8.1).[169] Triple points were observed in a number of cases (in *p*-methoxybenzoic acid, see Fig. 8.2). The results of an examination of the state diagram in which supercooled phases were observed are also of interest (see Figs. 8.3 and 8.4).[37,171]

Figure 8.8 shows the lines of phase transitions between the solid, smectic-A, nematic, and isotropic phases for 4-nitrophenyle 4′-alkoxybenzoates (n = 8, 10, and 12).[173] The pressure-induced region of the nematic phase is interesting.

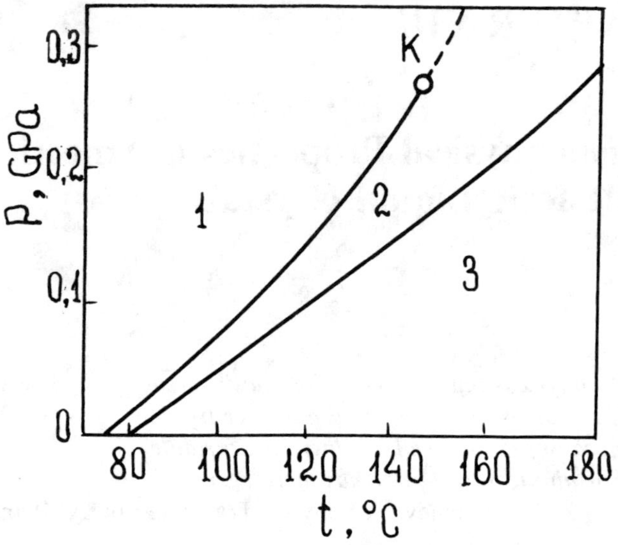

Figure 8.1. Pressure-temperature phase diagram of *n-p*-cyanobenzylidene-*p*-nonylaniline (CBNA)[169]; 1) SLC-A; 2) NLC; 3)IL. *K* is the critical point.

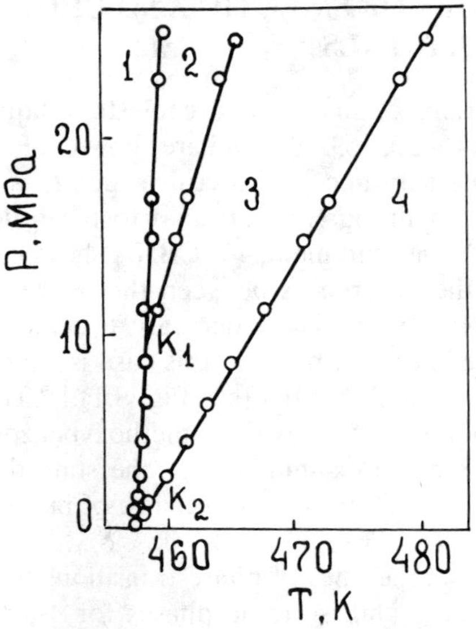

Figure 8.2. Phase diagram of *p*-methoxybenzoic acid[16,170]: 1) solid crystal; 2) SLC; 3) NLC; 4) IL. K_1 and K_2 are the triple points.

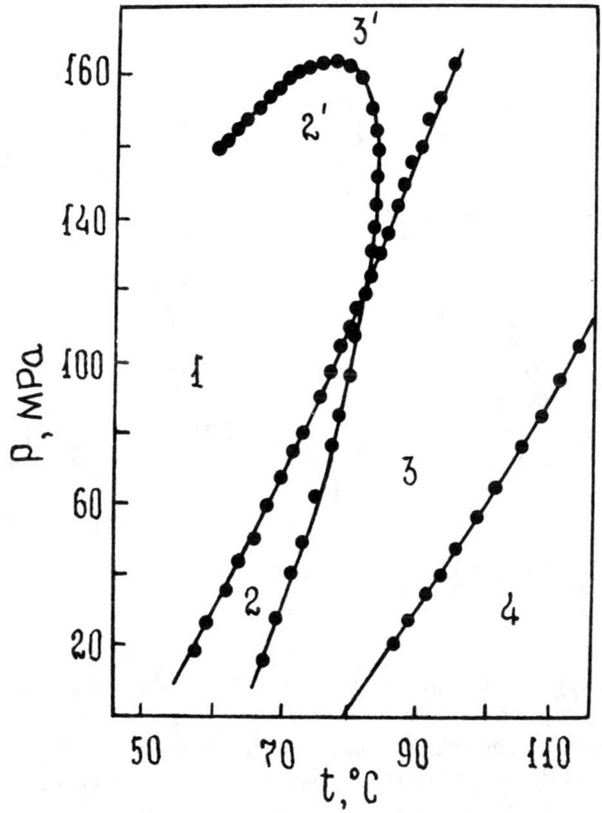

Figure 8.3. Phase diagram of 4-cyano-4'-octyloxybiphenyl (COOB)[37]: 1) solid crystal; 2) SLC-A 2') supercooled SLC-A; 3) NLC; 3') supercooled NLC; 4) IL.

The effect of pressure on strongly polar liquid crystals was studied for a large group of substances (80CB, CBOOA, CBNA, etc.).[174] The phase P–T diagrams and the layer thickness of SLC were given. Phase transitions in 5-*n*-hexyl-2-(4-*n*-pentyloxyphenyl)-pyridine were investigated by calorimetric and microscopic methods.[175] On the basis of these results the classification of smectic phases (C, F, G and H-phases) was proposed.

The temperatures of the transitions between isotropic liquid, nematic and smectic A and C phases of 13 of 4-alkoxy-*N*-(9-oxofluoren-2-ylmethylene)-anilines[176] and 12 of 4-alkoxy-*N*-(fluoren-2-ylmethylene)-anilines (AFYA) and 4-alkoxy-*N*-(2-naphthylmethylene)-anilines (ANMA)[177] were found experimentally. The thorough study of

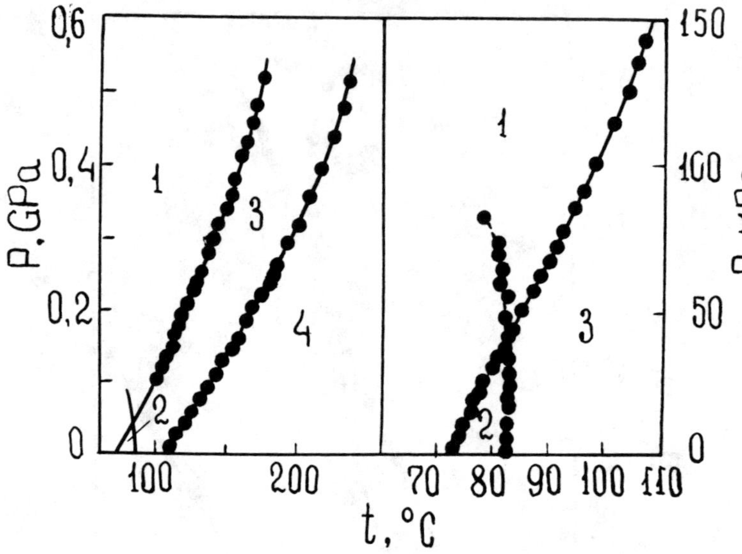

Figure 8.4. Phase diagram of 4-cyanobenzylidene-4′-*n*-octyloxyaniline (CBOOA)[171]: 1) solid crystal; 2) SLC-A; 3) NLC; 4) IL.

the topology of the phase diagrams for substances demonstrating nematic, smectic A and C and reentrant phases was given in papers 178, and 179. The conception of the universality of the phase transitions was discussed on this basis. It should be noted that the consideration of the phase diagrams was accompanied by the analysis of the molecular mechanisms causing phase transitions (molecular overlapping in reentrant nematic phase of OBBC, orientational melting of the end chains of molecules[180] etc.).

The generalization of the data published and the discussion of polymorphism and phase transitions in smectic liquid crystals (including A,C,H,K,G,I,J,F,B,E phases) were done in Refs. 181 and 182.

Cholesteric liquid crystals of practical significance (cholesteryl valerate, cholesteryl pelargonate, cholesteryl myristate, and cholesteryl oleate) have been investigated in detail up to a pressure of 0.5 GPa by N.A. Tikhomirova and A.V. Ginzberg,[172] who studied melting temperatures (DTA method) and temperatures of transitions between liquid phases (study of transparency). It was possible to observe the appearance of a new phase in cholesteryl valerate (at $P \simeq 0.85$ MPa).

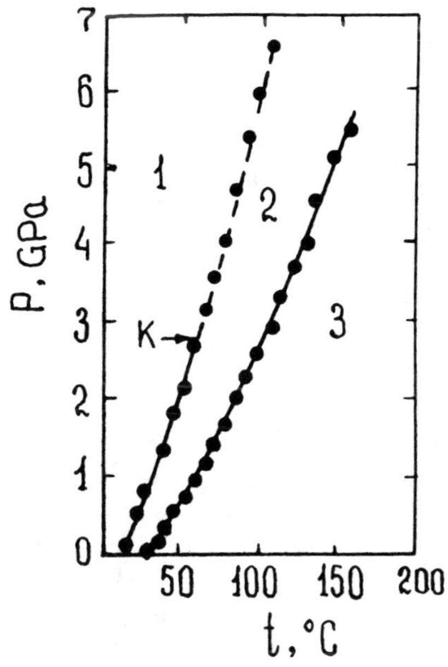

Figure 8.5. Phase diagram of cholesteryl oleyl-carbonate:[171] 1) SLC-A; 2) CLC; 3) IL. *K* is the tricritical point.

Figures 8.9 and 8.10 present results of experimental investigations of the temperatures of phase transitions between IL, CLC or NLC, and SLC-A as a function of the length of the alkyl chain. The dependence of transition temperatures on the number of carbon atoms in the *p*-*n*-alkoxyphenyl group of esters and on the halogen in *p*-halogenphenyl esters[21] is presented in Figs. 8.11 and 8.12.

N.A. Nedostup and V.V. Gal'tsev[183] investigated the effect of molecular structure on the temperatures of phase transitions and on the number of mesomorphic states in a homologous series of cholesterol esters and fatty acids. It was established that the temperature of IL–CLC transitions decreases with increasing alkyl chain length (Fig. 8.13), and an even–odd effect is also observed: transition temperatures of homologs with an odd number of carbon atoms in the alkyl chain are higher than for homologs with an even number of carbon atoms. The temperature of transitions into the solid crystalline state varies in a similar manner. A possible explanation is a periodic change in the polarizability of molecules and a corresponding change in the intermolecular interaction forces, inasmuch as the alkyl group in one

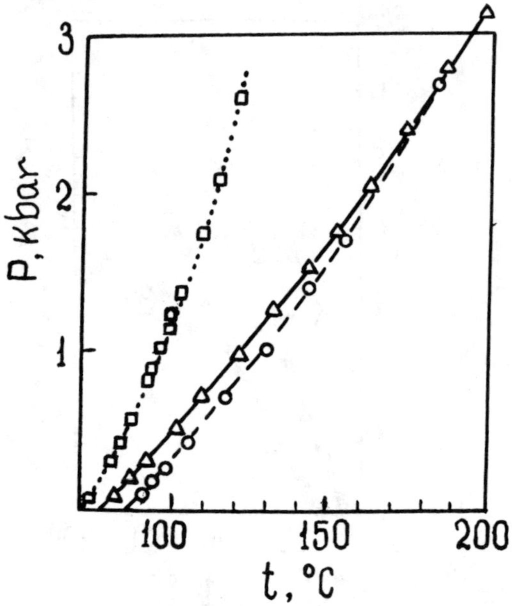

Figure 8.6. Phase diagram of cholesteryl nonanoate[171]: the dots are for the CLC-SLC-A transition (monotorpic transition); the solid line is the line of the solid-phase-CLC transition; the broken line is the SLC–IL transition.

case is attached to a chain in the direction of the long molecular axis and, in the other, at some angle to the axis.[36] This explanation agrees with the results of an investigation of the dielectric constant of substances in this series.[184]

Different experimental methods (polarizational microscopy, selective reflection of the light, optical rotational dispersion) were used for the study of the stability and phase diagrams of all three principal types of blue phases (I, II and III phases) in changing the temperature and helix pitch.[185] It was found that in increasing the pitch, phase III is the first to lose the stability, then phase II and, at last, phase I.

Many papers are concerned with the study of the density of SLC and CLC at atmospheric pressure.[36,186–188] Results obtained for a number of substances are presented in Figs. 8.14–8.19. A characteristic feature is the presence of a small jump in density with phase transitions between IL, SLC, and CLC. In some cases, the slope of the temperature dependence of the density also changes with the transition. The change in the jump of the specific volume with a transition into the solid crystalline state as a function of the number

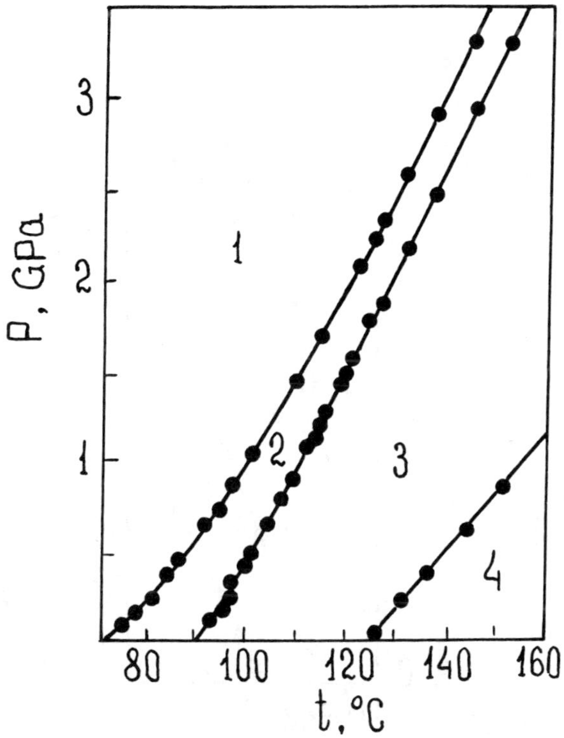

Figure 8.7. Phase diagram of 4,4′-heptyloxyazoxybenzene (HAB)[171]: 1) solid crystal; 2) SLC-C; 3) NLC; 4) IL.

of carbon atoms in the alkyl chain is presented in Fig. 8.20.[183] The even–odd effect is absent or is very weak here.

The temperature dependence of the specific volume of 4-cyano-benzylidene-4′-*n*-octyloxyaniline (CBOOA) accompany a decrease of the temperature was studied in Ref. 189 (Fig. 8.21); in addition, these data, together with the results of the study of scattering of x-rays, were successfully used to interpret the structure of SLC layers.

The experimental determination of the density was made for a number of mesogenic substances having the ability to form smectic liquid crystals: NOBA,[190] POBOA,[191] HOAB,[192] BBPAA,[193] HpBPA,[194] and HOBPA.[195] These measurements cover the cases of different mesophases (smectic A,B,C,G) and in some cases the results obtained enabled the character of phase transitions to be determined.

Considerably fewer investigations of the PVT properties of SLC and CLC have been performed. Cholesteryl valerate, butyrate, palmitate, stearate, myristate, and pelargonate have been studied. Some

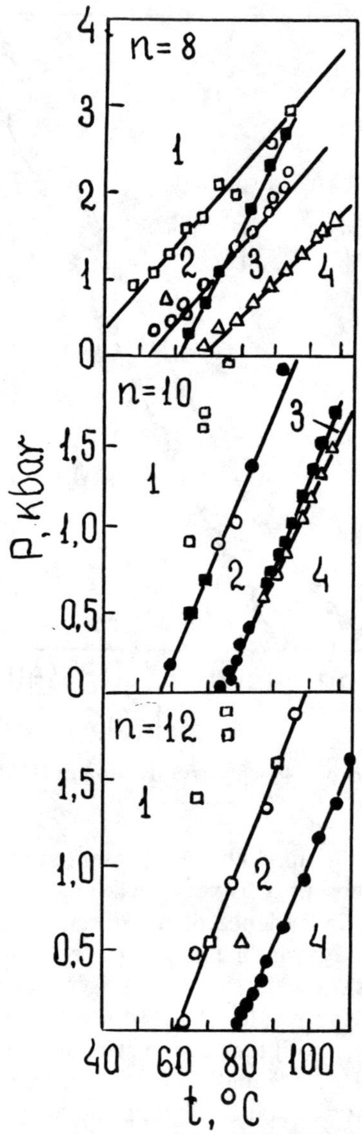

Figure 8.8. Lines of phase transitions of 4-nitrophenyl 4′-alkoxybenzoates with n = 8, 10, and 12[173]: 1) solid crystal, 2) smectic-A, 3) nematic, 4) isotropic liquid. The data for n = 10 demonstrate a pressure-induced nematic phase.

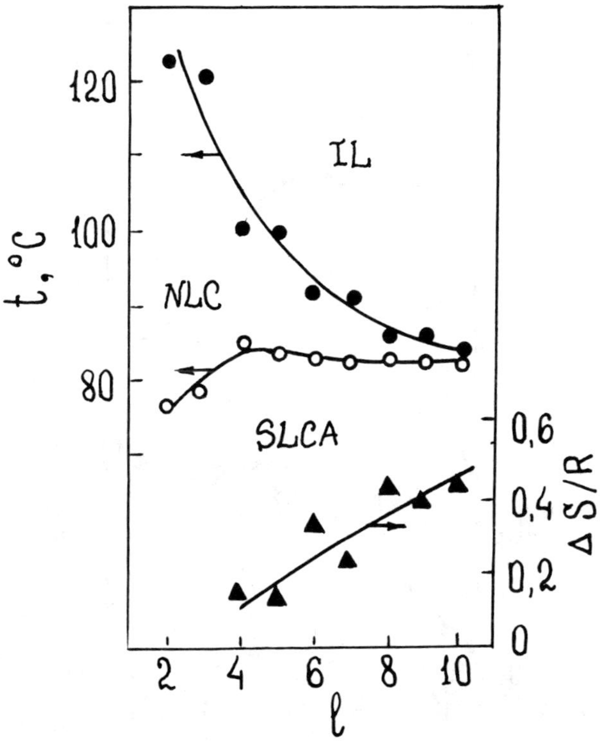

Figure 8.9. Temperatures and entropies of SLC-A–NLC transitions in 4-ethoxybenzal-4′-amino-*n*-alkyl-α-methylcinnamate as a function of the length of the alkyl chain.[3]

results of such a study by V.Ya. Baskakov, V.K. Semenchenko, N.A. Nedostup and V.M. Byankin are presented in Figs. 8.22–8.24 (Refs. 187, 196 and 197). It was possible to obtain interesting results from analysis of these data. For example, insignificant changes in the slope of the isotherms of cholesteryl valerate suggested the existence of a new phase of this substance not observed under ordinary conditions.[187]

8.2. CALORIC PROPERTIES AND THE VELOCITY OF SOUND

The caloric properties (heats of phase transitions) of a group of homologs of *n*-alkoxyazoxybenzene were studied by Arnold.[198] Some re-

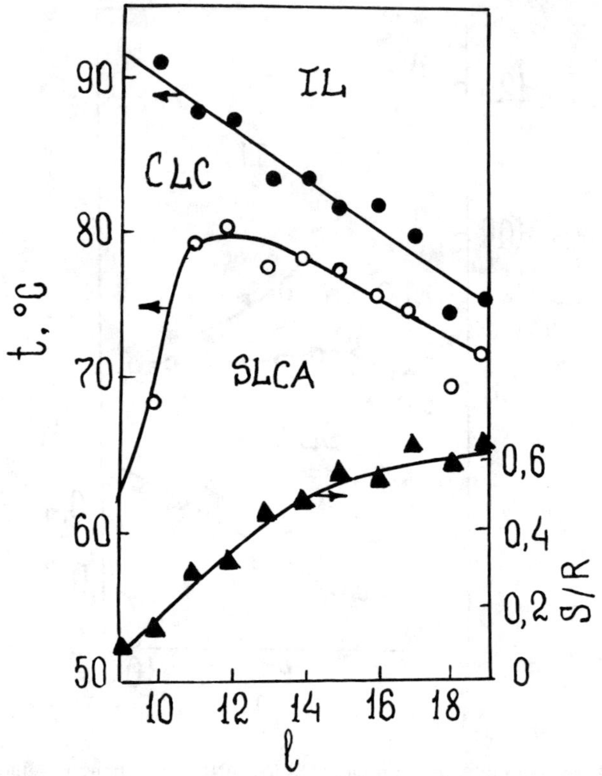

Figure 8.10. Temperatures and entropies of SLC-A–CLC transitions in cholesterol esters and saturated aliphatic acids as a function of the length of the alkyl chain.[3]

sults of these investigations concerning systems exhibiting SLC are presented in Table 8.1. The temperatures and heats of phase transitions in substances exhibiting CLC were studied in detail in Ref. 199 (Table 8.2).

Table 8.3 presents the temperatures and enthalpies of transitions of some materials exhibiting a cholesteric (and, in a number of cases a smectic-A also) phase.

Dependences of entropies of the transitions SLC-A–NLC and SLC-A–CLC on length of the alkyl chain were examined by McMillan (see Figs. 8.9 and 8.10).[3]

Experimental data on the thermal effects of phase transitions supplement calculation methods interesting for practice. The correlation between the jump value of the entropy ΔS and the ratio of the di-

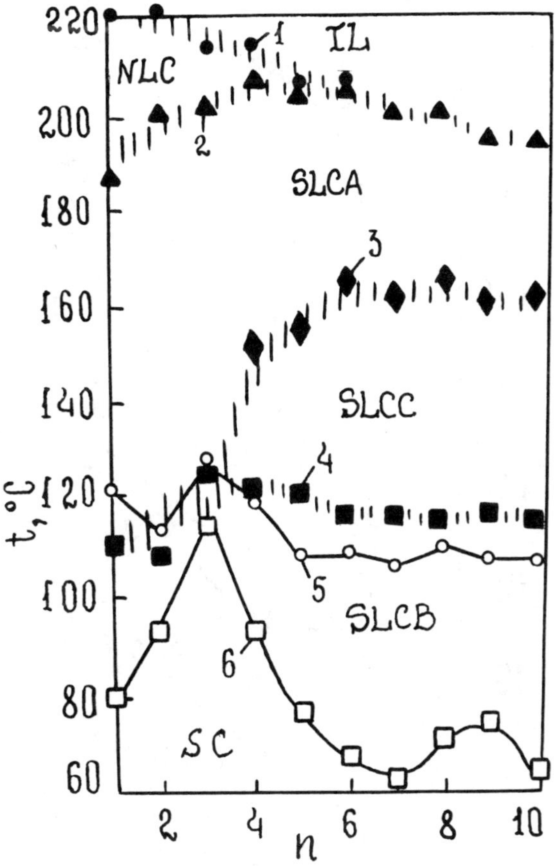

Figure 8.11. Transition temperatures as a function of the number of carbon atoms (*n*) in the *p*-*n*-alkoxyphenyl group of esters of 4′-*n*-octyloxybiphenyl-4-carboxylic acid:[21] 1) IL–NLC; 2) IL (or NLC)–SLC-A; 3) SLC-A–SLC-C; 4) SLC-A (or SLC-C)–SLC-B; 5) SLC-A (or SLC-B)–solid crystal; 6) SLC-B–solid crystal (SC).

electric relaxation times τ_2/τ_1[201] is among them. The calculation of the enthalpy of reentrant transitions was performed in paper 202.

Many investigations have been concerned with the heat capacities of SLC and CLC. The temperature dependence of the heat capacity for amyl ester of dodecyloxybenzal-*n*-aminocinnamatic acid (AEDOC), which forms a series of smectic mesophases, is characteristic (Fig. 8.25). The heat capacity is discontinuous at all points of the phase transitions. Temperature dependences of the heat capacities of other compounds are presented in Figs. 8.26 and 8.27.

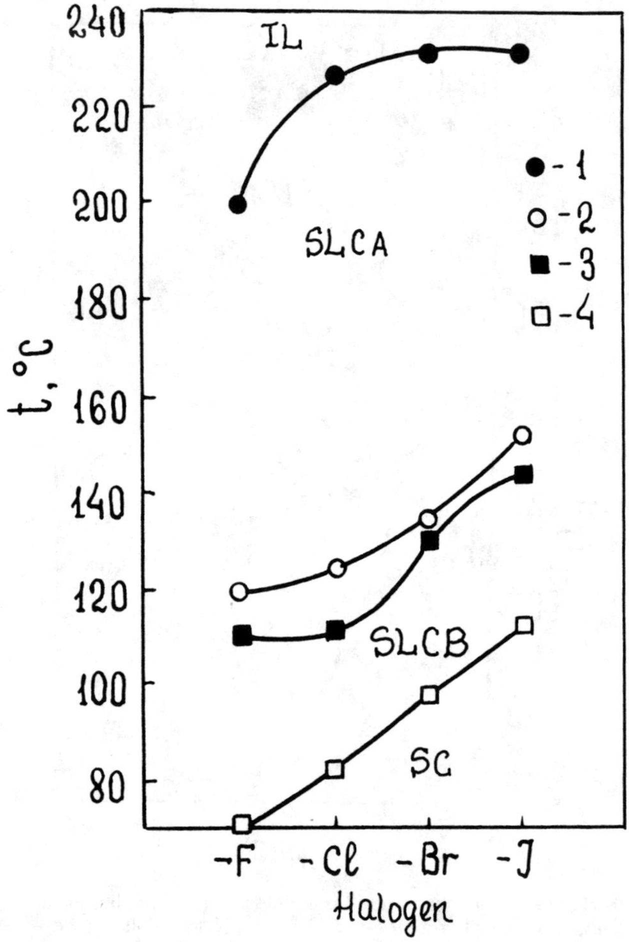

Figure 8.12. Dependence of the transition temperatures on the halogen in *p*-halogenophenyl esters of 4′-*n*-octyloxybiphenyl-4-carboxylic acid[21]: 1) IL–SLC-A; 2) SLC-A–solid crystal; 3) SLC-A–SLC-B; 4) SLC-B—solid crystal.

During the last years the liquid crystal heat capacity especially in the field of phase transitions between nematic and different smectic phases was intensively studied.[203–206] The heat capacity, tilt angles of molecules, polarization of the chiral smectic C phase were also studied.[207,208] The experimental data on heat capacity of the liquid crystalline blue phases (BP I, BP II and BP III) were obtained.[209]

Figure 8.15 shows that the temperature dependence of the adiabatic

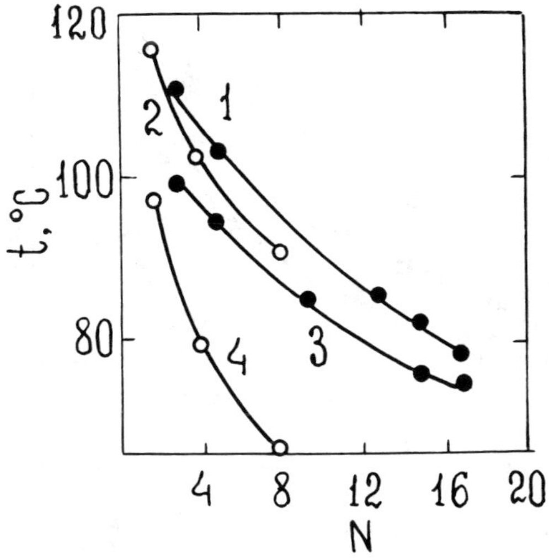

Figure 8.13. Change in the phase transition temperatures with increasing number of carbon atoms in the alkyl chain (N) in homologs of a series of cholesterol esters and fatty acids[183]: 1) IL–NLC transition, odd N; 2) same, even N; 3) liquid crystal–solid crystal transition, odd N; 4) same, even N.

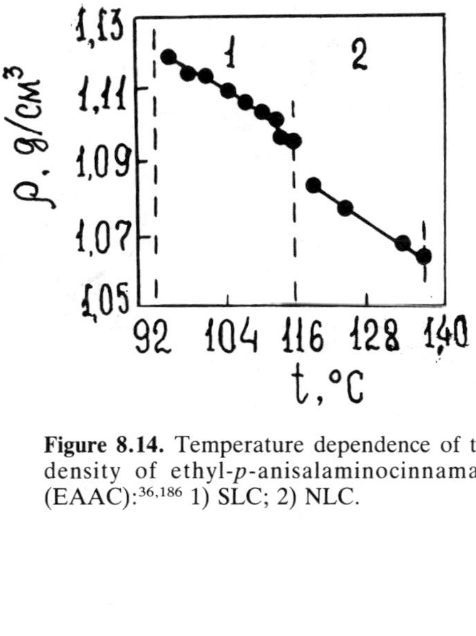

Figure 8.14. Temperature dependence of the density of ethyl-p-anisalaminocinnamate (EAAC):[36,186] 1) SLC; 2) NLC.

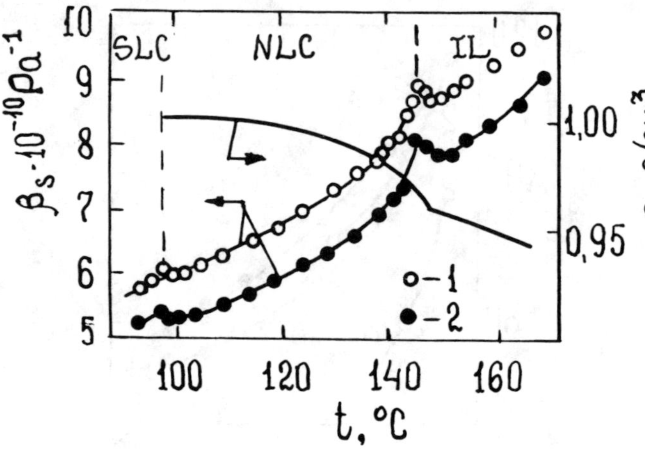

Figure 8.15. Temperature dependences of the adiabatic compressibility and density of HOBA[187]: at 3 MHz (1) and 60.5 MHz (2).

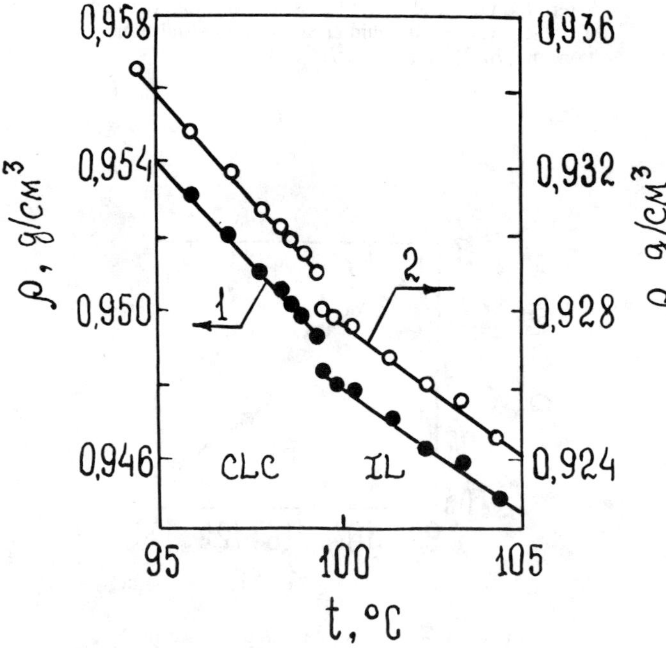

Figure 8.16. Temperature dependence of the density of cholesteryl valerate (1) and cholesteryl capronate (2).[188]

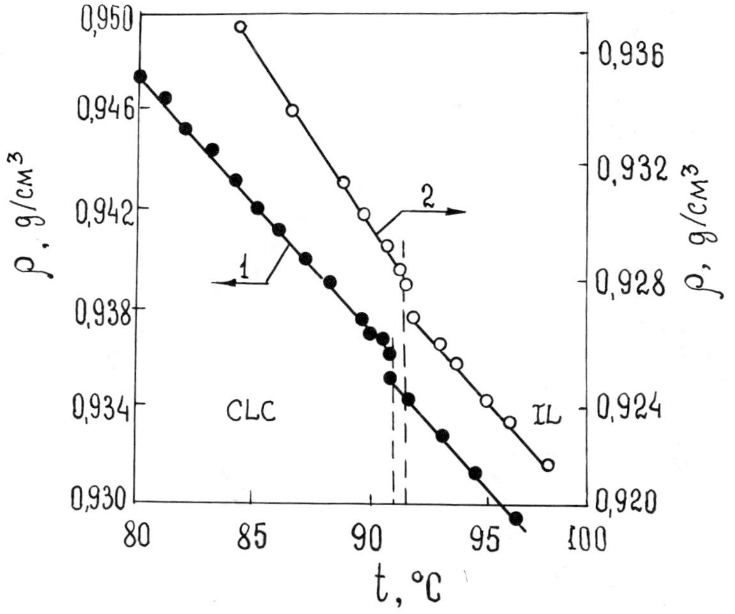

Figure 8.17. Temperature dependence of the density of cholesteryl pelargonate (1) and cholesteryl caprylate (2).[188]

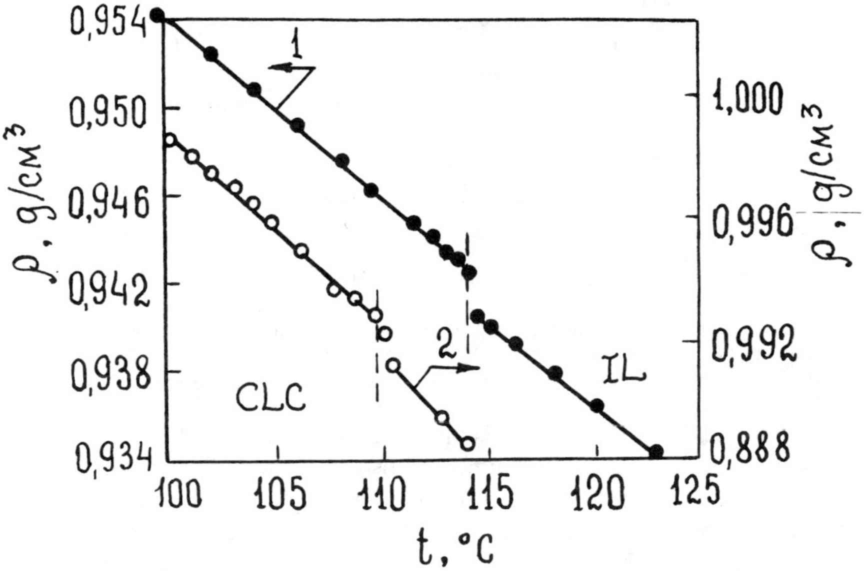

Figure 8.18. Temperature dependence of the density of cholesteryl propionate (1) and cholesteryl butyrate (2).[188]

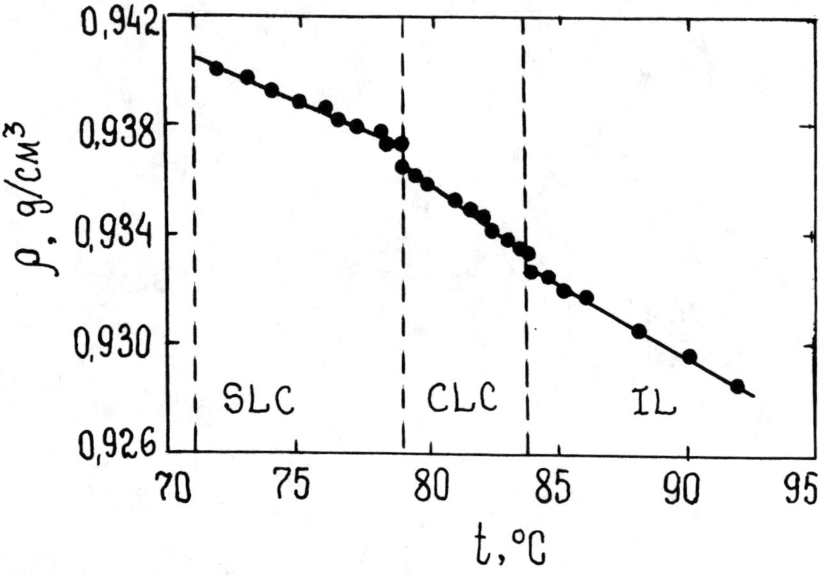

Figure 8.19. Temperature dependence of the density of cholesteryl myristate.[188]

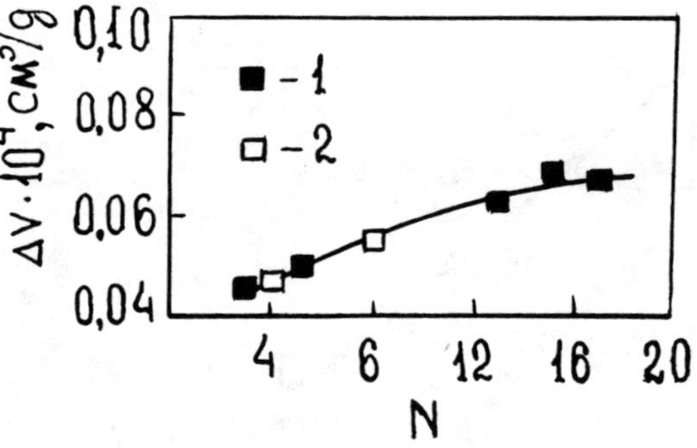

Figure 8.20. The change in the jump in the specific volume accompanying a CLC–solid crystal transition in a number of cholesterol esters and fatty acids as a function of the number of carbon atoms in the alkyl chain (*N*)[183]: 1) odd *N*; 2) even *N*.

compressibility of HOBA[187] is approximately linear with the exception of regions adjacent to phase transition points. The features of these dependences at the temperature of the SLC–NLC transition are considerably less distinct than at the temperature of the NLC–IL transition. These features are clearly manifested in cholesteryl stearate (Fig. 8.18).[210] Temperature and pressure dependences of the

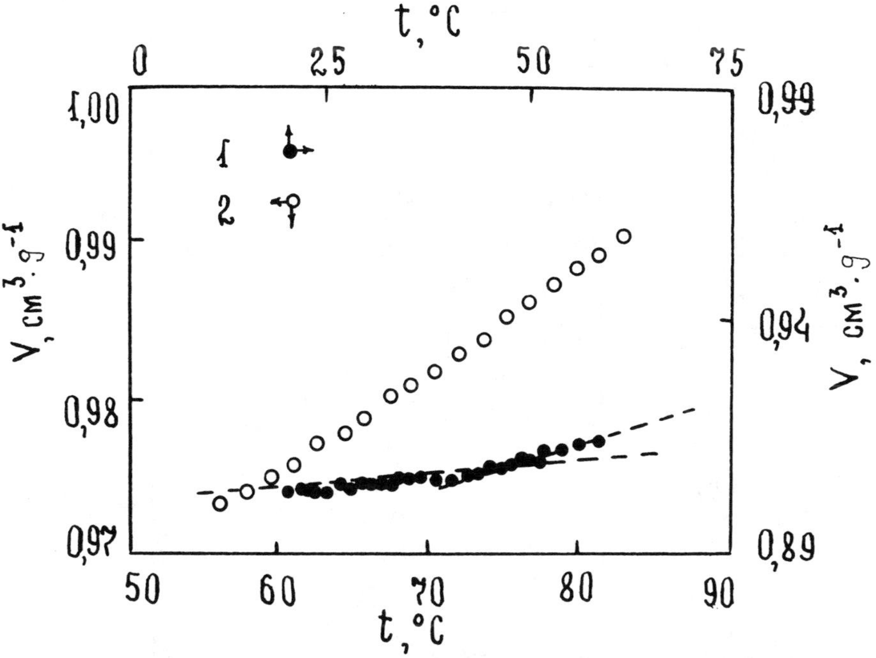

Figure 8.21. Change in the specific volume accompanying a decrease of the temperature in CBOOA[189]: 1) for a wide temperature range, 2) for the region of the smectic-A phase.

isothermal compressibility of cholesteryl stearate are presented in Figs. 8.29 and 8.30.

Temperature dependences of the velocity of sound in HOBA are presented in Fig. 8.31 for two frequencies. Approximately linear dependences are also characteristic here with the exception of regions near phase transition points. As in the case of adiabatic compressibility, singularities related to phase transitions are strongly manifested at the temperature of the NLC–IL transition.

The temperature dependences of the velocity of ultrasound and of ultrasonic absorption for a number of cholesteric liquid crystals are shown in Figs. 8.32–8.34 based on the data in Ref. 211. The singularities in these dependences, appearing at temperatures close to the transition temperatures, are discussed in the cited work from the viewpoint of pretransitional effects. Table 8.4 summarizes the information on the experiments on propagation of ultrasound in the region of phase transitions in liquid crystals.[212]

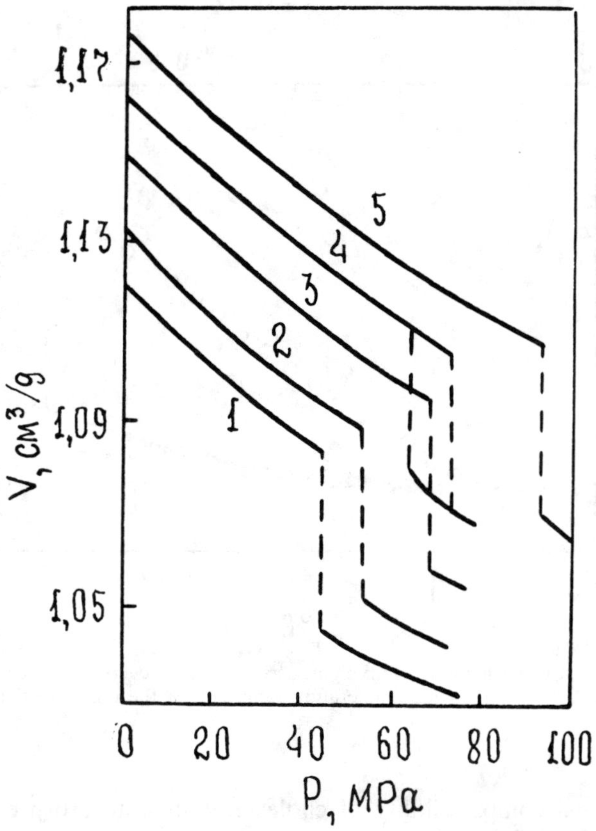

Figure 8.22. Isotherms of cholesteryl valerate[187]: 1) at 99°C; 2) 103°C; 3) 109°C; 4) 110°C; 5) 119°C.

The experimental measurements of the ultrasound velocity were also made for POBOA[213] and BBPAA.[214] Moreover, the damping of the sound in different liquid crystalline phases (see 215, for example) was studied.

8.3. SURFACE TENSION

Surface tension is an important property of liquid crystals. There are, however, very few reliable measurements of the surface tension of SLC and CLC, though NLC have been studied in relatively greater detail.[216]

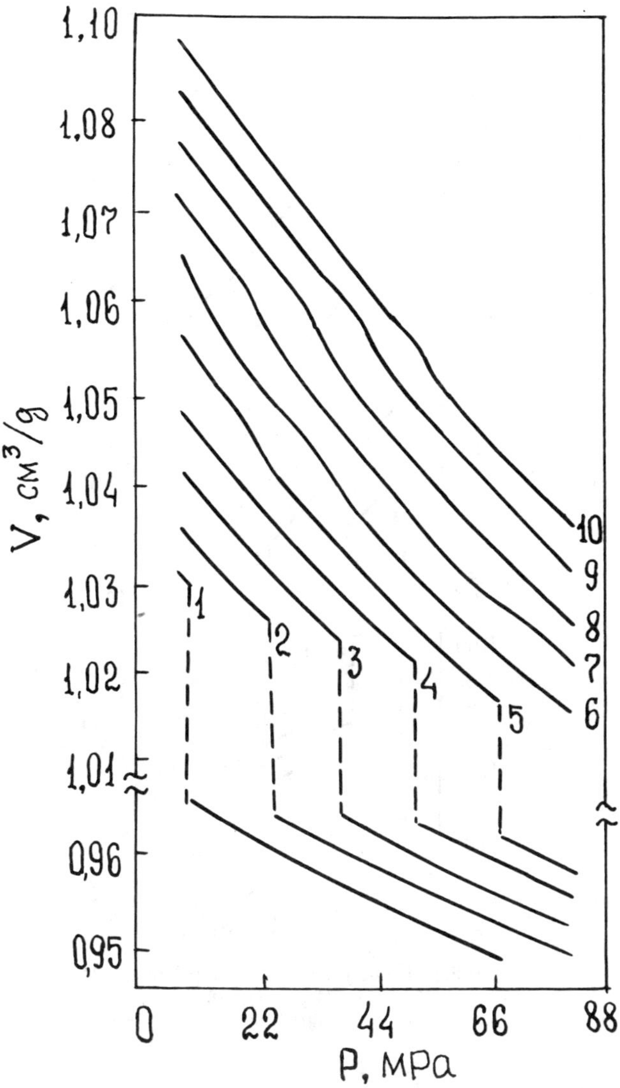

Figure 8.23. Isotherms of cholesteryl myristate[187]: 1) at 68°C; 2) 72°C; 3) 76°C; 4) 80°C; 5) 84°C; 6) 88°C; 7) 92°C; 8) 96°C; 9) 100°C; 10) 104°C.

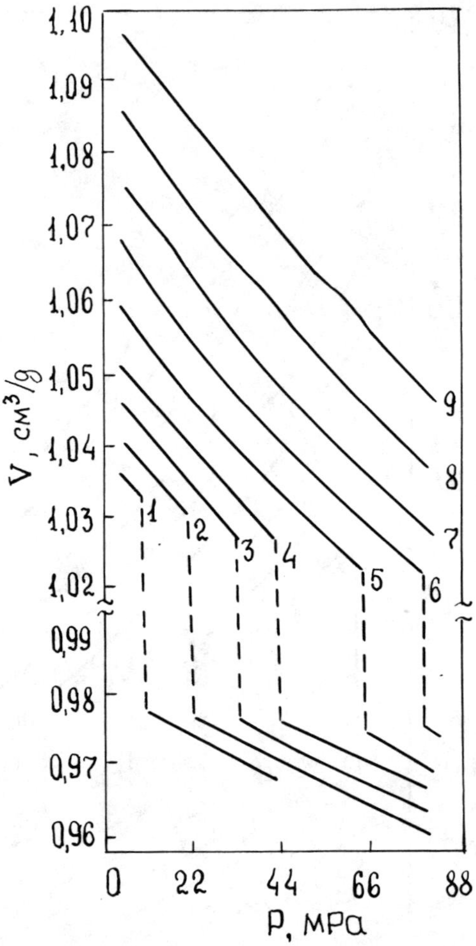

Figure 8.24. Isotherms of cholesteryl pelargonate[187]: 1) at 70°C; 2) 74°C; 3) 78°C; 4) 82°C; 5) 88°C; 6) 94°C; 7) 100°C; 8) 108°C; 9)116°C.

The surface tension of cholesteryl stearate was measured by Selle and Neumann.[217] Their results indicate the presence of a distinct maximum in the surface tension at the transition temperature (72–76°C). Churchill and Bailey,[218] who studied the surface tension of cholesteryl myristate (which has two mesophases, smectic and cholesteric) by a continuous method with different cooling rates, arrived at similar conclusions. It was established[216,218] that there is a high

Table 8.1
Heats of Fusion and Phase Transitions of a Group of Homologs of *n*-alkoxyazoxybenzene[198]

Alkyl Group	Molar Heat of Fusion, J/mole	Molar Heat of Phase Transitions, J/moles		
		SLC–NLC	NLC–IL	SLC–IL
Heptyl	40920	1592	1900	—
Octyl	42180	1181	2810	—
Nonyl	38170	1651	3560	—
Decyl	38580	2316	4930	—
Undecyl	41090	—	—	12110
Dodecyl	42060	—	—	13910

peak in the surface tension with a broad base (approximately 4–5°) that nearly encompasses the entire temperature range of the cholesteric phase. Comparison with viscosity curves indicates a qualitative similarity.

Krishnaswamy and Shashidhar[219] measured the surface tension of CBOOA in the region of existence of SLC, NLC, and IL (Fig. 8.35). Their goal was to verify the theory of surface tension of liquid crystals proposed by Croxton and Chandrasekhar.[220] Their analysis was based on the effects of the competition between the orientational order and the relative disorder (that occurs at a free surface) on surface tension. The results of this theory agree on the whole with the results of the molecular theory of NLC developed by Parsons and based on use of the Fowler–Kirkwood–Waff approximation. The positive slope of the temperature dependence of the surface tension of CBOOA in the SLC region and the change of positive slope to a negative one in the NLC region were observed, which are in complete agreement with the predictions of the theory. The slope again changes sign after the NLC–IL transition while a maximum occurs in the isotropic liquid region (see Fig. 8.35).

The results of measurements of the surface tension of CLC at some temperatures are presented in Table 8.5.

Figure 8.36 shows the temperature dependence of the surface tension of 4-*n*-octyloxy-4′-cyanobiphenyl,[225] which exhibits smectic A and C mesophases in addition to the nematic phase. These results agree well with the data obtained by the method of scattering of light by the surface waves and by the method of weighing of a drop of isotropic phase (at 353 K, $\sigma = 26.2$ dyne·cm^{-1}).

Anomalies of the temperature dependence of the surface tension

Table 8.2
Thermodynamic Properties of Members of a Homologous Series, Found in the Heating Regime[199]

		Solid Crystal–CLC Phase Transition			CLC–IL Phase Transition		
n	Ester	T, K	ΔH, J/mole	ΔS, J/mole · K	T, K	ΔH, J/mole	ΔS, J/mole · K
3	Propionate	373.0	21.5	57.6	385.9	0.35	0.91
3	Butyrate	373.1	24.1	64.6	384.7	0.48	1.25
5	Valerate	363.9	20.6	56.6	371.9	0.33	0.89
6	Capronate	370.4	34.1	92.1	372.6	0.23	0.62
9	Pelargonate	352.7	23.2	65.8	364.4	0.35	0.96
10	Caprylate	356.9	38.2	107.0	365.0	0.57	1.56
16	Palmitate	350.5	64.7	184.6	354.5	1.43	4.03

Table 8.3
The Temperatures (T, K) and enthalpies (ΔH, kJ/mole) of Transitions of Some Cholesterogenic Substances[200]

	Solid–LC		SLCA–CLC		CLC–IL	
Substance	T	ΔH	T	ΔH	T	ΔH
Cholesteryl propionate	372.8	21.8			387.3	0.67
Cholesteryl laurate	364.5	33.1	354.6	1.42	361.8	1.34
Cholesteryl myristate	344.2	47.6	352.3	1.76	357.8	1.51
Cholesteryl ethylcarbonate	356.9	21.1			378.9	0.67

for cholesteryl pelargonate bordering on glucerol were recently discussed.[226]

The very interesting and important practical problem of the effect of molecular orientation at the surface of a film or drop on the surface tension (this orientation can be changed by applying a magnetic field, for example) has not yet been studied although such experiments are planned.[219]

8.4. TRANSPORT PROPERTIES

At present the self-diffusion and viscosity (bulk and shear) coefficients are the only transport properties of SLC and CLC that have been

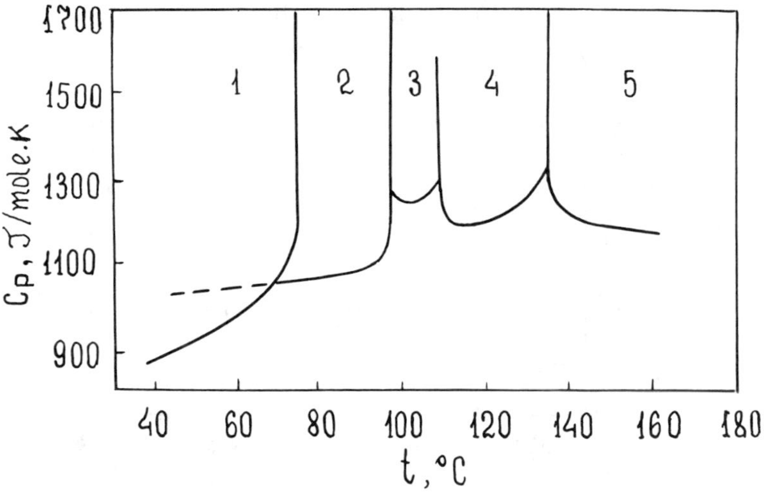

Figure 8.25. Temperature dependence of the molar heat capacity of amyl ester of dodecyloxybenzal-n-aminocinnamic acid (AEDOC)[14]: 1) solid crystal; 2) SLC-B; 3) SLC-C; 4) SLC-A; 5) IL.

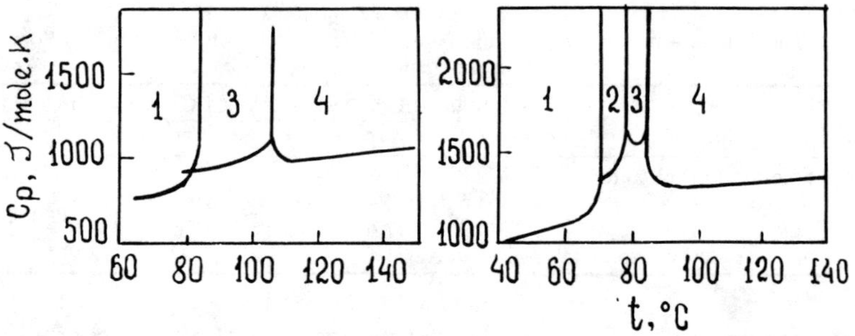

Figure 8.26. Temperature dependence of the molar heat capacity of cholesterol ester of carbonic (left) and myristate (right) acids[36]: 1) solid crystal; 2) SLC; 3) CLC; 4) IL.

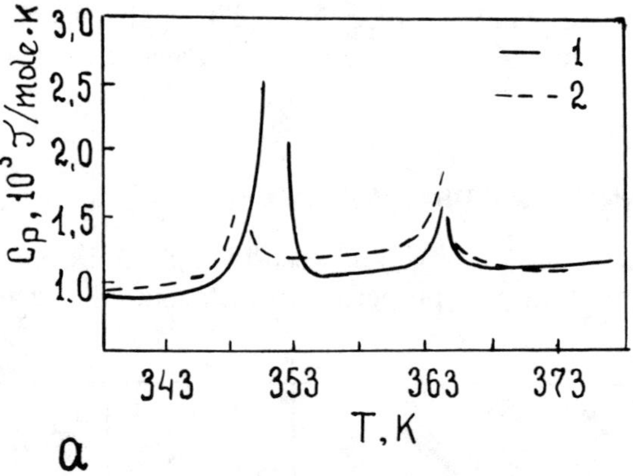

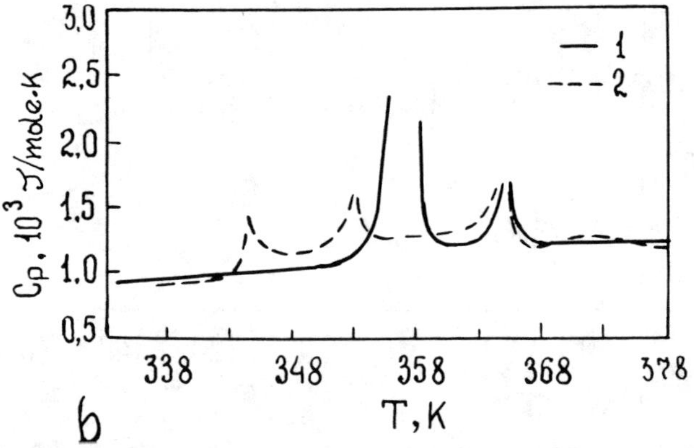

Figure 8.27. Temperature dependence of the molar heat capacity of cholesteryl pelargonate (a) and cholesteryl caprinate (b) with heating (1) and cooling (2).[187]

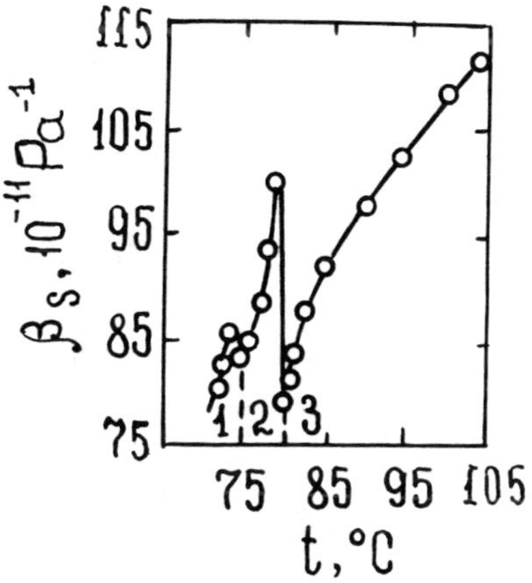

Figure 8.28. Temperature dependence of the adiabatic compressibility of cholesteryl stearate at atmospheric pressure[210]: 1) SLC; 2) CLC; 3) IL.

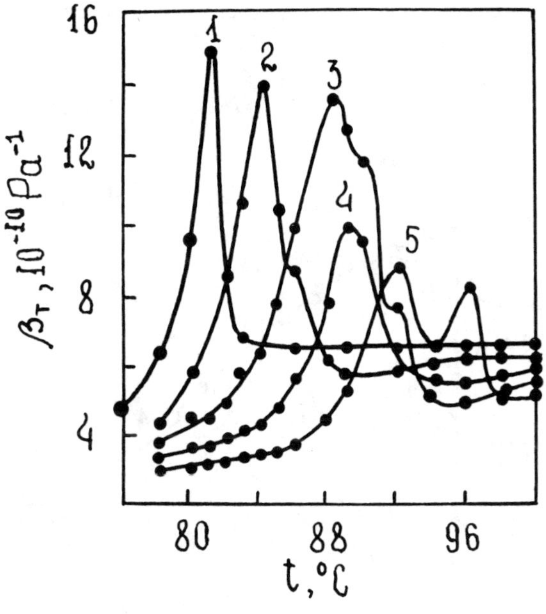

Figure 8.29. Isothermal compressibility of cholesteryl stearate[210] as a function of temperature along the isobars P = 11 MPa (1), P = 22 MPa (2); P = 33 MPa (3), P = 44 MPa (4), and P = 55 MPa (5).

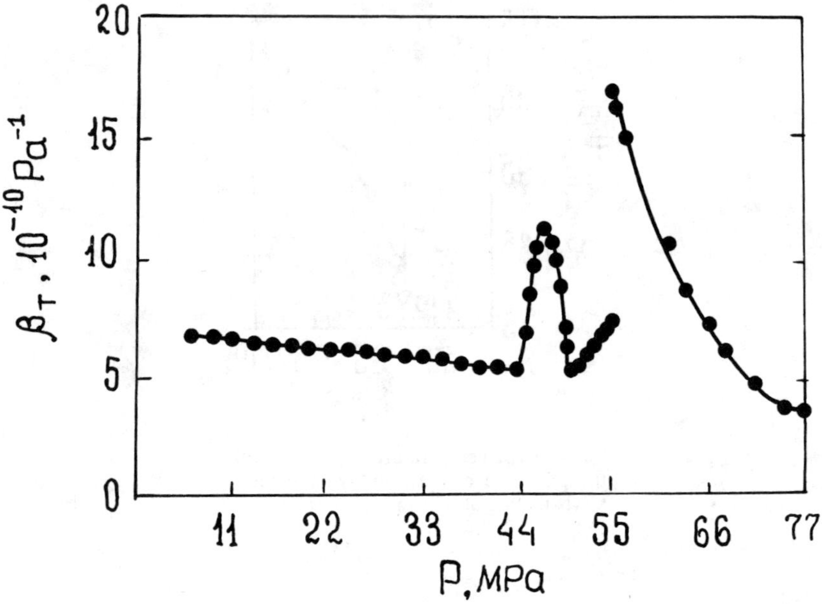

Figure 8.30. Isothermal compressibility of cholesteryl stearate as a function of pressure at 94°C.[210]

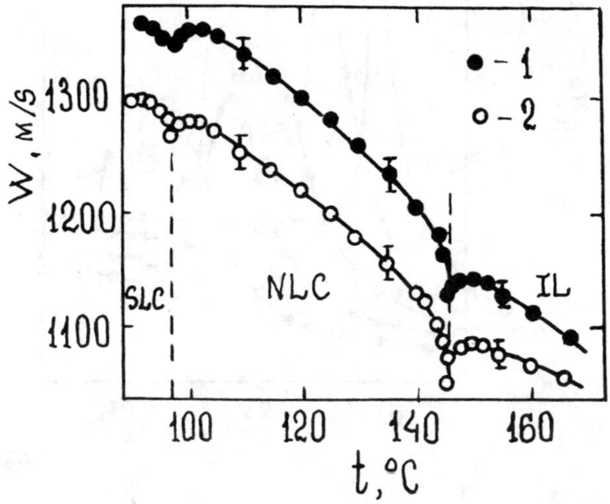

Figure 8.31. Temperature dependence of the velocity of sound at 3 MHz (1) and 60.5 MHz (2) in HOBA.[187]

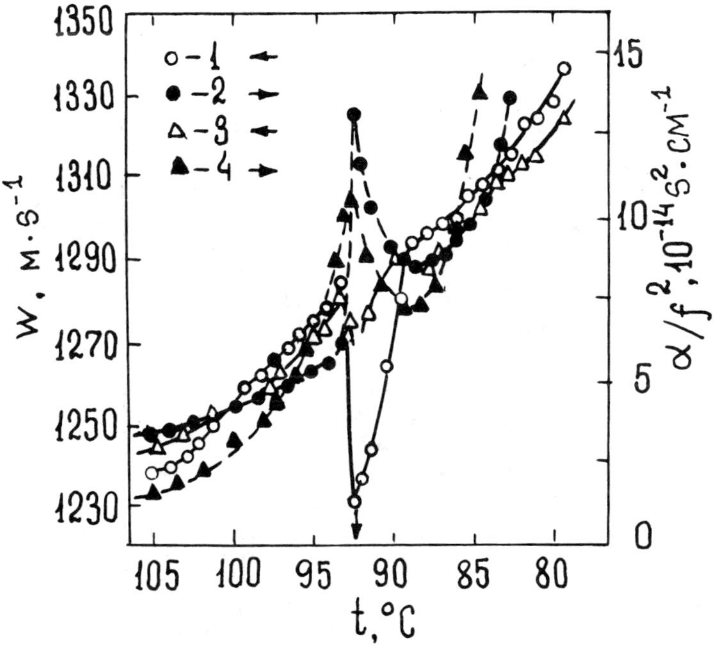

Figure 8.32. Temperature dependences of the velocities of ultrasound (open symbols) and of ultrasonic absorption (dark symbols) in cholesteryl nonanoate (CN) (circles) and cholesteryl decanoate (CD)[211]: 1) CN, velocity; 2) CN, absorption; 3) CD, velocity; 4) CD, absorption.

studied experimentally in detail. Only fragmentary information exists concerning the electrical conductivity of some CLC.[227,228] The computational–theoretical investigations are now being carried out only for self-diffusion. (Some estimates of viscosity, following Helfrich's theory, only concern interpretation of the order of magnitudes of experimental data.) The remaining kinetic characteristics (thermal conductivity, thermodiffusion coefficients, etc.) have been little studied.

Dctailed experimental investigations of the coefficients of self-diffusion of a number of substances forming smectic mesophases were carried out by Krüger *et al.*[143,229] The coefficients of self-diffusion of TBBA, which forms the smectic-A, -C, and -H mesophases, were studied in Ref. 143 (Fig. 8.37), and the coefficients of self-diffusion for a series of homologs of *p*-dodecanoyl-benzylidene-*p*′-aminoazobenzene were determined in Ref. 229. Dasannacharya[145] investigated

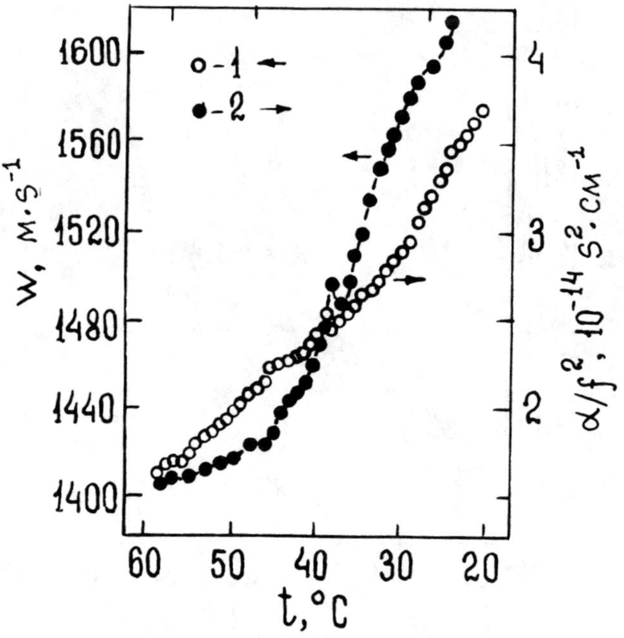

Figure 8.33. Temperature dependences of the velocity of ultrasound (1) and of ultrasonic absorption (2) in cholesteryl oleate.[211]

the coefficients of self-diffusion of EABAC. Diffusion of impurities in SLC-A was studied in Ref. 230.

The coefficient of self-diffusion of the smectic mesophases E, B, and A of partly deuterated *n*-butyl-4-(4′-phenylbenzylidene amino)-cinnamate (D-BPBAC) was determined in Ref. 231 by the method of incoherent quasielastic neutron scattering. It was found that the coefficient of self-diffusion in the A mesophase is of the order of $1 \cdot 10^{-6}$ cm²/s and is an order of magnitude smaller in the E and B phases. Simple rotational models cannot explain these experimental data.

Table 8.6 presents the apparent average coefficients of self-diffusion of smectic mesophases observed in D-BPBAC, in which the series of transitions has the form

$$\text{SC} \xrightarrow{350} \text{SLCE} \xrightarrow{380} \text{SLCB} \xrightarrow{445} \text{SLCA} \xrightarrow{479} \text{IL}$$

(the temperatures are presented in K). This information on self-

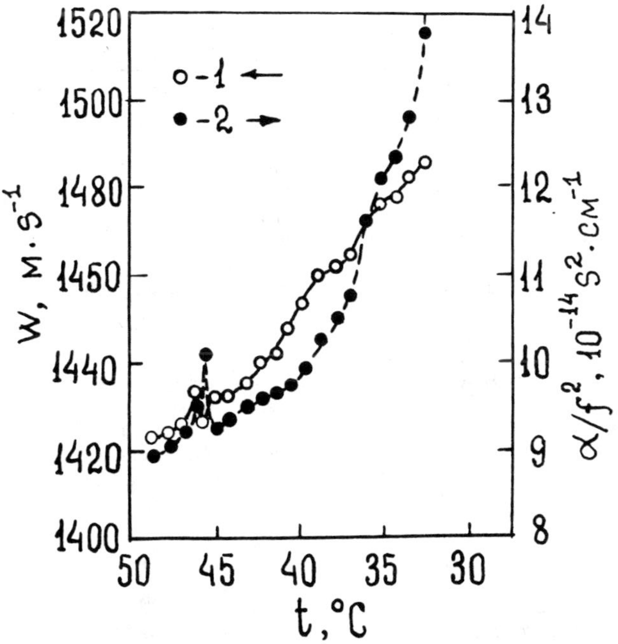

Figure 8.34. Temperature dependences of the velocity of ultrasound (1) and of the ultrasonic absorption (2) in cholesteryl oleate.[211]

diffusion was obtained by the method of incoherent quasi-elastic scattering of neutrons.[232]

The apparent average coefficients of self-diffusion of TBBA are presented in Fig. 8.38.[233] The large jump in the coefficient of self-diffusion of this material at the point of the SLC-H–SLC-C transition is interesting. The experimental data for the smectic-A phase of TBBA were compared with the computed values obtained based on the theory of nuclear spin relaxation, developed in application to translational diffusion in SLC-A.[234]

A qualitative discussion of the values obtained and their agreement with the computed data, in particular with data obtained by the molecular-dynamic method, is given in Sec. 7.5. A review of earlier measurements of the self-diffusion coefficients is given in Ref. 144.

A heat pulse method for the determination of thermal diffusivity of liquid crystals was proposed and used for the study of this property of 8CB.[235] The data obtained in this work are shown in Figure 8.39. It should be noted that for smectic A and nematic phases $D^t_\parallel/D^t_\perp >$

Table 8.4
Summary of Experimental Data on Propagation of Ultrasound in the Region of Phase Transitions in Liquid Crystals*

Type of Transition	Studied Substances	Type of Experiment**	Frequency, MHz
SLC A–NLC	CBOOA BBOA EMBAC	Sound velocity, sound absorption (oriented and non-oriented samples)	2–36; 220; 560; 840
SLC C–NLC	HAB	The same (oriented samples)	12
CLC–IL	CB, CA, CF, CH, CLa, CN, CD, CO, CL, CM, CV, CP	The same (generally non-oriented samples)	0.5–15
SLC A–IL	DEAB	Sound velocity (oriented samples)	2.5; 12; 20
SLC B–SLC A	EMBAC	Sound velocity, sound absorption (oriented samples)	4.12
SLC A–CLC	CD, CN	The same	2–15; 560; 1000
CLC–NLC***	Mixture CC-PAA	Sound absorption at different concentrations and temperatures	3

*The detail bibliography of these experimental works is given in Reference[212].
**For CLC–NLC transition the dependence of sound absorption on field intensity is determined; for other transitions sound velocity and sound absorption were measured as temperature dependence.
***This transition was induced by external field.

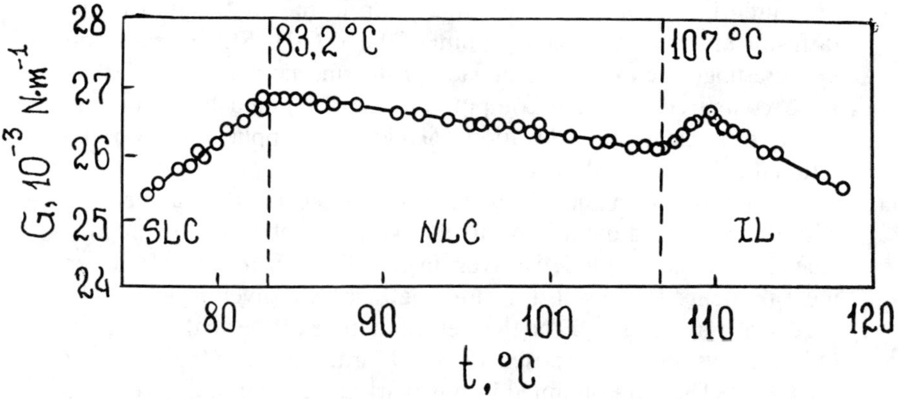

Figure 8.35. Surface tension of CBOOA in smectic, nematic, and isotropic phases.[219]

Table 8.5
Surface Tension σ (dyne · cm^{-1}) of Some Cholesteric Liquid Crystals*

Substance	t, °C	σ	Method of Measurement	Reference
Cholesteryl propionate	90	26.0	Wilhelmy's plate	221
Cholesteryl butyrate	90	25.9	—"—	221
Cholesteryl pentanoate	90	25.6	—"—	221
Cholesteryl hexanoate	90	25.5	—"—	221
Cholesteryl decanoate	88	28.2	Maximum bubble pressure	222
Cholesteryl tetradecanoate	82	23.9	Wilhelmy's plate	218
Cholesteryl octadecanoate	75	24.9	—"—	223

*Summary of the data is given in Reference[224].

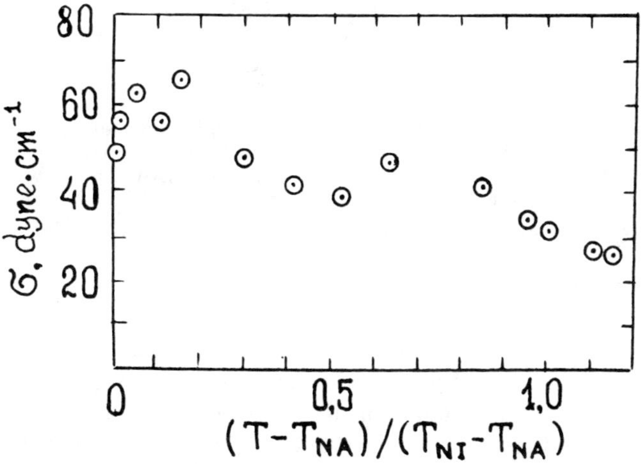

Figure 8.36. Surface tension of 4-*n*-octyloxy-4′-cyanobiphenyl[225]: T_{NA} = 352,5 K, T_{NI} = 339,8 K, T_{AC} = 327,7 K.

1, while for coefficients of normal diffusion of smectic A phase $D_{\parallel}/D_{\perp} < 1$ is usually observed.

Measurements of self-diffusion in the cholesteric liquid crystalline phase by means of NMR-method were performed by using isotrope ^{13}C.[236] The values of self-diffusion changed within (3–20)·10^{-12} m^2/s versus the temperature, and activation energy was about 40 kJ/mole which is close to values characteristic of NLC.

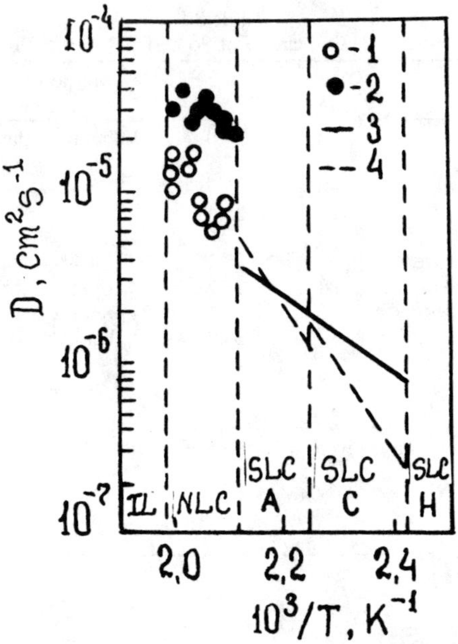

Figure 8.37. Self-diffusion coefficients of TBBA[143]: 1) D for NLC; 2) D for NLC; 3) D for SLC-A and SLC-C; 4) D for SLC-A and SLC-C.

As noted in Sec. 7.6, the viscous properties of SLC and CLC generally do not satisfy Poiseuille's and Newton's laws. Thus, the measured viscosity coefficients depend on the rate of shear and are anomalously large compared with the viscosity of NLC or isotropic liquids. The results obtained in Ref. 237 and presented in Fig. 8.40 demonstrate these features. In particular, it is evident that the coefficient of effective (apparent) viscosity of cholesteryl myristate depends on the rate of shear (with the high temperature range ex-

Table 8.6
Apparent Average Coefficients of Self-Diffusion (D) of D-BPBAC[232]

Phase	Temperature, K	D, 10^{-7} cm$^2\cdot$s^{-1}
SLC A	469	13 ± 2
SLC A	455	9.5 ± 1
SLC B	439	≲ (0.4 ± 0.1)
SLC E	363	≲ (0.4 ± 0.1)
SLC E	333	≲ (0.4 ± 0.1)

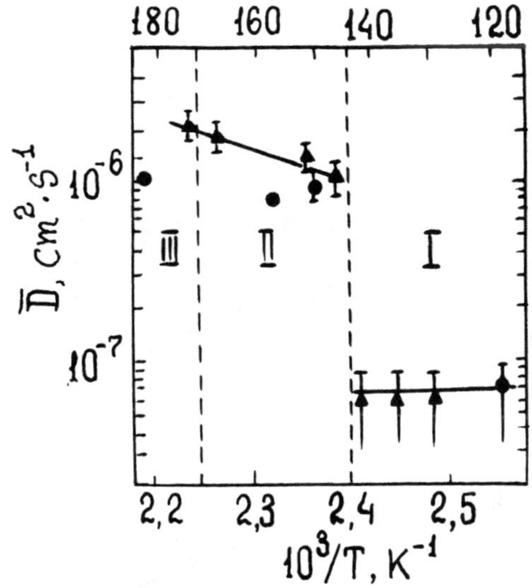

Figure 8.38. Values of the average coefficient of translational diffusion of TBBA:[233] I) SLC–H, II) SLC–C, III) SLC–A.

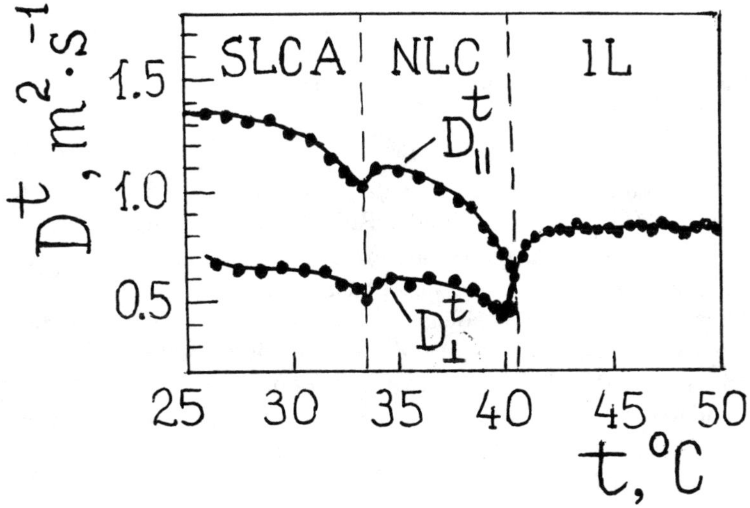

Figure 8.39. Temperature dependence of the thermal diffusivity of 8CB for smectic A phase (SLCA), nematic phase (NLC) and isotropic liquid (IL).[235]

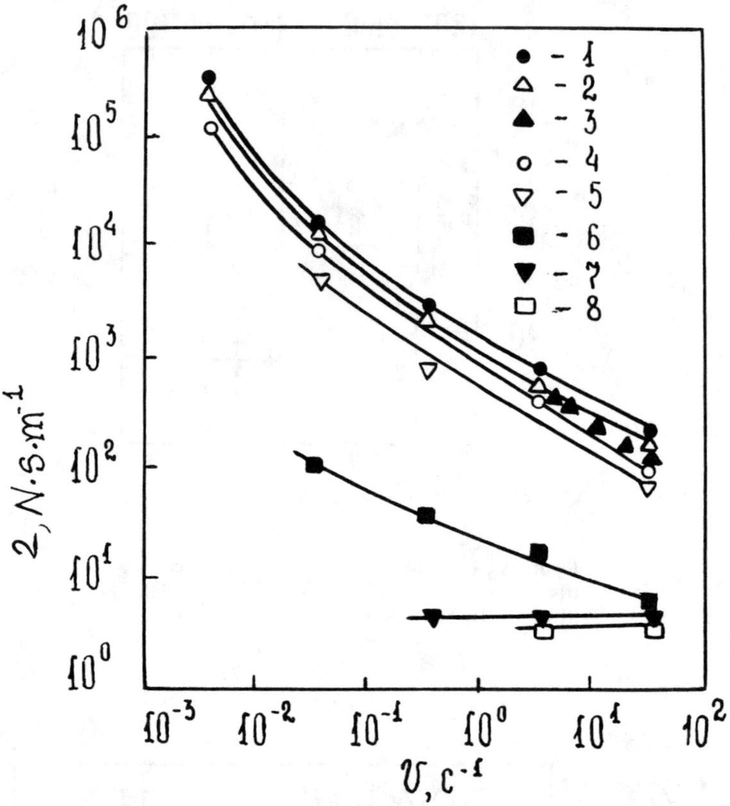

Figure 8.40. Viscosity of chlesteryl myristate as a function of the rate of shear at 65°C (1), 70°C (2), 74°C (3), 75°C (4), 77°C (5), 79°C (6), 84°C (7), and 90°C (8).[7,237]

cepted). An analysis of experimental data on the viscosity of cholesteryl acetate[238] (Fig. 8.41) presented in Ref. 16 leads to similar conclusions.

A. A. Tabidze[239] studied the effective and dynamic viscosities of the smectic and cholesteric mesophases of cholesteryl caprinate, cholesteryl palmitate, and cholesteryl myristate over a wide temperature range (Fig. 8.42). The value of the activation energy of SLC (estimated from the slope of temperature dependence of viscosity) was found to be constant within 10% over a wide range of rates of shear.[187] The dynamic viscosity of three cholesteryl esters was studied by the impedance method in Ref. 240. Detailed discussion of these results is presented in Ref. 187.

The values of the shear and bulk viscosities (the latter can be found

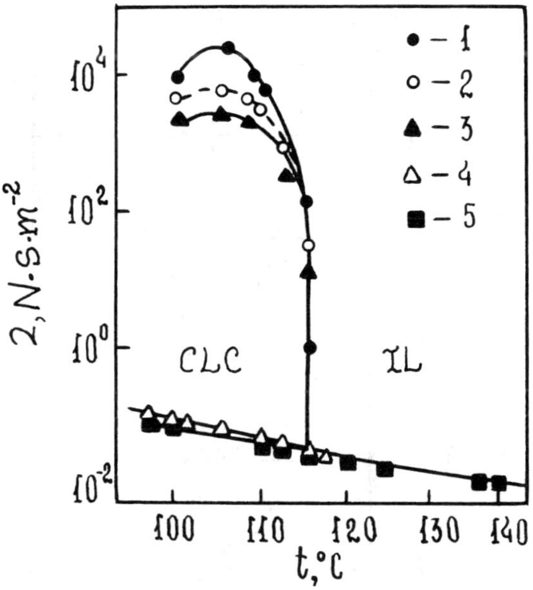

Figure 8.41. Temperature dependence of the viscosity of chlesteryl acetate with the following rates of shear:[16,238] 1) 10 s^{-1}; 2) 50 s^{-1}; 3) 100 s^{-1}; 4) 10^4 s^{-1}; 5) high rates of shear and an isotropic liquid.

from data on the velocity and absorption of sound[36]) are compared in Fig. 8.43. Constancy of the ratio η_V/η for regions of the cholesteric and isotropic phases is characteristic and is related to the activation energies of flow and orientational rotation being equal in these phases.

The coefficients of viscosity of cholesteryl stearate, cholesteryl myristate, cholesteryl valerate, cholesteryl caprylate, and the viscosity coefficient of the nematic *N*-(*p*-ethoxybenzylidene)-*p*-*n*-butylaniline (EBBA) were studied in Ref. 228 (Fig. 8.44). These data characterize the relative change in viscosity with the transition from the isotropic to the cholesteric phase. The value of the results, however, is limited because they were obtained for a constant rate of shear. A discussion of problems related to the viscous properties of SLC and CLC, as well as some quantitative data are presented in Refs. 241–244.

The results of studies of the singularities in the temperature dependence of the viscosity near the point of the SLC-A–SLC-B transition are of great interest. This study was performed for butyloxybenzylidene-octylaniline (BBOA or 40.8),[245] which exhibits the

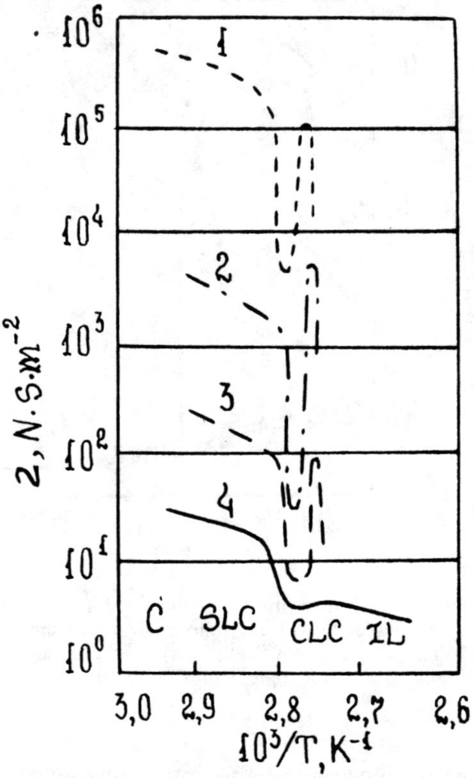

Figure 8.42. Temperature dependence of the viscosity of cholesteryl myristate for the following rates of shear:[187,239] 1) 0.0034 s^{-1}; 2) 0.34 s^{-1}; 3) 34.0 s^{-1}; 4) dynamic viscosity.

following sequence of transitions (the transitions temperatures are indicated in *K*):

$$\text{SC} \xrightarrow{310.9} \text{SLCB} \xrightarrow{322.4} \text{SLCA} \xrightarrow{336.9} \text{NLC} \xrightarrow{352.0} \text{IL}.$$

The heat of the SLC-A–SLC-B transition is equal to 2.09 kJ/mole. The results of this study are presented in Fig. 8.45. It is evident that in the direct proximity to the point of the transition the "critical" viscosity η_c is satisfactorily described by the expression:

$$\eta_c = \eta_o \exp\left[b/(T - T_c)^n\right] \tag{8.1}$$

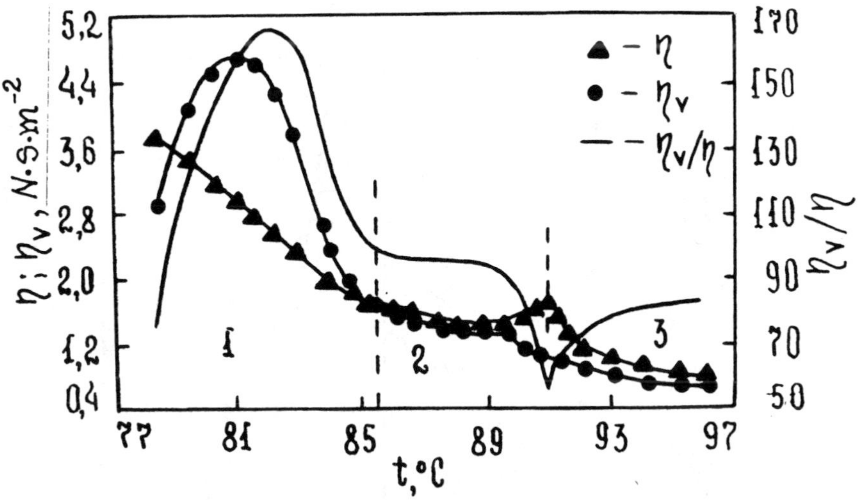

Figure 8.43. Shear η and bulk η_V viscosities and the ratio η_V/η for cholesteryl caprilate[36]: 1) SLC; 2) CLC; 3) IL.

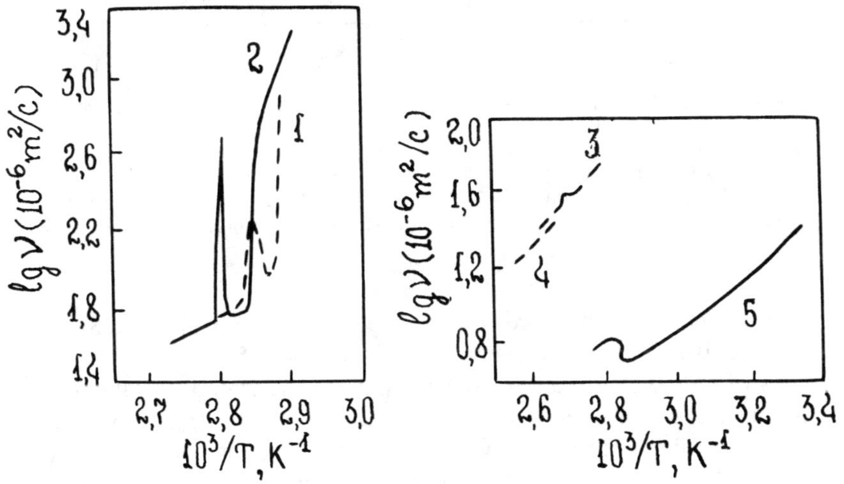

Figure 8.44. Temperature dependences of the kinematic viscosity of liquid crystals:[228] 1) cholesteryl stearate; 2) cholesteryl myristate; 3) cholesteryl valerate; 4) cholesteryl caprilate; 5) EBBA.

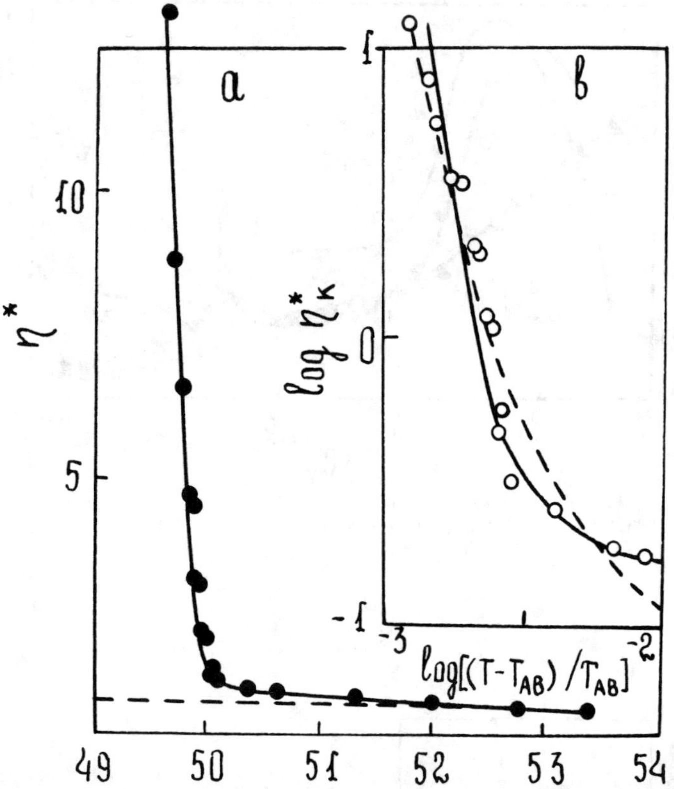

Figure 8.45. a) Temperature dependence of the reduced viscosity $\eta^* = \eta_0/\eta_1$, where η_1 is the viscosity at 53.4°C in the vicinity of the SLC–A–SLC–B transition of butyloxybenzylidene-octylaniline (BBOA). The broken line corresponds to Arrhenius' classical law with an activation energy of 46.1 kJ/mole. b) The logarithmic dependence of the reduced critical viscosity $\eta_c^* = \eta_c/\eta_1$. The solid and broken lines were calculated based on Eq. (8.1) with $\eta_0/\eta_1 = 0.05$ and 0.15, respectively (for n equals to 0.8 and 2.0).[245]

The theoretical models previously proposed for describing the SLC-A–SLC-B transition are also discussed in Ref. 245.

In recent years attention has been paid to the experimental study of the viscosity of smectic and cholesteric liquid crystals. The method of the capillary viscometer was used for the viscosity measurement of 8OCB and compounds of homologous series of 5-n-alkyl-2-(4′-isotiocyanotophenyl) dioxane-1,3 demonstrating the smectic mesophase.[246] The viscosity measurement of hexatic smectic B phase of 65OBC was performed by means of the method of stress relaxation.[247]

It was possible to explain the obtained results on the basis of the model for the free dislocations movement. The relaxation method was also used for the observation of the infiltration effect in a cholesteric liquid crystal.[248] The viscosity of blue phases BPI, BPII and BPIII was measured.[249]

Thermal-conductivity studies near the smectic A-hexatic B transition in a liquid crystal compound 65CBC was performed by means of the method of the periodic heating at 20 mK $< T - T_c <$ 3K.[250] The obtained data indicate the decrease of thermal conductivity in the vicinity of the critical point. The critical index of the thermal-conductivity proved to be equal to 0.51 ± 0.04, while the critical index of the heat capacity is 0.60 ± 0.03.

The features of the molecular movement of the smectic B,E,G,H phases,[251] smectic F phase of PBHA,[252] and cholesteric phase of saturated acyl cholesteryl esters[253] were studied on the basis of different experimental data (temperature dependence of the time of the spin-lattice relaxation, measurements by the method of spin label, temperature dependences of NMR spectrum). Theoretical models of molecular dynamics of SLC and CLC were discussed in these works.

References for Part II

CHAPTER V

1. A. L. Tsykalo, Teplofizicheskie svoistva nematicheskikh zhidkikh kristallov. Obzornaya informatsiya, No. 6, Inst. Vys. Temp. Akad. Nauk, Moscow (1978) [English translation: *Thermophysical Properties of Nematic Liquid Crystals*. General Information, Harwood Academic Publishers, New York (1982)].
2. K. K. Kobayashi, *Phys. Lett. A*, **31**, 125 (1970).
3. W. L. McMillan, *Phys. Rev. A*, **4**, 1238 (1971).
4. W. L. McMillan, *Phys. Rev. A*, **8**, 1921 (1973).
5. R. B. Meyer and W. L. McMillan, *Phys. Rev. A*, **9**, 899 (1974).
6. W. J. A. Goossens, *Phys. Lett. A*, **31**, 413 (1970).
7. P. G. de Gennes, *The Physics of Liquid Crystals*, Clarendon Press, Oxford (1974).
8. M. J. Stephen and J. P. Straley, *Rev. Mod. Phys.*, **46**, 617 (1974).
9. V. M. Chaikovskii and I. G. Chistyakov in: Sb. dokladov 1 nauchnoi konferentsii po zhidkim kristallam (Contributions to the 1st Scientific Conference on Liquid Crystals), Ivanovo (1972), p. 301.
10. E. I. Kovshev, L. M. Blinov, and V. V. Titov, *Usp. Khim.*, **46**, 753 (1977) [*Russ. Chem. Rev.*, **46**, 395 (1977)].
11. I. G. Chistyakov, *Zhidkie kristally* (*Liquid Crystals*), Nauka, Moscow (1966), p. 127.
12. L. M. Blinov, *Usp. Fiz. Nauk*, **114**, 67 (1974) [Sov. Phys. Usp. 17, 658 (1975)].
13. L. N. Bardukov, B. I. Bel'tsov, A. A. Bronnikova, and I. P. Igoshin in: *Zhidkie kristally* (*Liquid Crystals*), Ivanovo (1978), p. 45.
14. A. A. Bronnikova in: Sb. dokladov 1 nauchnoi konferentsii po zhidkim kristallam (Contributions to the 1st Scientific Conference on Liquid Crystals), Ivanovo (1972), p. 112.
15. I. G. Chistyakov and L. S. Shabyshev. *ibid.*, p. 43.
16. S. Chandrasekhar, *Rep. Prog. Phys.*, **39**, 613 (1976).
17. G. G. Maidachenko, R. P. Smirnov, T. E. Kalinnikova, and L. N. Kotovich in: *Zhidkie Kristally* (*Liquid Crystals*), Ivanovo (1978), p. 132.
18. V. G. Tishchenko, M. M. Fetisova, and R. M. Cherkashina, *ibid.*, p. 111.
19. V. G. Tishchenko and R. M. Cherkashina in: *Kholestericheskie Zhidkie Kristally* (*Cholesteric Liquid Crystals*), Novosibirsk (1976), p. 26.
20. G. W. Gray and J. W. Goodby, *Mol. Cryst. Liq. Cryst.*, **37**, 157 (1976).
21. G. W. Gray and D. G. McDonnell, *Mol. Cryst. Liq. Cryst.*, **37**, 189 (1976).
22. A. L. Tsykalo, *Dokl. Akad. Nauk SSSR*, **249**, 1348 (1979) [*Sov. Phys. Dokl.*, **24**, 1002 (1979)].
23. B. I. Halperin and D. R. Nelson, *Phys. Rev. Lett.*, **41**, 121 (1978).
24. D. E. Moncton and R. Pindak, *Phys. Rev. Lett.*, **43**, 701 (1979).
25. W. H. de Jeu and J. A. De Poorter, *Phys. Lett. A*, **61**, 114 (1977).
26. G. G. Natale, *J. Acoust. Soc. Am.*, **63**, 1677 (1978).
27. J. W. Goodby and G. W. Gray, *Mol. Cryst. Liq. Cryst.*, **49**, 217 (1979).
28. H. Sackmann and D. Demus, *Mol. Cryst.*, **2**, 81 (1966).
29. H. Sackmann and D. Demus, *Fortschr. Chem.*, **12**, 349 (1969).
30. D. Demus, G. Kunicke, J. Neelsen and H. Sackmann, *Z. Naturforsch.*, **A23**, 84 (1968).

31. S. Diele, P. Brand and H. Sackmann, *Mol. Cryst. Liq. Cryst.*, **17**, 163 (1972).
32. D. Coates, K. J. Harrison and G. W. Gray, *Mol. Cryst. Liq. Cryst.*, **22**, 99 (1973).
33. J. Doucet, A. M. Levelut, M. Lambert, L. Liebert and L. Strzelecki, *J. Phys. (Fr.)*, **15**, C13 (1975).
34. D. Demus, S. Diele, M. Klapperstuck, V. Link and H. Zaschke, *Mol. Cryst. Liq. Cryst.*, **15**, 161 (1971).
35. A. de Vries and D. L. Fishel, *Mol. Cryst. Liq. Cryst.*, **16**, 311 (1972).
36. A. P. Kapustin, *Elektroopticheskie i akusticheskie svoistva zhidkikh kristallov* (*Electrooptical and Acoustical Properties of Liquid Crystals*), Nauka, Moscow (1973), p. 232.
37. P. A. Cladis, R. K. Bogardus, W. B. Daniels, and E. N. Taylor, *Phys. Rev. Lett.*, **39**, 720 (1977).
38. G. S. Ranganath. Pramana, *J. Phys.*, **27**, 299 (1986).
39. A. L. Tsykalo and A. D. Bagmet in: *VI Vsesoyuz. konf. po teplofizicheskim svoistavam veshchestrv* (*6th All-Union Conference on the Thermophysical Properties of Substances*), Minsk (1978), p. 108.

CHAPTER VI

40. J. G. Kirkwood and E. Monroe, *J. Chem. Phys.*, **9**, 514 (1941).
41. J. G. Kirkwood, *Phase Transformations in Solids*, John Wiley and Sons, Inc., New York (1951), p. 387.
42. P. J. Wojtowicz, *RCA Rev.*, **35**, 388 (1974).
43. T. R. Taylor, S. L. Arora, and J. L. Fergason, *Phys. Rev. Lett.*, **25**, 722 (1970).
44. L. Doucet, A. M. Levelut, and M. Lambert, *Phys. Rev. Lett.*, **32**, 301 (1974).
45. D. Guillon, P. Seurin, and A. Skoulios, *Mol. Cryst. Liq. Cryst.*, **51**, 149 (1979).
46. D. Guillon, P. Finn, W. B. Daniels, and A. C. Griffin, *Mol. Cryst. Liq. Cryst. Lett.*, **64**, 93 (1980).
47. Ali H. Al-Mowali, *Z. Phys. Chem. (BRD)*, **141**, 23 (1984).
48. J. J. Benattar, B. Deloche, and J. Charvolin, *J. Phys. (Fr.)*, **46**, 2179 (1985).
49. S. Dumrongrattana and C. C. Huang, *Phys. Rev. Lett.*, **56**, 464 (1986).
50. Y. Ouchi, T. Uemura, H. Takezoe, and A. Fukuda, *Jap. J. Appl. Phys.*, **24**, 893 (1985).
51. A. Saupe, *Ann. Rev. Phys. Chem.*, **24**, 441 (1973).
52. N. L. Kramarenko, I. V. Kurnosov, Yu. V. Naboikin, and V. G. Tishchenko in: *Zhidkie Kristally* (*Liquid Crystals*), Ivanovo (1976), p. 118.
53. L. N. Lisetskii, A. V. Tolmachev, and V. G. Tishchenko, *Pis'ma Zh. Eksp. Teor. Fiz.*, **27**, 205 (1978) [*JETP Lett.*, **27**, 191 (1978)].
54. A. L. Tsykalo, *Zh. Fiz. Khim.*, **53**, 2528 (1979) [*Russ. J. Phys. Chem.*, **53**, 1143 (1979)].
55. A. Wulf, *J. Chem. Phys.*, **59**, 1487 (1973).
56. P. N. Keating, *Mol. Cryst. Liq. Cryst.*, **8**, 315 (1969).
57. P. N. Keating, (*Liquid Crystals*), Gordon and Breach, New York (1976), p. 279.
58. B. Böttcher, *Chem. Z.*, **96**, 214 (1972).
59. P. G. de Gennes, *Solid State Commun.*, **10**, 753 (1972).
60. P. V. Vigman and V. M. Filev, *Zh. Eksp. Teor. Fiz.*, **69**, 1466 (1975) [*Sov. Phys. JETP*, **42**, 747 (1976)].

61. B. W. Van der Meer and G. Vertogen, *Phys. Lett. A*, **59**, 279 (1976).
62. B. W. Van der Meer, G. Vertogen, A. J. Dekker, and J. G. J. Ypma, *J. Chem. Phys.*, **65**, 3935 (1976).
63. R. Pindar and T. H. John, *Phys. Lett. A*, **59**, 277 (1976).
64. Y. R. Lin-Liu, Yu Ming Shih, Chia-Wei Woo, and H. T. Tan, *Phys. Rev. A*, **14**, 445 (1976).
65. H. Stegemeyer, *Ber. Bunsenges, Phys. Chem.*, **78**, 860 (1974).
66. L. N. Lisetskii, B. L. Timan, V. G. Tishchenko, and O. D. Kolotii, *Fiz. Tverd. Tela*, **19**, 3461 (1977) [*Sov. Phys. Solid State*, **19**, 2023 (1977)].
67. T. Harada and P. Crooker, *Mol. Cryst. Liq. Cryst.*, **30**, 79 (1975).
68. Y. R. Lin-Liu, Yu Ming Shih, and Chia-Wei Woo, *Phys. Rev. A*, **15**, 2550 (1977).
69. S. Chandrasekhar and B. R. Rahana, *Mol. Cryst. Liq. Cryst.*, **35**, 109 (1976).
70. P. Pollmann and H. Stegemeyer, *Chem. Phys. Lett.*, **20**, 87 (1973).
71. H. Stegemeyer and K. J. Mainusch, *Chem. Phys. Lett.*, **16**, 38 (1972).
72. A. Wulf, *J. Chem. Phys.*, **60**, 3994 (1974).
73. H. Kozawaguchi and M. Wada, *Mol. Cryst. Liq. Cryst.*, **45**, 55 (1978).
74. L. M. Blinov, *Elektro- i magnito-optika zhidkhikh kristallov* [*Electro- and Magneto-Optics of Liquid Crystals*], Nauka, Moscow (1978), p. 384.
75. P. Adamski, and A. Dylik-Fromiec, *Mol. Cryst. Liq. Cryst.*, **35**, 337 (1976).
76. P. J. Collins and J. R. McColl, *Solid State Commun.*, **28**, 997 (1978).
77. C. E. Tarr, M. E. Field, and L. R. Whalley, *Mol. Cryst. Liq. Cryst.*, **37**, 353 (1976).
78. A. V. Tolmachev, V. G. Tishchenko, and L. N. Lisetskii, *Fiz. Tverd. Tela*, **19**, 1886 (1977) [*Sov. Phys. Solid State*, **19**, 1105 (1977)].
79. A. S. Sonin, A. V. Tolmachev, and V. G. Tishchenko, *Kristallografiya*, **21**, 1164 (1976) [*Sov. Phys. Crystallogr.*, **21**, 673 (1976)].
80. E. B. Priestley, *RCA Rev.*, **35**, 144 (1974).
81. S. Jen, N. A. Clark, P. S. Pershan, and E. B. Priestley, *Phys. Rev. Lett.*, **31**, 1552 (1973).
82. L. N. Lisetskii and V. G. Tishchenko in: *Kholestericheskie Zhidkie kristally (Cholesteric Liquid Crystals)*, Novosibirsk (1976), p. 14.
83. C. Lehmann, *Z. Phys. Chem.*, **56**, 750 (1906).
84. F. P. Price and J. H. Wendorff, *J. Phys. Chem.*, **75**, 2839 (1971).
85. H. Arnold and P. Roediger, *Z. Phys. Chem.*, **283**, 239 (1968).
86. K. Bergmann and H. Stegemeyer, *Ber. Bunsgenges. Phys. Chem.*, **82**, 1309 (1978).
87. K. Bergmann and H. Stegemeyer, *Z. Naturforsch.*, **A34**, 251 (1979).
88. *Ibid*, 253.
89. *Ibid*, 255.
90. V. A. Kizel and V. V. Prokhorov, *Zh. Exp. Teor. Fiz.*, **87**, 450 (1984).
91. R. Barbet-Massin, P. E. Cladis and P. Pieranski, *Phys. Rev. A: Gen. Phys.*, **30**, 1161 (1984).
92. R. Barbet-Massin and P. Pieranski, *P. Phys. Lett.* (*Fr.*), **45**, 799 (1984).
93. Matthew A. Marcus, *Phys. Rev. A: Gen. Phys.*, **30**, 1109 (1984).
94. P. Pieranski, R. Barbet-Massin and P. E. Cladis, *Phys. Rev. A: Gen. Phys.*, **31**, 3912 (1985).
95. V. A. Belyakov, D. I. Demikhov, V. E. Dmitrienko and V. K. Dolganov, *Zh. Exp. Teor. Fiz.*, **89**, 2035 (1985).
96. F. Porsch and H. Stegemeyer, *Chem. Phys. Lett.*, **125**, 319 (1986).

CHAPTER VII

97. F. T. Lee, H. T. Tan, and C. W. Woo, *Phys. Lett. A*, **48**, 68 (1974).
98. H. Arnold and P. Roediger, *Z. Phys. Chem. (Leipzig)*, **239**, 283 (1968).
99. J. G. J. Ypma and G. A. Vertogen, *J. Phys. (Paris)*, **37**, 557 (1976).
100. R. G. Priest, *Mol. Cryst. Liq. Cryst.*, **37**, 101 (1976).
101. P. G. de Gennes, *C. R. Acad. Sci. Ser. B*, **274**, 758 (1972).
102. R. Wise, D. Smith, and W. Doane, *Phys. Rev. A*, **7**, 1366 (1973).
103. A. Wulf, *Phys. Rev. A*, **17**, 2077 (1978).
104. A. Wulf, *Mol. Cryst. Liq. Cryst.*, **47**, 225 (1978).
105. H. Arnold and P. Roediger, *Z. Phys. Chem. (Leipzig)*, **231**, 407 (1966).
106. H. Hervet, F. Tolino, A. I. Dianoux, and R. E. Lechauer, *Phys. Rev. Lett.*, **34**, 451 (1975).
107. G. S. Attard and G. R. Luckhurst, *Chem. Phys. Lett.*, **117**, 523 (1985).
108. E. P. Sokolova, *Vestnik Leningradskogo gosudarstvannogo universiteta (Leningrad State University News)*, **1**, 28 (1987).
109. A. Kloczowski and J. Stecki, *Mol. Phys.*, **55**, 689 (1985).
110. F. Dowell, *Phys. Rev. A: Gen. Phys.*, **31**, 2464 (1985).
111. Y. Drossinos and D. Ronis, *Phys. Rev. A: Gen. Phys.*, **33**, 589 (1986).
112. W. J. A. Goossens, *Europhys. Lett.*, **3**, 341 (1987).
113. J. Katriel *et al.*, *Phys. Rev. A: Gen. Phys.*, **28**, 3037 (1983).
114. G. F. Kventsel, G. R. Luckhurst and H. B. Zewdie, *Mol. Phys.*, **56**, 589 (1985).
115. H. Stegemeyer and H. Finkelmann, *Chem. Phys. Lett.*, **23**, 227 (1973).
116. H. Hanson, A. J. Dekker, and F. Van der Woude, *J. Chem. Phys.*, **62**, 1941 (1975).
117. H. Finkelmann and H. Stegemeyer, *Ber. Bunsenges. Phys. Chem.*, **78**, 869 (1974).
118. T. C. Lubensky, *J. Phys. Chem. Solids.*, **34**, 365 (1973).
119. F. P. Price and C. S. Bak, *Mol. Cryst. Liq. Cryst.*, **29**, 225 (1975).
120. E. I. Dzyaloshinskii, S. G. Dmitriev, and E. I. Kats, *Zh. Eksp. Teor. Fiz.*, **68**, 2335 (1975) [*Sov. Phys. JETP*, **41**, 1167 (1975)].
121. A. G. Khachaturyan, *J. Phys. Chem. Solids*, **36**, 1055 (1975).
122. W. J. A. Goossens, *Liq. Cryst.*, **1**, 521 (1986).
123. T. R. Taylor, J. L. Fergason and S. L. Arora, *Phys. Rev. Lett.*, **24**, 359 (1970); **25**, 722 (1970).
124. W. Helfrich and C. S. Oh, *Molec. Cryst. Liq. Cryst.*, **14**, 289 (1971).
125. V. M. Filev, *Pizma Zh. Exp. Teor. Fiz.*, **43**, 523 (1986).
126. D. C. Wright and N. D. Mermin, *Phys. Rev. A: Gen. Phys.*, **31**, 3498 (1985).
127. V. A. Belyakov and V. E. Dmitrienko, *Usp. Fiz. Nauk.*, **146**, 369 (1985).
128. H. Stegemeyer, Th. Blümel, K. Hiltrop, H. Onusseit and F. Porsch, *Wiss. Beitr. M.*-Luther Univ. Halle-Wittenberg, **17**, 64 (1986).
129. P. Pieranski and P. E. Cladis, *Phys. Rev. A: Gen. Phys.*, **35**, 355 (1987).
130. P. Pieranski and P. E. Cladis, *Europhys. News.*, **17**, 113 (1986).
131. B. Pansu, R. Dandoloff, and E. Dubois-Viollette, *J. Phys. (Fr.)*, **48**, 297 (1987).
132. A. L. Tsykalo and A. D. Bagmet, *Kristallografiya*, **21**, 1101 (1976) [*Sov. Phys. Crystallogr.*, **21**, 637 (1976)].
133. A. L. Tsykalo and A. D. Bagmet, *Fiz. Tverd. Tela*, **20**, 1326 (1978) [*Sov. Phys. Solid State*, **20**, 762 (1978)].
134. A. L. Tsykalo and A. D. Bagmet, *Mol. Cryst. Liq. Cryst.*, **46**, 111 (1978).

135. A. L. Tsykalo, *Zh. Tekh. Fiz.*, **50**, 54 (1980) [*Sov. Phys. Tech. Phys.*, **25**, 28 (1980)].
136. A. L. Tsykalo and A. D. Bagmet, *Kristallografiya*, **25**, 568 (1980).
137. A. L. Tsykalo, *Teplofizicheskie svoistva smekticheskikh i kholestericheskihk zhidkikh kristallov* (*Thermophysical Properties of Smectic and Cholesteric Liquid Crystals*). Obzornaya informatsiya, N 6. Inst. Vys. Temp., Akad. Nauk, Moscow (1980).
138. A. L. Tsykalo, *Molec. Cryst. Liq. Cryst.*, **129**, 409 (1985).
139. B. J. Berne and P. Pechukas, *J. Chem. Phys.*, **56**, 4213 (1972).
140. G. J. Krüger, H. Spiesecke, and R. Weiss, *Phys. Lett. A*, **51**, 295 (1975).
141. Tadashi Akahane, Masumi Kondoh, Kenji Hashimoto and Masahiro Nakagawa, *Jap. Appl. Phys.*, **26**, 1000 (1987).
142. B. Quentrec, *Phys. Rev. A*, **12**, 282 (1975).
143. G. J. Krüger and R. Weiss, *J. Phys.*, **38**, 353 (1977).
144. T. E. Faber and G. R. Luckhurst, *Ann. Rep. Prog. Chem. A*, **72**, 31 (1975/1976).
145. B. A. Dasannacharya, *Proc. Nucl. Phys. Solid State Phys. Symp.* (Calcutta), **C18**, 338 (1975).
146. A. L. Tsykalo and A. D. Bagmet, *Acta Phys. Pol. A*, **55**, 111 (1979).
147. L. Kivelson, *Mol. Phys.*, **28**, 321 (1974).
148. A. Stroobants, H. N. W. Lekkerkerker and D. Frenkel, *Phys. Rev. Lett.*, **57**, 1452 (1986).
149. A. L. Tsykalo, E. E. Nyankina and A. D. Bagmet, *Zh. Tekh. Fiz.*, **50**, 2365 (1980).
150. R. Alben, *Solid State Commun.*, **13**, 1783 (1973).
151. S. Torza and P. Cladis, *Phys. Rev. Lett.*, **32**, 1406 (1974).
152. R. G. Priest, *J. Phys. (Paris)*, **36**, 437 (1975).
153. A. Kloczkowski and J. Stecki, *Mol. Phys.*, **55**, 1223 (1985).
154. E. I. Katz and V. V. Lebedev, *Physica. A*, **135**, 601 (1986).
155. W. J. A. Goossens, *J. Phys. (Fr.)*, **46**, 1411 (1985).
156. C. C. Huang and S. Mumrongattana, *Phys. Rev. A: Gen. Phys.*, 34, 5020 (1986).
157. L. D. Landau and E. M. Lifshitz, *Statisticheskaya fizika*, Nauka, Moscow (1976), p. 584 [English translation: *Statistical Physics*, Pergamon Press, New York (1980)].
158. D. A. Badalyan and A. G. Khachaturyan in: *Vtoraya konferentsiya sotsialisticheskikh stran po zhidkim cristallam* (2nd Conference of Socialist Countries of Liquid Crystals), Solnechnyi Bereg, Bulgaria (1977), p. 93.
159. R. B. Meyer and T. C. Lubensky, *Phys. Rev. A*, **14**, 2307 (1976).
160. F. T. Lee, H. T. Tan, and C. W. Woo, *Phys. Lett. A*, **48**, 68 (1974).
161. S. A. Pikin, "Macroscopic theory of the stability of the structure of liquid crystals," Doctoral Dissertation, Institute of Crystallography, Academy of Sciences of the USSR, Moscow (1978).
162. Lech Longa, *J. Chem. Phys.*, **85**, 2974 (1986).
163. S. Somasekhara, R. Shashidhar and B. R. Ratna, *Phys. Rev. A: Gen. Phys.*, **34**, 2561 (1986).
164. H. Kresse, A. Wiegeleben and B. Gajewska, *Z. Phys. Chem.*, **266**, 193 (1985).
165. W. Helfrich, *Phys. Rev. Lett.*, **23**, 372 (1969).
166. P. G. de Gennes, *Phys. Fluids*, **17**, 1645 (1975).
167. Ronald Y. Dong, *J. Magn. Res.*, **66**, 422 (1986).
168. E. I. Katz and V. V. Lebedev, *Zh. Exp. Teor. Fiz.*, **91**, 878 (1986).

CHAPTER VIII

169. T. J. McKee and J. R. McColl, *Phys. Rev. Lett.*, **34**, 1076 (1975).
170. S. Chandrasekhar, *Proc. Int. Liq. Cryst. Conf.*, Bangalore, 1973; Pramana Supplement 1, p. 117.
171. P. H. Keyes, H. T. Weston, W. J. Lin, and W. B. Daniels, *J. Chem. Phys.*, **63**, 5006 (1975).
172. N. A. Tikhomirova and A. V. Ginzberg, *Kristallografiya*, **22**, 155 (1977) [*Sov. Phys. Crystallogr.*, **22**, 87 (1977)].
173. P. E. Cladis, D. Guillon, W. B. Daniels, and A. C. Griffin, *Mol. Cryst. Liq. Cryst.*, **56**, 89 (1979).
174. Prasad S. Krishna, S. Pfeiffer, G. Heppke and R. Shashidhar, *Z. Naturforsch.*, **A40**, 632 (1985).
175. A. Mädicke, D. Demus, A. I. Pavluchenko, V. V. Titov and N. I. Smirnova, *Cryst. Res. and Technol.*, **21**, 111 (1986).
176. Philip J. Bullock, David J. Byron, David J. Harwood, Robert C. Wilson and Adrian M. Woodward, *J. Chem. Soc. Perkin Trans.*, **12**, 2121 (1984).
177. David J. Byron, Robert C. Wilson, Philip P. Baker, Ighinas Danilewicz, Gian S. Hayer and Julian M. Wyer, *Ibid.*, **V2**, 297 (1985).
178. Akira Takase and Sakumitsu Sakagami, *Jap. J. Appl. Phys.*, **23**, 1532 (1984).
179. R. Shashidhar, *Physica. BC*, **139–140**, 609 (1986).
180. E. M. Averyanov, P. V. Adomenas, V. A. Zhuikov and V. Ya. Zyryanov, *Pizma Zh. Eksp. Teor. Fiz.*, **43**, 89 (1986).
181. F. Hardouin, *Physica. A*, **140**, 359 (1986).
182. H. Sackmann, *Wiss. Beitr. M.*-Luther-Univ. Halle Wittenberg, **17**, 43 (1986).
183. N. A. Nedostup and V. V. Gal'tsev, *Zh. fiz. Khim.*, **51**, 214 (1977) [*Russ. J. Phys. Chem.*, **51**, 121 (1977)].
184. S. Chadrasekhar and B. Bahadur, *J. Chim. Phys.-Phys.-Chim. Biol.*, **70**, 605 (1973).
185. Th. Blümel, Peter J. Collings, H. Onusseit and H. Stegemeyer, *Chem. Phys. Lett.*, **116**, 529 (1985).
186. D. Vorlander, *Phys. Z.*, **31**, 428 (1930).
187. A. P. Kapustin, Eksperimental'nye issledovaniya zhidkikh kristallov (Experimental Investigations of Liquid Crystals), Nauka, Moscow (1978), p. 368.
188. S. N. Mochalin and P. P. Pugachenich in: Sb. Dokladov I nauchnoi konferentsii po zhidkim kristallam (Contributions to the 1st Scientific Conference on Liquid Crystals), Ivanovo (1972), p. 200.
189. D. Guillon and A. Skoulios, *Mol. Cryst. Liq. Cryst.*, **51**, 149 (1979).
190. N. V. S. Rao and V. G. K. M. Pisipati. *Z. Naturforsch.*, **A40**, 466 (1985).
191. V. G. K. M. Pisipati, N. V. S. Rao, P. V. Datta Prassad and P. R. Alapati, *Z. Naturforsch.*, **A40**, 472 (1985).
192. N. V. S. Rao, V. G. K. M. Pisipati and D. Saran, *Phase Transit.*, **4**, 275 (1984).
193. K. R. K. Rao, J. V. Rao, P. Venkatacharyulu and V. Baliak, *Acta Phys. Pol.*, **A70**, 541 (1986).
194. J. V. Rao, K. R. K. Rao, L. V. Chondary and P. Venkatacharyulu, *Cryst. Res. and Technol.*, **21**, 1245 (1986).
195. K. R. K. Rao, J. V. Rao and P. Venkatacharyulu, *Acta Phys. Pol.*, **A69**, 261 (1986).

196. V. Ya. Baskakov and V. K. Semenchenko, *Pis'ma Zh. Eksp. Teor. Fiz.*, **17**, 580 (1973) [*JETP Lett.*, **17**, 414 (1973)].
197. V. K. Semenchenko, V. M. Byankin, and V. Ya. Baskakov, *Kristallografiya*, **20**, 187 (1975) [*Sov. Phys. Crystallogr.*, **20**, 111 (1975)].
198. H. Arnold, *Z. Chem.*, **4**, 211 (1964).
199. E. I. Bukhman, "Experimental investigation of the thermophysical properties of cholesteric liquid crystals," Candidate's Dissertation, Odessa State University, Odessa (1976).
200. H. Arnold and P. Roediger, *Z. Phys. Chem. (Leipzig)*, **239**, 283 (1968).
201. H. Kresse, A. Wiegeleben and B. Gajewska, *Z. Phys. Chem.*, **266**, 193 (1985).
202. J. O. Indekeu, A. Nihat Berker, C. Chiang and C. W. Garland, *Phys. Rev. A: Gen. Phys.*, **35**, 1371 (1987).
203. M. A. Anisimov, V. P. Voronov., A. O. Kyl'kov and F. Kholmuradov, *Pisma Zh. Eksp. Teor. Fiz.*, **41**, 248 (1985).
204. T. Pitehford, G. Nounesis, S. Dumrongrattana, J. M. Viner and C. C. Huang, *Phys. Rev. A: Gen. Phys.*, **32**, 1938 (1985).
205. G. Nounesis, C. C. Huang, T. Pitchford, E. Hobbie and S. T. Lygerwall, *Ibid*, **35**, 1141 (1987).
206. G. W. Garland and M. E. Huster. *Ibid*, **35**, 2365 (1987).
207. S. Dumrongrattana, C. C. Huang, G. Nounesis, S. C. Lien and J. M. Viner, *Ibid*, **34**, 5010 (1986).
208. S. Dumrongrattana, G. Nounesis, J. J. Stofko and P. A. Arimilli, *Ibid*, **35**, 1460 (1987).
209. R. N. Kleiman, D. J. Bishop, R. Pindak and P. Taborek, *Ibid*, **53**, 2137 (1984).
210. N. A. Nedostup and V. V. Gal'tsev, *Zh. Fiz. Khim.*, **50**, 2949 (1976).
211. S. K. Kor and S. K. Pandey, *J. Chem. Phys.*, **64**, 1333 (1976).
212. G. G. Natale, *J. Acoust. Soc. Am.*, **63**, 1677 (1978).
213. V. G. K. M. Pisipati, N. V. S. Rao, P. V. Datta Prassad and P. R. Alapati, *Z. Naturforsch.*, **A40**, 472 (1985).
214. K. R. K. Rao, J. V. Rao, P. Venkatacharyulu and V. Baliak, *Acta Phys. Pol.*, **A70**, 541 (1986).
215. D. Collin, J. L. Gallani and P. Martinoty, *Phys. Rev. Lett.*, **58**, 254 (1987).
216. S. N. Mochalin, in: *SB. dokladov 1 nauchnoi Konferentsii po zhidkim kristallam* (Contribution to the 1st Scientific Conference on Liquid Crystals), Ivanovo (1972), p. 321.
217. P. J. Selle and A. W. Neumann, *Kol. Zsch. Polym.*, **2**, 222 (1968).
218. D. Churchill and L. W. Boiley, *Mol. Cryst. Liq. Cryst.*, **7**, 285 (1969).
219. S. Krishnaswamy and R. Shashidhar, *Mol. Cryst. Liq. Cryst.*, **38**, 353 (1977).
220. C. A. Croxton and S. Chandrasekhar, *Proc. Int. Liq. Cryst. Conf.*, Bangalore, 1973; Pramana Supplement 1, p. 237.
221. A. W. Neumann and P. J. Selle, *Z. Physik. Chem. (Frankfurt)*, **65**, 19 (1969).
222. F. K. Gorskii and N. M. Sakevich, *Kristallografiya*, **12**, 586 (1968).
223. A. W. Neumann, L. J. Klementowski, and R. W. Springer, *J. Coll. Int. Sci.*, **41**, 538 (1972).
224. I. Haller, *Progress in Solid-State Chem.*, **10**, 103 (1975).
225. L. Leger and A. Martinet, *J. Phys. (France)*, **37**, C3 (1976).
226. O. D. Lavrentovich and L. N. Tarahan, *Zh. Tekhnicheskoi Fiz.*, **56**, 2071 (1986).
227. K. Yoshino, K. Yamashiro, and Y. Inuishi, *J. Appl. Phys.*, **13**, 1471 (1974).

228. A. S. Babaev, A. I. Bogdanov, and V. I. Stafeev, *Elektrokhimiya*, **13**, 1235 (1977) [*Sov. Electrochem.*, **13**, 1054 (1977)].
229. G. J. Kruger and H. Spiesecke, and R. van Steenwinkel, *Mol. Cryst. Liq. Cryst.*, **40**, 103 (1977).
230. Kwo-Sun Chu, K. Ailawadi, and D. S. Moroi, *Mol. Cryst. Liq. Cryst.*, **38**, 45 (1977).
231. R. M. Richardson, A. J. Leadbetter, C. J. Carlile, and W. S. Howells, *Mol. Phys.*, **35**, 1697 (1978).
232. R. M. Richardson, A. J. Leadbetter, C. J. Carlile, and W. S. Howells, *Molec. Phys.*, **35**, 1697 (1978).
233. J. Janik, *Wiss. Beitr. M.* Luther-Univ., Halle-Wittenberg, **4**, 43 (1976).
234. S. Zumer and M. Vilfan, *J. Molec. Struct.*, **46**, 475 (1978).
235. Tadashi Akahane, Masumi Kondoh, Kenji Hashimoto and Masahiro Nakagawa, *Jap. J. Appl. Phys.*, **26**, 1000 (1987).
236. R. Stannarius and H. Schmiedel, *J. Magn. Res.*, **65**, 1 (1985).
237. K. Sakamoto, R. Porter, and J. Johnson, *Mol. Cryst. Liq. Cryst.*, **8**, 443 (1969).
238. R. S. Porter, E. M. Barrall, and J. F. Johnson, *J. Chem. Phys.*, **45**, 1452 (1966).
239. A. A. Tabidze, Investigation of the viscosity of liquid crystals, Candidate's Dissertation, Pedagogical Institute, Moscow Oblast', Moscow (1975).
240. J. F. Dyro and P. D. Edmonds, *Mol. Cryst. Liq. Cryst.*, **25**, 75 (1974); *ibid*, **29**, 263 (1975).
241. R. S. Porter and J. F. Johnson, *J. Appl. Phys.*, **34**, 55 (1963).
242. C. G. Kartha and R. K. L. Padmini Agakwal, *J. Phys. Soc. Jpn.*, **28**, 470 (1970).
243. P. Martinoty and S. Candau, *Mol. Cryst. Liq. Cryst.*, **14**, 243 (1971).
244. F. K. Gorskii and M. E. Mikhlin in: *Kristallizatsiya i fazovye prevrashcheniya* (Crystallization and Phase Transformations), Nauka i technika, Minsk (1971), p. 5.
245. S. Bhattacharya and S. V. Lether, *Phys. Rev. Lett.*, **42**, 458 (1979).
246. K. Czupryński, R. Dabrowski, J. Baran, A. Zywociński and J. Przedmojski, *J. Phys.* (*Fr.*), **47**, 1577 (1986).
247. P. Oswald, *J. Phys.* (*Fr.*), **47**, 1279 (1986).
248. N. Scaramuzza, F. Simoni, R. Bartolino and G. Durand, *Phys. Rev. Lett.*, **53**, 2246 (1984).
249. R. N. Kleiman, D. J. Bishop, R. Pindak and P. Taborek, *Phys. Rev. Lett.*, **53**, 2137 (1984).
250. G. Nounesis, C. C. Huang and J. W. Goodby, *Phys. Rev. Lett.*, **56**, 1172 (1986).
251. Ronald Y. Dong, *J. Magn. Res.*, **66**, 422 (1986).
252. Ali H. Al-Mowali, *Z. Phys. Chem.* (*BRD*), **141**, 23 (1984).
253. David H. Croll, Donald M. Small and James A. Hamilton, *J. Chem. Phys.*, **85**, 7380 (1986).

PART III.

Chiral Nematics and Liquid Crystals in Supercooled States

I would build my house of crystal,
With a solitary friend,
Where the cold cracks like a pistol
And the needle stands on end.
Robert Frost, "In the Long Night"

Commercial production of many useful instruments and devices, display models of which have been known for a long time, is in many ways held up by the lack of sufficiently stable liquid-crystalline materials required for their operation. For example, until comparatively recently, esters of cholesterol and other steroids formed the only group of cholesteric liquid crystals. However, widespread use of these materials in thermometers, information display systems, and other devices was impossible owing to the low chemical, electro-, and photochemical stability of these compounds. Until recently, it was impossible to observe and make use of liquid-crystalline properties in a very interesting and practically important range of low and cryogenic temperatures, since under these conditions the well-known liquid crystals transformed into solids. The situation changed radically only in recent years due to the fact that new types of liquid-crystalline substances—chiral nematics—were obtained and studied.

In what follows, we examine the basic results obtained in this new and important area. The features of the molecular structure of chiral nematics are elucidated and their optical and thermodynamic properties are discussed. The behavior of liquid-crystalline substances subjected to rapid cooling is also examined. The capability of mesogenic substances to retain their liquid-crystalline properties in the supercooled states could open up new areas of application of liquid crystals at very low temperatures, which is not only of theoretical interest but also of great practical interest for refrigeration technology, cryobiology, and medicine.

CHAPTER IX.

Properties of Chiral Nematics

He was a winter wind,
Concerned with ice and snow,
Dead weeds and unmated birds,
And little of love could know.
Robert Frost, "Wind and Window Flower"

The instability of liquid-crystalline materials often retards their widespread application. This refers, in particular, to cholesteric-type liquid crystals: esters of cholesterol and other derivatives of optically active steroids. For this reason, substances referring to another type of compound are of great interest. We are talking here about the so-called chiral nematic liquid crystals (chiral nematics): non-steroid chiral compounds, belonging to the same classes as the nematic liquid crystals. Although the structure of the molecules of chiral nematics is similar to the structure of molecules of ordinary NLC, these molecules exhibit chirality, leading to the appearance of twisted structures characteristic of CLC.[1]

Chiral nematics are characterized by their high stability, as well as the possibility of practical application in a wide temperature range. Analogously to the derivatives of cholesterol, the pitch of the helix in these substances changes with temperature, which gives rise to a change in the wavelength of selectively reflected light. For this reason, chiral nematics can be widely used as thermometers. Chiral nematics exhibit interesting electrooptical effects. Many of these substances form, at well defined temperatures, a smectic C phase and they include substances which exhibit ferroelectrical properties.[2] A very significant factor here is that this class includes substances that exhibit liquid-crystalline properties at very low temperatures (-30°C and lower). This determines the role of liquid crystals of this type in the solution of practical problems in refrigeration technology, cryobiology, and cryomedicine.

In the early work concerning the synthesis and study of chiral nematogens the object of study was the optically active MBBA[3,4]

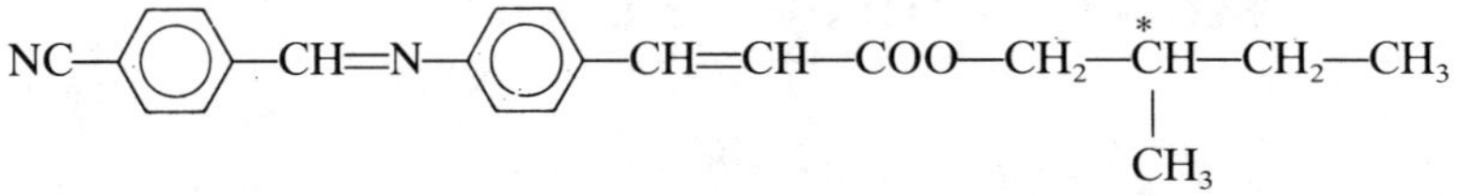

(here the asterisk indicates the chiral center). Later, compounds containing a terminal substituent with an asymmetric carbon atom,[5] as well as compounds in which chirality appeared as a result of the conformational factors[6] or substitution of one of the hydrogen atoms bound with the carbon atom by deuterium[7] were described and studied. Table 9.1 shows the structural formulas and transition temperatures of some representative liquid crystals of this type.

For the chiral nematogen and racemate with the same compounds, the temperatures of the transition into the isotropic liquid are usually practically identical. The insignificant difference could be due to the additional gain in energy in connection with the presence of helical ordering[1]:

$$\Delta T_{is} \approx (K_{22} q_0^2 / 2L) T_{is},$$

where L is the heat of transition, K_{22} is the twist elastic constant, $q_0 = 2\pi/h$, and h is the pitch of the helix. The quantity ΔT_{is} is usually much smaller than the change in the transition temperature caused by the presence of impurities. For example, a study of the thermodynamic properties of mixtures of optically active EBMBA and its racemate[14] showed that for any composition of the mixtures, the temperature of the isotropic transition lies in the range 332.8–333.5 K. Example 1 in Table 9.1 indicates that the melting temperatures of the racemate and of the optically active isomer likewise coincide.

Analysis of the dependence of transition temperatures for transitions into the isotropic liquid on the structure of molecules with terminal substituents, containing an n-alkyl chain, has led to the conclusion that branching of the chain appreciably lowers the temperature of this transition.[1] Based on the results of the study of phase transition temperatures in a number of chiral azometines it was established[1,11] that as the distance between the asymmetric atom and the framework of the molecule increases, the thermal stability of the smectic and cholesteric mesophases increases in a zig-zag manner.

Table 9.1
Some Examples of Chiral Nematic Liquid Crystals[1]

NN	Structure Formulae	Temperatures of Transitions, K Solid–LC–IL		Reference
1	$CH_3CH_2\overset{*}{C}H(CH_3)CH_2OOC{-}C_6H_4{-}N{=}CH{-}C_6H_4{-}CH{=}N{-}C_6H_4{-}COOCH_2\overset{*}{C}H(CH_3)C_2H_5$	396	457	8
		(racemate and isomer)		
2	$C_2H_5O{-}C_5H_4{-}\underbrace{N{=}N}_{O}{-}C_6H_4{-}C_5H_{11}{-}H$	309	384	9
3	$CH_3O{-}C_6H_4{-}CH{=}N{-}C_6H_4{-}O\underset{\underset{O}{\Vert}}{C}{-}C_5H_{11}{-}H$	356	376	10
4	$CH_3O{-}C_6H_4{-}CH{=}N{-}C_6H_4{-}O\underset{\underset{O}{\Vert}}{C}{-}\overset{*}{C}_5H_{11}$	314	351	11
5	$C_2H_5O{-}C_6H_4{-}\underbrace{N{=}N}_{O}{-}C_6H_4{-}\overset{*}{C}_5H_{11}$	277	349	12
6	$C_4H_9O{-}C_6H_4{-}N{=}N{-}C_6H_4{-}C_5H_{11}{-}H$	316	339	13
7	$C_4H_9O{-}C_6H_4{-}N{=}N{-}C_6H_4{-}\overset{*}{C}_5H_{11}$	309	312	12

The possibility of creating substances with the required magnitude of the pitch of the helix is of great importance. The required pitch can be obtained by mixing levo- and dextrorotary forms of the same compound (which has no effect on the thermodynamic properties) or by adding a nematic substance (solvent) to the chiral nematogen. The dependence of the pitch of the helix on the concentration of the dissolved chiral nematogen usually has the form[1,15]

$$h = 1/2\ \beta x,$$

where β is the "twist" force, and x is the concentration of the chiral nematogen. As far as the temperature dependence of the pitch is concerned, it can both increase and decrease with increasing temperature.[1]

The study of chiral nematogens, which belong to a series of azo- and azoxy-compounds, containing one or two substituents with asymmetric carbon atom,[12] has led to the conclusion that h is virtually independent of the chemical nature of the main part of the molecule, but increases as the number of carbon atoms in the n-alkyl chain increases from one to seven, and as the length of the chain increases further, the pitch decreases. When one asymmetric chain is replaced by another, the pitch of the helix changes considerably (by a factor of 2–4).[2] An increase in the number of optically active centers in the molecule decreases the pitch.[16] The results of a study of the relationship between the direction of rotation of the helix and the structure of the chiral nematogens[12] are examined in Ref. 1.

During the last few years, significant new progress has been achieved

Table 9.2
Some Properties of Compositions I and II[17]

		Temperatures of Phase Transitions, K*,**			Melting Heat of the Most Stable Solid Crystal	Helix Pitch
Substance	n	SC–IL	IL–CLC	CLC–SLCA	(kJ/mole)	mcm***
I	1	277	(243)	(219)	6.3	0.2
I	2	282	(259)	(251)	13	0.3
I	3	301	(263)	(253)	25	0.4
II	1	326.7	(282) ± 3†	—	15	1.5
II	3	330.7	(308) ± 3†	—	30	1.0

*SC—solid crystal, IL—isotropic liquid, CLC—cholesteric liquid crystal, SLCA—smectic liquid crystal A.

**Temperatures of monotropic transitions are given in brackets.

***Obtained by extrapolation of measurements in 10%-solutions of 4-n-pentyl-4′-cyanobiphenyl.

†Virtual transition.

in the synthesis, study, and application of chiral nematics. The problem consisted of creating additives to NLC, which are used in electrooptical displays, based on the twist effect and on the field-controlled cholesteric-nematic phase transition and, in addition, the new substances had to have the same chemical, electro-, and photochemical stability as the basic component—NLC. As a result, it was possible to obtain an entire collection of chiral nematics with a wide range of required properties (transition temperatures, pitch of the helix, dielectric properties, etc.).[12,17–19] We shall briefly examine these results, turning our attention to the materials that are suitable for work at low temperatures.

Table 9.2 shows some thermodynamic properties of two types of chiral nematics, synthesized by Gray and McDonnell[17]

I 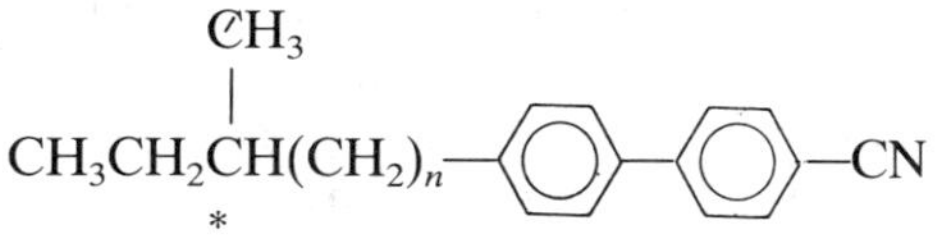

II 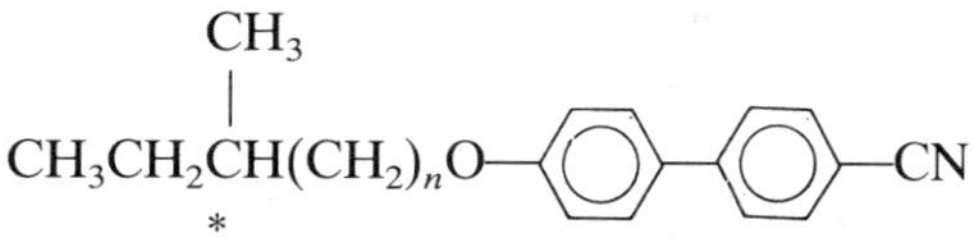

Here *n* varies from one to three. From the data in Table 9.2 it is evident that these substances have low melting temperatures and low transition temperatures for other transitions and demonstrate the monotropic cholesteric and smectic A phases. The suitable thermodynamic and optical characteristics of these materials, as well as of chiral nematics with the following molecular structure

III 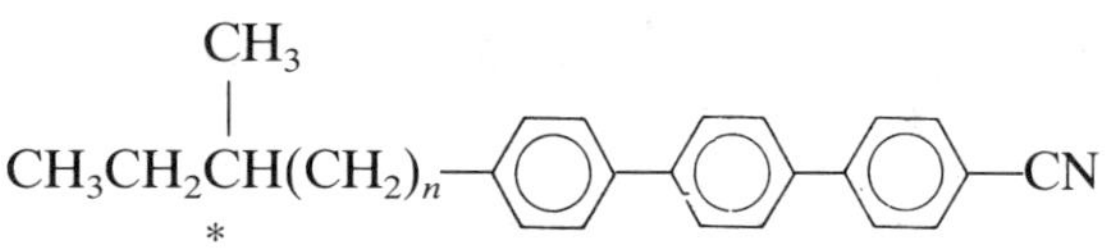

IV 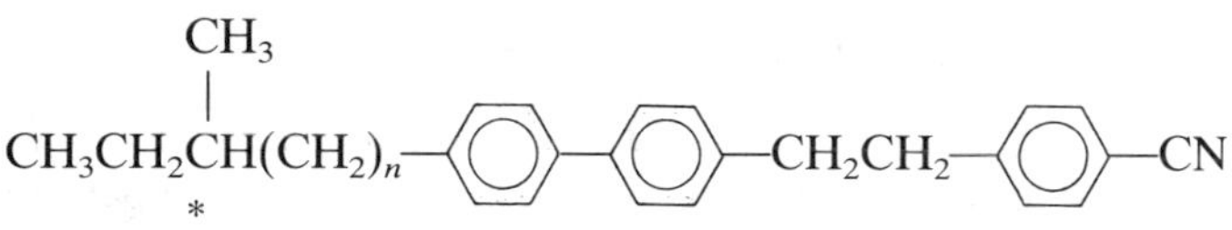

make them very promising for use in electro-optics.

Substances consisting of these molecules can be used both to obtain cholesteric mixtures with a relative large pitch of the helix, suitable for displays, and to create mixtures with a pitch that gives selective reflection of light in the visible range. Examples of such mixtures, with an indication of the phase transition temperatures, as well as the color-temperature characteristics of the selected reflection are presented in Table 9.3. These mixtures retain their color properties unchanged for a long time and give a high constant intensity of color without any contrivances and special devices.

Table 9.4 presents some of the properties of compounds III and IV.[17] An interesting property of the substance IV with $n = 1$ is the small pitch of the cholesteric helix. This substance can be used in combination with the substance *I* ($n = 1$) in order to obtain a wide selection of cholesteric mixtures with a pitch of 1–2 μm for use in displays. Substances consisting of molecules having this type of structure can be used in mixtures with one another and with other chiral nematics in order to extend the selection of thermochromic mixtures.

Chiral nematics can also be obtained based on complex esters, which have now been well studied and are quite stable. Below, we present the structure formulas of some compounds of this type, which are promising for practical applications.[17]

V 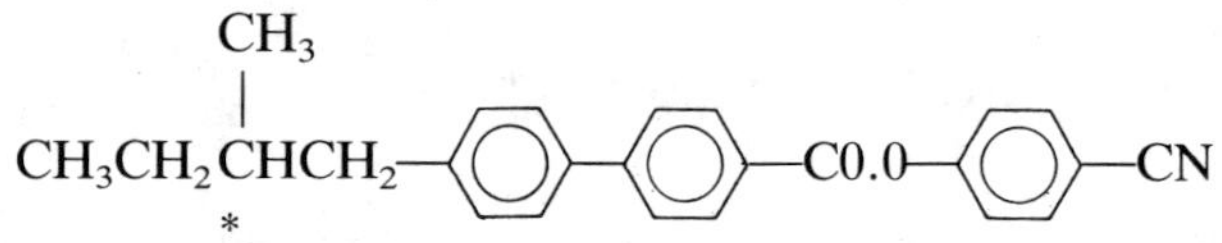

VI 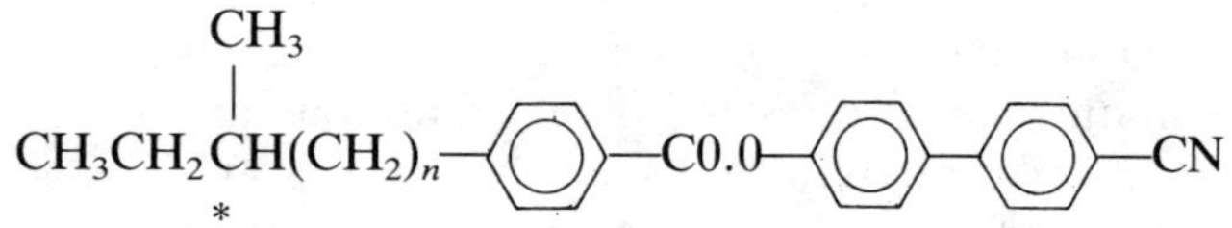

VII 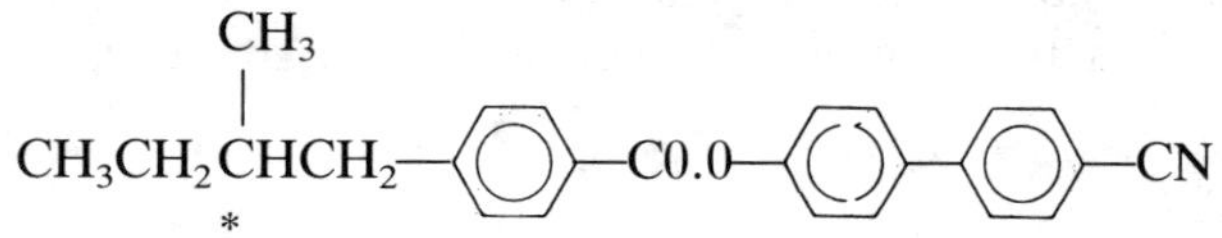

VIII 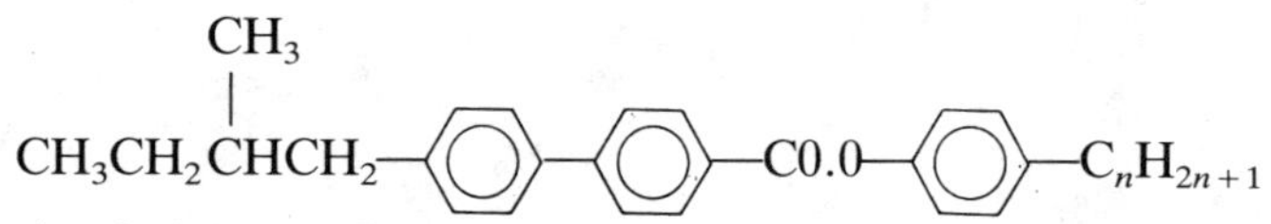

Here n varies from 1 to 3. These substances are characterized by positive values of the dielectric anisotropy and comparatively high transition temperatures. Some of the physical properties of separate representatives of substances of this type are summarized in Table 9.4.

The group of chiral esters includes compounds which are capable of forming, together with a cholesteric phase, an optically active

Table 9.3

Temperatures of Transitions (K) and Thermooptical Characteristics of Some Mixtures of Chiral Nematics[17]*

Character of Phase Transition or Color of Selective Reflection	Number and Mixture Content			
	1	2	3	4
	(I, n = 2)—86% wt. (III, n = 2)—14% wt.	(I, n = 3)—74% wt. (I, n = 1)—14% wt. (III, n = 3)—12% wt.	(I, n = 3)—80% wt. (IV, n= 1)—20% wt.	(I, n = 3)—80% wt. (VI, n = 1)—20% wt.
SLCA–CLC	260	278.0	270	263
Red		278	270	263
Pink	261			
Yellow	261.5	281	270.5	
Green	263	282	271	264
Turquoise	264.5	285	271.5	265
Blue	267	287	272.2	266
CLC–IL	277	290.7	288	292

*See footnotes for Table 9.2.

Table 9.4
Some Properties of Compounds III–VII[17]*

Compound	n	Temperatures of Phase Transitions, K: SC–SLCA or SC–CLC	SLCA–CLC	CLC–IL	Melting Heat of the Most Stable Solid Crystal, kJ/mole	Helix pitch, mcm
III	2	430	—	475	16	3.7
III	3	393	436	459	14	2.8
IV	1	364.9	—	376.6	24	1.5
IV	2	364.8	—	384.0	26	4.5
V	—	372.7	—	469.1	30	1.6
VI	1	369	—	483	22	1.6
VI	2	372.4	—	478	19	4.7
VII	—	350.8	411	463.6	19	4.7

*See footnotes for Table 9.2.

smectic chiral phase C, which has a helicoidal twisted structure with a pitch of the order of 0.2 μm. The structural formula of substances of this type has the following form

1X 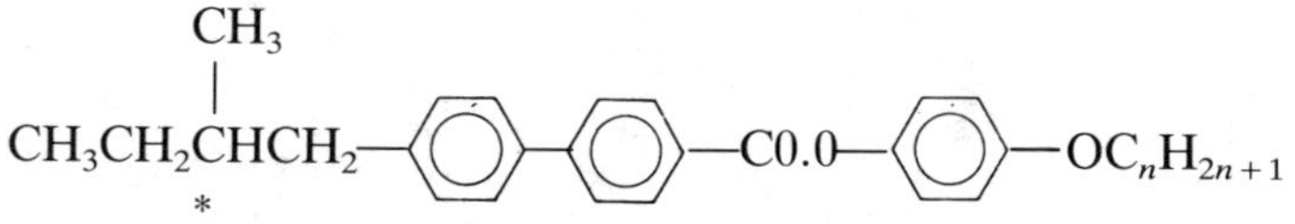

Table 9.5 presents the phase transition temperatures and heats of fusion of such compounds (n = 4–10). Figure 9.1 shows the dependence of the transition temperatures of these substances on the number of carbon atoms in the alkyl chain of the esters.

Figure 9.2 shows the temperature dependence of the pitch of the helix and the wavelength of selectively reflected light for compound IX (n = 6). It is evident that, in this case, the wavelength of reflection increases with temperature. Figure 9.3 shows the temperature dependence of the pitch of the helix and the wavelength of the selectively reflected light for one of the mixtures of compounds IX. The change in the pitch of the helix with an increase in temperature has a complicated character here in the region of the smectic C phase (decrease followed by an increase), and a sharp peak is observed in this dependence at the smectic C–cholesteric transition.[17] It should be noted that for individual substances of this type the color range of the smectic chiral phase can be very small, but it can be extended by forming appropriate mixtures.

Based on the complex esters, materials with two chiral centers, which have, in connection with this feature, a very small pitch of the

Table 9.5
Some Properties of Compounds IX[17]*

n	Temperatures of Phase Transitions, K: SC–CSC or SC–CLC	CSC–CLC	CLC–IL	Melting Heats of the Most Stable Solid Crystal, kJ/mole
4	354.9	—	442.5	21
5	353.0	(343.2)	437.7	29
6	342.0	353.4	436.7	26
7	353.6	360.5	430.9	30
8	349	361.8	428.6	45
9	352	362.7	420.2	43
10	350.0	366.6	421.7	48

*CSC—chiral smectic C, other abbreviations see footnotes for Table 9.2.

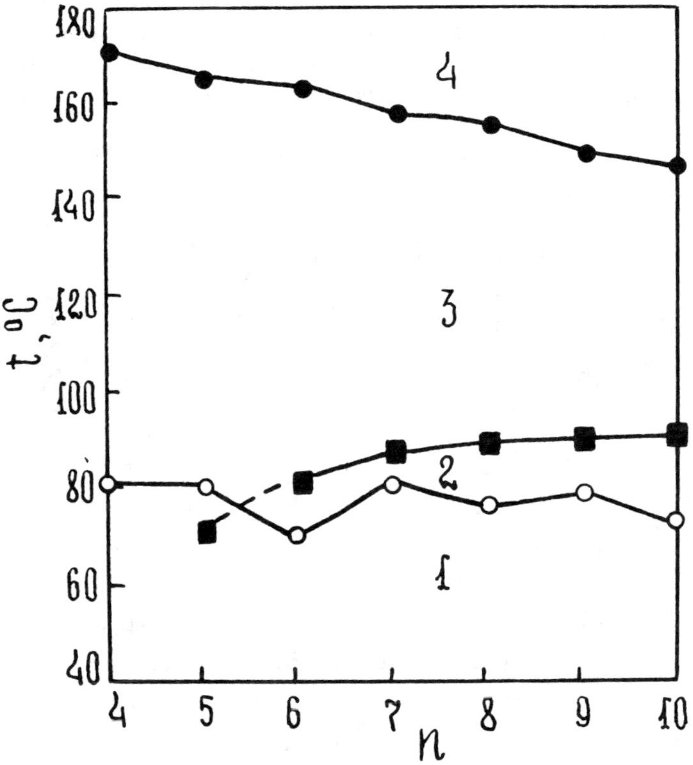

Figure 9.1. Dependence of temperatures of transitions upon the quantity of carbon atoms n in alkyl chain of ester for 4-n-alkoxyphenyl esters of 4′-(2″-methylbutyl)biphenyl-4-carboxylic acid[17]: 1—solid; 2—chiral smectic C; 3—cholesteric phase; 4—isotropic liquid.

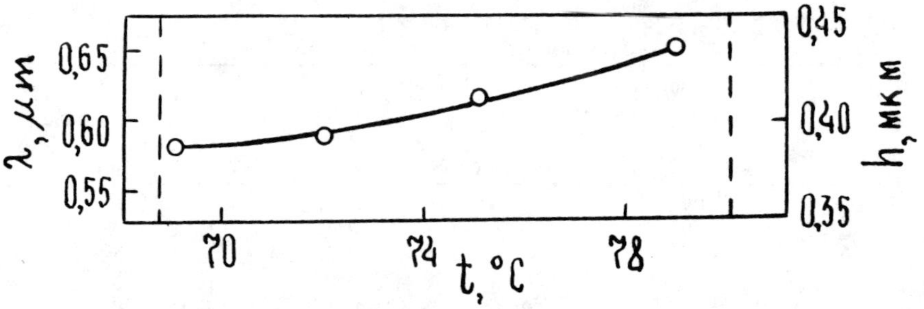

Figure 9.2. Temperature dependence of wavelength of selectively reflected light λ and helix pitch *h* in the range of chiral smectic C phase of substance IX.[17]

cholesteric helix of about 0.1 μm,[18] can also be synthesized. Such substances can be used as trace additives to the nematic materials in order to obtain cholesteric mixtures with the required pitch, which can be successfully used in electro-optical devices. Such substances with a high "twist force" are especially useful in thermography, in-

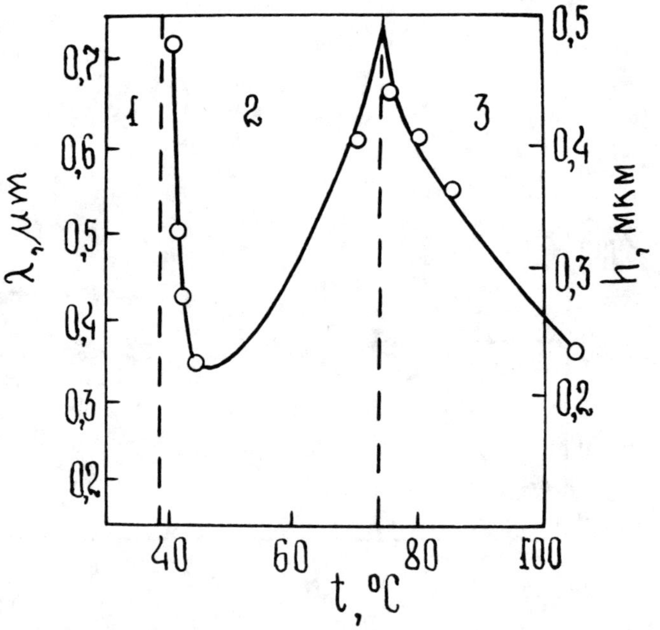

Figure 9.3. Temperature dependence of wavelength of selectively reflected light λ and helix pitch in the range of chiral smectic C phase and cholesteric phase for mixture of esters IX: 1—solid; 2—chiral smectic C phase; 3—cholesteric phase (regions of phases are divided by a dashed line).[17]

asmuch as only 25–35% of this "force" is required in order to obtain a pitch guaranteeing selective reflection in the visible part of the spectrum. For this reason, it is easy to prepare thermochromic mixtures based primarily on nematic materials. An example of such a mixture is the following four-component composite material.[18]

This mixture exhibits phase transitions S–C and C–IL at temperatures of 18.5 and 43.0°C, respectively, and has a color range between 23.5 and 35.0°C, exhibiting a change in the color of the selective reflection in this range from red to blue.

Commercial mixtures, developed on the basis of research in this area,[17–19] on which basis it is possible to obtain thermochromic materials that have a working temperature range with color variation in the interval from 0.5 to 5°C at temperature levels from −30°C to 40°C and higher, have found great practical application in recent years. Figure 14.19 shows a typical phase diagram of such a mixture and the concentration dependence of the color interval.

The results of modern experimental and theoretical studies of anisotropy of physical properties of liquid crystals with induced spiral structures were discussed in the review.[20] Optical properties of hiral nematic and hiral smectic liquid crystals (Bragg's reflection, electrooptical effects, abnormal light absorption, dispersion of eigenwaves) were studied especially thoroughly. Regularities of the spiral structure induction were defined on the basis of investigation of more than 30 unmesomorphic optically active admixtures.

CHAPTER X.

Liquid Crystals in Supercooled (Metastable) States

Then the mortal coldness of the soul like death itself
comes down;
It cannot feel for others' woes, it dare not dream its own;
That heavy chill has frozen o'er the fountain of our tears,
And though the eye may sparkle still, 'tis where the ice
appears.

George Gordon Byron, "Stanzas for Music"

It is well known that metastable states are not completely stable states of the system consisting of a large number of particles, capable of undergoing a first-order phase transition.[21] Although the system is stable relative to small (continuous) changes in the thermodynamic parameters, it exhibits an instability when a competing phase appears in it. Thermodynamically, this is due to the existence under fixed conditions of at least two minima in the thermodynamic potential and, in addition, the absolutely stable state of the system corresponds to the lowest of these, while the other minima correspond to the metastable states. Since the transition into the stable phase requires overcoming a certain potential barrier, such metastable states can exist for a long time, inasmuch as, by themselves, they are stable, appearing internally in equilibrium according to all other indications.[21] However, "frozen" nonequilibrium glassy-type states, when molecular rearrangements, related with continuous change in the starting phase, are inhibited and sharply slowed down due to the high viscosity and weak role of the thermal motion of particles of the rapidly cooled starting system, can also exist.

It is usually difficult to overheat appreciably the solid phase above the point of the phase transition, due to the ease with which a high-temperature more disordered (for example, liquid) phase can appear on the interface, which is in a state of tension. But, the formation of a solid phase from the liquid phase accompanying cooling of the latter requires that high energy barriers, related with the decay of the molecular complexes existing in the liquid and formation of a crystalline lattice from the molecules, must be overcome.[22] If the high-temperature phase contains molecular clusters, whose structure differs considerably from the subunits of the low-temperature phase, then this substance admits large overcoolings and can be transformed into a glass-like state.[23,24]

Observations show that most organic liquids do not crystallize, but rather vitrify. This is apparently related with the decrease in the mobility of the molecules, and suppression of their rotational and translational motions. For this reason, as the temperature decreases, the frequency of nucleation in such substances at first increases and then drops sharply, while the nuclei of the new phase that appear during the cooling process cannot grow.[21] This indicates the existence of a high activation barrier for the transition into the crystalline phase, inasmuch as translations of molecules, related with a change in the structure of the substance, require preliminary rearrangement of their immediate environment.

The crystallization of supercooled liquids is accompanied by two processes: formation of nuclei of the new phase and their development and growth, related with the interaction between the nuclei and the parent phase and with one another. Random insoluble impurities, which can have different initiating capabilities, usually serve as the "seed." If such centers or active impurities are absent, then crystallization begins on spontaneous nuclei of homogeneous origin. This is also valid in the case of mesogenic substances, although, in the latter case, these effects are even more strongly manifested. On the basis of the kinetic theory of formation of the crystalline phase accompanying cooling of isotropic organic liquids,[21,25] it is possible to estimate the most important characteristics of the process of supercooling of liquid crystals (frequency of nucleation, size of critical nuclei and the number of molecules in them, maximum supercooling, delay time, etc.).

While working with different types of liquid crystals, supercooled states of these substances are often observed. This is due to the nature of the phase diagrams, which often demonstrate the presence of an entire series of intermediate liquid-crystalline phases, some of which correspond to narrow temperature intervals for the equilibrium state.

In addition, the relatively high viscosity of liquid crystals, especially at low temperatures, also increases the probability for the appearance of nonequilibrium states.

Let us discuss, following Ref. 22, the thermodynamic interpretation of such metastable states. Viewing the thermodynamic potential ϕ, as a function of temperature T, pressure P, and concentration x, the transition points of first-order phase transitions can in general be defined as the points of intersection of the potential curves of different phases $\phi(T, P, x)$ with fixed P and x (Fig. 10.1a). This interpretation is justified by the fact that by avoiding the appearance of a new phase in an experiment, it is in principle possible to investigate the potential of the other phase for values of T, P, and x corresponding to regions of its metastable existence.

It is well known that there exists a type of SLC (bimolecular smectics with two density waves) which under high pressures form an A phase only as a metastable state, which is explained by the strong repulsion of the polar parts of molecules as they approach one another.[22,26] As far as the supercooled nematic phase of such substances is concerned, it can exist in a range of temperatures of the order of 40 K, which is usually related with the presence of large smectic fluctuations in the nematic phase (cybotactic clusters), whose existence has been observed by many experimenters.[27–29] This more complicated case is examined in Fig. 10.2, which shows the temperature dependences of the potentials of different phases for different values of the pressure. At low pressures (a), the reversible phase transitions NLC ↔ SLC-A and SLC-A ↔ SC occurs at the points T_{NA} and T_{AS}.

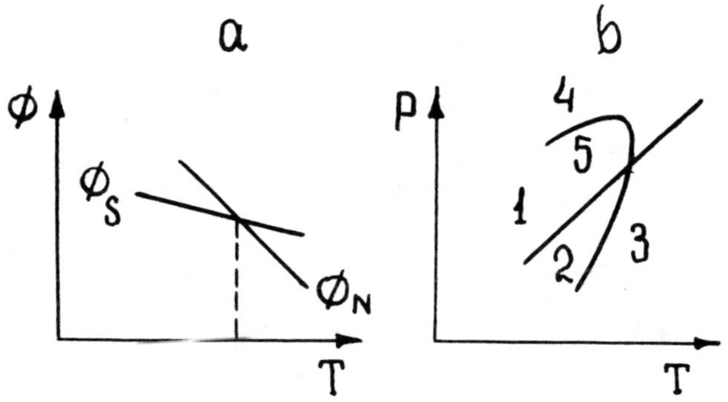

Figure 10.1. Solidification of NLC at constant pressure (a): ϕ_S and ϕ_N—thermodynamic potentials of solid (S) and nematic (N) phases;[22] phase diagram of supercooled LC (b): 1—solid; 2—SLC A; 3— NLC; 4—supercooled NLC; 5—supercooled SLC A.[22,26]

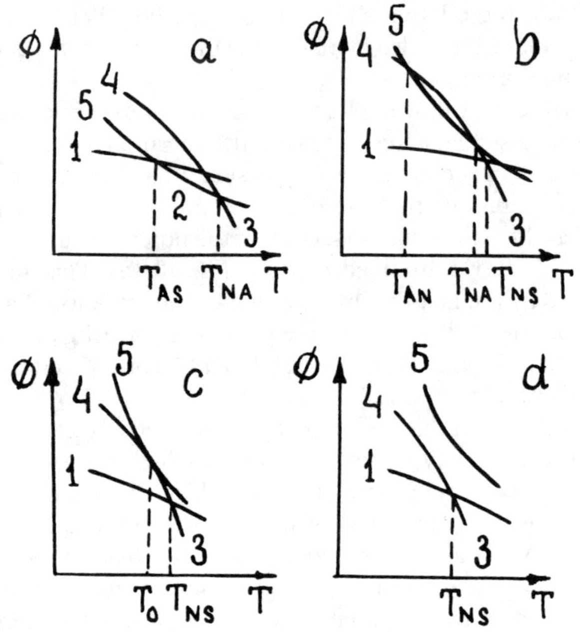

Figure 10.2. Thermodynamic potentials of NLC, SLC A and solid at different temperatures and pressures:[22] 1—solid crystal; 2—SLC A; 3—NLC; 4—supercooled NLC; 5—supercooled SLC A. a—phase transitions NLC↔SLC A↔SC at low pressures; b—phase transitions NLC↔SC and NLC (supercooled)↔SLC A (supercooled)↔NLC (supercooled) in conditions of supercooling at interjacent values of pressure; c—critical values of pressure ($P = P_o$); d—phase SLC A at high pressures is absent ($P \geqslant P_o$).

As the pressure increases, the curve of the potential of SLC-A is displaced upwards, so that the SLC-A phase can exist only under the conditions of the supercooled phase of NLC (b). For some critical pressure $P = P_0$, the curves of the potentials of NLC and SLC-A touch at a single point $T = T_0$ (c), and for $P > P_0$, the region of the metastable existence of SLC-A vanishes (d).

Since at the point (P_0, T_0) the thermodynamic potentials of the NLC and SLC-A phases and their first derivatives with respect to temperature are equal, for P less than or equal to P_0 these potentials can be expanded with respect to the small deviations $(P - P_0)$ and $(T - T_0)$ and the following condition for the equality of these potentials at these points can be obtained:

$$\left[\frac{\partial\phi_{NLC}(P_0, T_0)}{\partial P} - \frac{\partial\phi_{SLCA}(P_0, T_0)}{\partial P}\right](P - P_0) + \frac{1}{2}\left[\frac{\partial^2\phi_{NLC}(P_0, T_0)}{\partial T^2} - \frac{\partial^2\phi_{SLCA}(P_0, T_0)}{\partial T^2}\right](T - T_0)^2 = 0,$$

where $\partial\phi_{SLCA}/\partial P > \partial\phi_{NLC}/\partial P$, $\partial^2\phi_{SLCA}/\delta T^2 > \partial^2\phi_{NLC}/\partial T^2$ at $P = P_0$ and $T = T_0$. For this reason, the temperatures of the transitions NLC (supercooled) → SLC-A (supercooled) and SLC-A (supercooled) → NLC (supercooled) are determined by the equation[22]

$$P = P_0 - \gamma(T - T_0)^2, \qquad \gamma > 0,$$

which corresponds to the phase diagram observed experimentally[26] (Fig. 10.1b). It is important to emphasize that this equation has two solutions for transition temperature with $P < P_0$ and reflects the successive appearance of NLC (supercooled), SLC (supercooled) and again NLC (supercooled) phases as the temperature decreases.

If the transitions between the liquid-crystalline phases are close to a second-order phase transition, they can be described based on the L. D. Landau's expansion for the thermodynamic potential, referred to unit mass of the material[30]:

$$\phi = a|\psi|^2 + C|\psi|^4 + d|\psi|^6,$$

where $c < 0$, $d > 0$, $a = a'[P - P^* + \gamma(T - T_0)^2]$, $\qquad a' > 0$.

This indicates the absolute instability of the SLC-A phase for $P > P_0 = P^* + C^2/4da'$ and the existence of NLC (supercooled) → SLC-A (supercooled) transitions for $P < P_0$ at the points

$$T_{NA,AN} = T_0 \pm \sqrt{(P_0 - P)\gamma}.$$

This approach to the description of the states examined was developed by Clark.[31] It is interesting to note that in accordance with the experiment, the critical value of the pressure P_0 increases with increasing parameter $C^2/4da'$, characterizing the hysteresis phenomena (including the entropy jump).

The appearance of a high-symmetry phase at low temperatures is a rare phenomenon.[22] Low-symmetry modifications SLC-H and -G appear much more often as a result of supercooling of the high-temperature phase.[32]

The formation of nuclei of crystals in supercooled mesomorphic phases of cholesterol derivatives was studied theoretically in Ref. 33. For the case of crystallization from supercooled smectic and cholesteric mesophases of cholesterol esters and its binary mixtures, the classical theory of nucleation of crystals accompanying supercooling of isotropic liquids was used. The energy of the interface between the phases, depending on the length of the alkyl chain, type of mesophase, and concentration of the mixture, was estimated. It turned out that for the systems examined it was an order of magnitude lower than for ordinary isotropic systems.

The general analysis of nucleation of crystals in supercooled liquids (glasses) is performed in Ref. 34. In the paper cited, Kalinina *et al.* proposed a method for determining the surface energy at the nucleus-glass interface. A condition for a stationary rate of nucleation of crystals is formulated and it is shown that in determining the surface energy it is necessary to include its dependence on the degree of ordering of the nuclei, which decreases with the size of the nuclei. A theoretical estimate of the thermodynamic potential, which is useful for different glasses with not too low (measurable) rate of nucleation of crystals, is given.

The propagation of transverse acoustic waves in cholesteric glasses, obtained by rapid cooling of CLC, was studied theoretically by Rapini.[35] The coefficient of rotation of the polarization plane and the dependence of the frequency on the wave vector are determined for the case when the direction of propagation of sound coincides with the axis of the cholesteric helix. It is noted that there is an analogy between the behavior of acoustic and electromagnetic waves. Nakagizi *et al.*[15] proposed that at frequencies greater than the inverse time for establishing local equilibrium, the behavior of the CLC corresponds to the behavior of a glass-like state, and this can explain the disagreement between the experimental data on propagation of acoustic waves in CLC at high frequencies and the theory, observed in a number of other papers.

The method of molecular dynamics, successfully used previously to investigate the metastable supercooled states of isotropic liquids,[36,37] is also very promising for studying supercooled states of liquid crystals. Apparently, the methods of computer simulation will in the future find broad application not only in studying the structure and properties of supercooled liquid-crystalline mesophases, but also to study the "shock" cooling, heating, and appearance and development of nuclei of new phases. Such an investigation in application to cholesteric liquid crystal was performed in Ref. 38. The use of the techniques of molecular dynamics modeling permitted not only "observing" in the computer "experiment" metastable supercooled states of CLC but also determining the qualitative dependence of the pitch of the helix of such a cholesteric supercooled film on temperature with gradual "heating." The nature of this dependence is confirmed by observations[39] and is shown in Fig. 10.3. It is remarkable that for supercooled states, the temperature dependence of the pitch has a qualitatively different character than the usual dependences $h(T)$ for CLC—derivatives of cholesterol.

Strong supercooling of cholesteryl hydrogenphthalate was observed by Tsuji *et al.*[40] A special technique for studying small pitches of the cholesteric helix, based on shock freezing of CLC to a temperature of liquid nitrogen and application of scanning electron microscopy, were developed by Voss *et al.*[41] Microphotographs of vitrified CLC are presented in Ref. 41 and the pitch of the helical structure of the

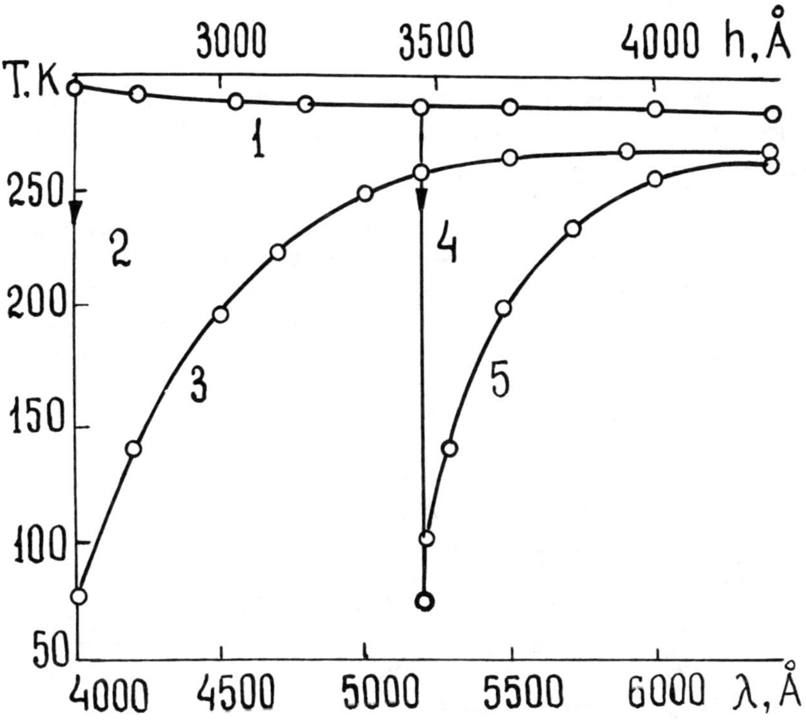

Figure 10.3. Dependences of temperature on wavelength of selectively reflected light (helix pitch) for mixture: cholesteryl oleate (61.5% wt.), cholesteryl pelargonate (26.0% wt.) and cholesteryl valerate (12.5%):[39] 1—curve of equilibrium states; 2 and 4—processes of fast cooling up to cryogenic temperature; 3 and 5—processes of warming (metastable states).

supercooled mixture cholesteryl nonanoate—cholesteryl chloride at 170 K, which turned out to equal 150 nm, is determined. A vitrified layer of NLC with a twisted structure was also obtained by rapid cooling in liquid nitrogen.[42]

A detailed study of metastable states in substances capable of forming nematic and smectic phases was performed in Refs. 43–47. We shall examine the data obtained for TBBA by Dubini *et al.*[45]

Using the method of differential scanning calorimetry, it was possible to investigate the two metastable phases of TBBA, existing at relatively low temperatures.[45] The phase diagram of this substance, refining the results in Ref. 43, was obtained:

$$\begin{array}{ccccccccccccc} \text{SC} & \overset{230}{\longleftrightarrow} & \text{SC} & \overset{386}{\longrightarrow} & \text{SLC B} & \overset{417}{\longleftrightarrow} & \text{SLC C} & \overset{444}{\longleftrightarrow} & \text{SLC A} & \overset{472}{\longleftrightarrow} & \text{NLC} & \overset{509}{\longleftrightarrow} & \text{IL} \\ \text{(IX)} & & \text{(VIII)} & & \text{(V)} & & \text{(IV)} & & \text{(III)} & & \text{(II)} & & \text{(I)} \\ & 332 \pm 3.5 & \uparrow & & \updownarrow\ 362 & & & & & & & & \\ & & \text{(VII)} & \underset{348}{\overset{353}{\rightleftarrows}} & \text{SLC E} & & & & & & & & \\ & & & & \text{(VI)} & & & & & & & & \end{array}$$

Here, the temperatures of the corresponding transitions (in *K*) are indicated near the arrows and the Roman numerals enumerate the phases observed. The scheme presented contains two metastable phases VI and VII. These phases were studied using x-ray scattering by Doucet *et al.*[43] It turned out that phases VI and V have similar short range structural order, and a considerable difference is observed between phases VI and VII. Experiments on quasielastic scattering of neutrons[44] showed that in the phase V, the molecules rotate uniformly around their long axis with characteristic time 10^{-11} s, while in the phase VI, jump-like reorientations of molecules around the long axis approximately with the same characteristic time are most probable. The transition from phase VI to phase VII represents freezing of these jumps. Although the nature of the phases VI and VII cannot be considered as finally established, the fact that these metastable phases exist in itself is reliably proven[43–47] and there are reasons for believing that these phases are smectic phases.[43] It is remarkable that the supercooling effect was observed in a number of cases.[45]

Table 10.1 shows the measured values of the changes in enthalpy and entropy of the phase transitions according to data in Ref. 45, which agree well with the results in Ref. 44. The disagreement between the changes in enthalpy for the transition VIII–V (17.3 kJ mole^{-1}) and the sum of changes in the enthalpies of transitions V–VI, VI–VII, and VII–VIII (13.4 kJ mole^{-1}) is interesting. The reason for this disagreement is not yet clear.

Table 10.1.
Changes of enthalpy and entropy in phase transitions in TBBA[45]*

Type of transition	VIII → V	V → VI	VI → V	VI → VII
Δ H, kJ·mole^{-1}	17.3 ± 0.6	0.63 ± 0.13	0.59 ± 0.13	0.19 ± 0.04
Δ S, kJ·mole^{-1} K^{-1}	44.83 ± 1.59	1.72 ± 0.42	1.68 ± 0.38	0.54 ± 0.13

Type of transition	VII → VI	VII → VIII	V → IV	IV → III
Δ H, kJ·mole^{-1}	0.17 ± 0.04	12.6 ± 1.7	3.02 ± 0.19	0
Δ S, kJ·mole^{-1} K^{-1}	0.50 ± 0.01	37.7 ± 5.4	7.29 ± 0.29	0

*Designations of phases (III–VIII) see in text.

A considerable supercooling of the smectic phase of CBOOA was observed by Guillon *et al.*[48]

An interesting investigation of unstable solid modifications of mesogenic substances, obtained by rapid cooling of an isotropic liquid state, was conducted by methods of Raman and infrared spectroscopy and scanning calorimetry by Kirov *et al.*[49] In particular, it was established that if rapid cooling is used, then the scheme of phase transitions in EBBA Sc $\xrightarrow{309K}$ NLC $\xrightarrow{350K}$ IL can be added to as follows.[49]

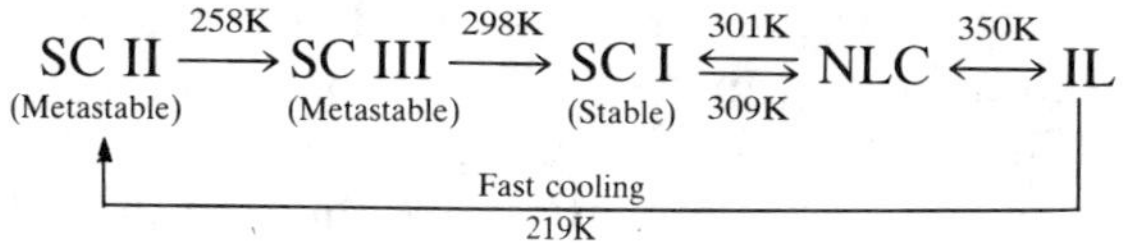

Here SC is the solid. Thus, with rapid cooling, the transition into the solid state in this substance occurs only at 219 K (instead of 301 K with slow cooling), which is much lower than the temperatures of transitions between the solid modifications SC III–SC I and SC II–SC III. The enthalpies and entropies of transitions, as well as their energy barriers are determined for MBBA, EBBA, and CBPA in Ref. 49, and an excellent review of data on metastable supercooled phases is presented. However, since these data concern, primarily, the solid state, their analysis falls outside the scope of this book.

Freeze-fracture electron-microscope observation of 4-(2-methylpentylphenyl)-4′-hexyloxybenzoate, cooled from cholesteric and from three blue phases (BP I, II and III) was performed.[50]

A special theory of destruction of cholesteric liquid crystals' surfaces during their abrupt deep freezing was developed by a group of authors.[51] The supposition that the energy of formation and growth of cracks are anisotropic and proportional to the order parameter lies in the basis of this theory. A numerical modeling of the texture of a quickly frozen cholesteric liquid crystal and the structure of a blue phase was carried out.

It may be expected that in the near future supercooled states of liquid crystals will be used to solve a broad range of practical problems. However, the methods of commercial preparation of liquid-crystalline specimens, capable of deep supercooling, have not yet been developed. Nevertheless, the fundamental directions along these lines are clear. Since crystallization of supercooled liquids almost always occurs on heterogeneous nuclei, and in addition the role of "seeds" is usually played by particles of random insoluble impurities, the specimens must be carefully cleaned. It is also known that small liquid drops (10–100 μm) are supercooled on the average much more than massive specimens,[21] and in addition this effect cannot be explained by the specific action of the curved surface of the interface.[21] At sufficiently high dispersity, most drops will not contain particles that initiate crystallization of the starting mesomorphous phase. Since the probability for such an impurity particle to occur in a drop of liquid crystal equals n_0/n, where n_0 is the number of particles, n is the number of drops, and in addition $n >> n_0$, this probability decreases as n increases compared with n_0. For example, if $n_0 \simeq 10^6$ cm^{-3}, then for a drop of diameter 27 μm, on the average, only one drop out of hundred will contain an active particle.* Here it is important that the effect of the particle initiating crystallization in the case of such a dispersed medium extends only to the corresponding drop. Of course, with fragmentation, the surface of the small drops must be clean. The effect of fragmentation can be examined in greater detail based on the so-called "isokinetic" relation.[52]

*This example is taken from Ref. 21.

References for Part III

CHAPTER IX

1. B. M. Bolotin and M. V. Loseva in: *Zhidkie kristally* (*Liquid Crystals*), Khimiya, Moscow (1979), pp. 24–33.
2. Yu. B. Amerik and B. A. Krentsel', *Khimiya zhidkikh kristallov i mezomorfiykh polimernykh sistem* (*Chemistry of Liquid Crystals and Mesomorphic Polymer Systems*), Nauka, Moscow (1981).
3. G. Friedel, *Ann. Physique*, **19**, 5273 (1922).
4. E. M. Barrall, R. S. Porter, and J. F. Johnson, *Mol. Cryst.*, **3**, 299 (1968).
5. L. Verbit, R. L. Tuggey, and A. R. Pinhas, *Mol. Cryst. Liq. Cryst.*, **30**, 201 (1975).
6. C. Mioskowski, J. Bourgwignon, and S. Candau, *Chem. Phys. Lett.*, **38**, 456 (1976).
7. G. W. Gray, *Mol. Cryst. Liq. Cryst.*, **21**, 161 (1973).
8. D. Coats, K. J. Harrison, and G. W. Gray, *Mol. Cryst. Liq. Cryst.*, **22**, 99 (1973).
9. R. Steinstrasser, *Pohl L. Tetrahedron Lett.* (1971), p. 1921.
10. H. M. Rosenberg and M. P. Serve, *J. Chem. Engrg.*, **16**, 496 (1971).
11. J. A. Castellano, C. S. Oh, and M. T. McCaffrey, *Mol. Cryst. Liq. Cryst.*, **27**, 417 (1974).
12. M. Tsukamoto, T. Ohtsuka, and K. Marimoto, *Jap. J. Appl. Phys.*, **14**, 1307 (1975).
13. R. Steinstrasser and L. Pohl, *Z. Naturforsch. B*, **26**, 577 (1971).
14. D. Dolphin, Z. Muljian, J. Cheng, *J. Chem. Phys.*, **58**, 413 (1973).
15. T. Nakagizi, H. Kodama, and K. K. Kobayashi, *Phys. Rev. Lett.*, **27**, 564 (1971).
16. B. H. Klandermann, *J. Amer. Chem. Soc.*, **97**, 1585 (1975).
17. G. W. Gray and D. G. McDonnell, *Mol. Cryst. Liq. Cryst.*, **37**, 189 (1976).
18. G. W. Gray and D. G. McDonnell, *Mol. Cryst. Liq. Cryst.*, **48**, 37 (1978).
19. D. M. Makow, *Applied Optics*, **19**, 1274 (1980).
20. G. S. Chilaya, *Fizicheskie svoistva zhidkikh kristallov s inducirovannoi spiralnoi strukturoi* (*Physical Properties of Liquid Crystals with Induced Spiral Structure*), Metsniereba. Tbilisi (1985).

CHAPTER X

21. V. P. Skripov, *Metastabil'naya zhidkost'* (*Metastable liquid*), Nauka, Moscow (1972).
22. S. A. Pikin, *Strukturnye prevrashcheniya v zhidkikh kristallakh* (*Structural Transformations in Liquid Crystals*), Nauka, Moscow (1981).
23. A. Ubbelode, *Plavlenie i kristallicheskaya struktura* (*Melting and Crystal Structure*), Mir, Moscow (1969).
24. G. Tamman, *Stekloobraznoe sostoyanie* (*The Glassy State*), ONTI, Moscow-Leningrad (1935).
25. V. I. Danilov, *Stroenie i kristallizatsiya zhidkosti* (*Structure and Crystallization of Liquids*), Izd. AN USSR, Kiev (1956).
26. P. E. Cladis, R. K. Bogardus, W. B. Daniels, and G. N. Taylor, *Phys. Rev. Lett.*, **39**, 720 (1977).
27. A. De Vries, *Mol. Cryst. Liq. Cryst.*, **10**, 31 (1970).

28. W. L. McMillan, *Phys. Rev. A*, **6**, 936 (1972).
29. W. L. McMillan, *Phys. Rev. A*, **7**, 1673 (1979).
30. L. D. Landau and E. M. Lifshitz, *Statisticheskaya fizika* (*Statistical Physics*), Nauka, Moscow (1976).
31. N. A. Clark, *J. Phys.* (*France*) *Coll.*, **40**, C3-345 (1979).
32. G. W. Gray and J. W. Goodby, *Ann. Phys.*, **3**, 123 (1978).
33. J. M. Pochan and H. W. Gibson, *J. Amer. Chem. Soc.*, **94**, 5573 (1972).
34. A. M. Kalinina, V. M. Fokin, and V. N. Filipovich, *Fizika i khimiya stekla*, **2**, 298 (1976).
35. A. Rapini, *C. r. Acad. Sci.*, **275**, B701 (1972).
36. V. G. Baidakov, A. E. Galashev, and V. P. Skripov, *Fizika nizkikh temperatur*, **2**, 957 (1976).
37. A. E. Galashev and V. P. Skripov, *Ibid*, **6**, 1106 (1980).
38. A. L. Tsykalo, *Zhurnal fizicheskoi khimii*, **56**, 2279 (1982).
39. A. L. Tsykalo, *Zhurnal tekhnicheskoi fiziki*, **51**, 1553 (1981).
40. K. Tsuji, M. Sorai, and S. Seki, *Bull. Chem. Soc. Jap.* (1971).
41. J. Voss and B. Voss, *Z. Naturforsch. A*, **31**, 1661 (1976).
42. J. O. Kessler and E. P. Raynes, *Phys. Lett. A*, **50**, 335 (1974).
43. J. Doucet, A. M. Levelut, and M. Lambert, *Phys. Rev. Lett.*, **32**, 301 (1974).
44. J. M. Schnur, J. P. Sheridan, and M. P. Fontana, *Pramana Supple.*, **1**, 175 (1975).
45. B. Dubini, S. Melone, M. G. Ponsi-Boss, S. Rustichelli, and F. Rustichelli, *Ann. Phys.*, **3**, 375 (1978).
46. Z. Luz and S. Meiboom, *J. Chem. Phys.*, **59**, 275 (1973).
47. M. P. Fontana and S. Bini, *Phys. Rev.*, **14**, 4 (1976).
48. D. Guillon, P. Seurin, and A. Skoulios, *Mol. Cryst. Liq. Cryst.*, **51**, 149 (1979).
49. N. Kirov, M. P. Fontana, and F. Cavatorta, *Mol. Cryst. Liq. Cryst.*, **54**, 207 (1979).
50. J. A. N. Zasadzinski, S. Meiboom, M. J. Sammon and D. W. Berreman, *Phys. Rev. Lett.*, **57**, 364 (1986).
51. D. W. Berreman, S. Meiboom, J. A. Lasadzinski and M. J. Sammon, *Phys. Rev. Lett.*, **57**, 1737 (1986).
52. D. N. Hollomon and D. Turnball, *Uspekhi fiziki metallov, I.* (*Progress in the Physics of Metals*), Metallurgizdat, Moscow (1956) p. 304.

PART IV.

Liquid Crystals Consisting of Disc-Like Molecules

"Who cares about you?" said Alice. . . .
"You're nothing but a pack of cards!"
Lewis Carroll, *Alice's Adventures in Wonderland*

In ordinary SLC-A, the density distribution function is periodic in one direction. However, the great variety of liquid-crystalline substances includes substances for which this function is periodic in two directions. Examples of such systems were discovered by Chandrasekhar and his coworkers. Studying substances consisting of molecules with a flat diskotic shape, they came to the conclusion that two-dimensional hexagonal ordering of the centers of mass of molecules in the absence of translational order in a direction perpendicular to the planes of the molecules exists. It turned out that these systems have a number of very interesting peculiarities, concerning their properties, structure, and phase transitions. The theory of such liquid crystals, which is only beginning to be developed, promises to open many exciting pages.

It became clear at the same time that the ideas of orientationally ordered "diskotic" phases can be very useful in an unexpectedly customary area, and they have already been encountered a long time ago. We are talking about "pitch" phases, which play an important role in metallurgy in obtaining coke and which exist at quite high temperatures. The isotropic melt of such compounds as lignin, petroleum pitch, and others can transform into a liquid crystal (carbonized mesophase), after which coking appears. For this reason, the conclusions and results obtained in the studying of liquid crystals formed by disc-like molecules (LC-D) can also be useful for studying and controlling coking, graphitization, making of carbon filaments, increasing the strength of materials, and solving other important practical problems.[1–3]

In this part of the book, we examine the features of the molecular structure and the structure of LC-D and the basic results obtained in the theoretical investigation of these interesting systems and phase transitions in them. The results of computer modeling of LC-D are of considerable interest. It is interesting that the technique of modeling, developed and proven in the case of ordinary liquid crystals, consisting of elongated molecules, can be easily extended to LC-D and useful results can be obtained. This method permits obtaining unique information on such important characteristics as the features of translational and rotational motions of disc-like molecules, coefficients of self-diffusion, correlation times, and other data, which it is difficult or impossible to obtain by other methods. A great deal of attention is devoted to experimental data on the properties of these as yet little-studied liquid crystals.

CHAPTER XI.

The Molecular Structure and Structural Features of Discotic Liquid Crystals and the Foundations of the Theory of These Substances

There are more things in heaven and earth, Horatio,
Than are dreamt of in your philosophy.
William Shakespeare, *Hamlet* (Act I, Scene V)

11.1. CHARACTERISTICS OF THE MOLECULAR STRUCTURE AND STRUCTURE OF LC-D

The structural formulas of some substances forming liquid crystals of the type being studied are presented in Table 11.1. As will be clear from what follows, the term "disc-like" molecules is essentially arbitrary. Although molecules of this type have a flat shape, their large cross section can differ considerably from a circle. Diskotic liquid crystals can consist not only of monomeric but also of polymeric molecules.

The first investigations, which permitted arriving at the conclusion that a new class of liquid crystals, formed not by elongated but by flat molecules, exists, concerned benzene-hexa-*n*-alkanoates (1, Table 11.1).[4–6,14] The method of x-ray scattering established not only the presence of a liquid-crystalline mesophase, but also according to the sizes of the scattering rings, that the structure of the liquid crystals corresponds to hexagonal packing (Fig. 11.1a).

According to the data in Ref. 4, the Bragg parameters, corresponding to the three inner rings of the x-ray diffraction pattern, are equal to 20.1, 11.7, and 10.1 Å for octanoate. If *a* is the lattice

Table 11.1
Structure formulae of molecules of some diskotic liquid crystals

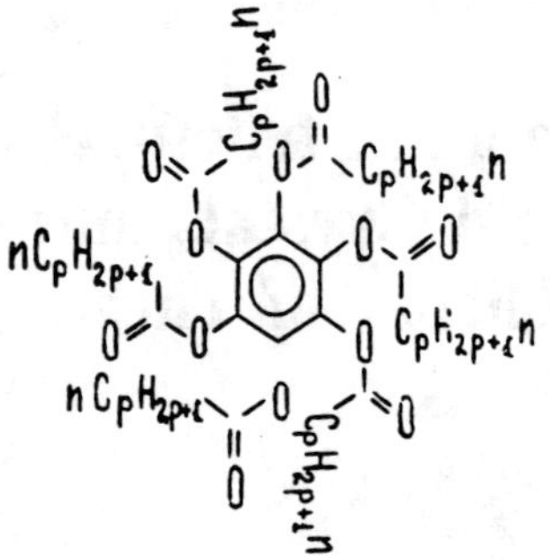

1. Benzene-hexa-*n*-alkanoates (BHA), $n = 5–8$[6], $n = 4–9$[4,5]

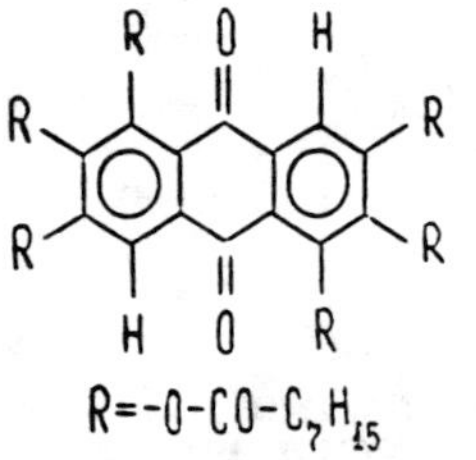

2. Hexa-*n*-octanoate of rufigallol (RHO)[7]

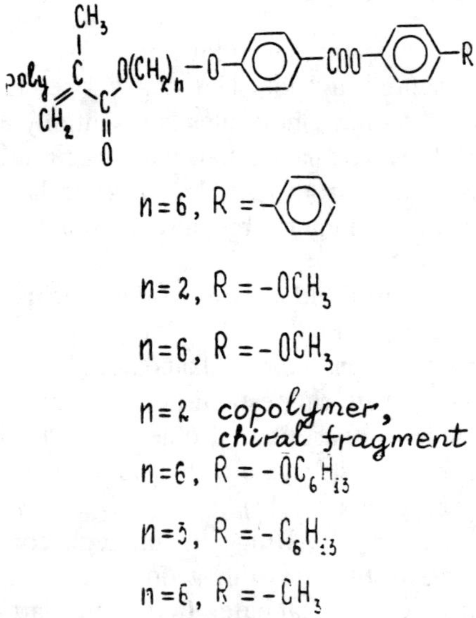

3. Polymeric mesogenic systems[8]

Table 11.1 (*continued*)

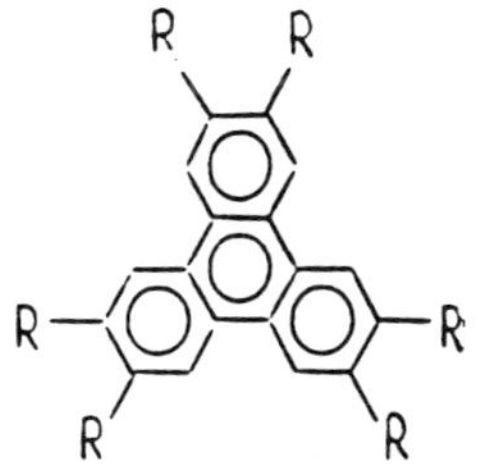

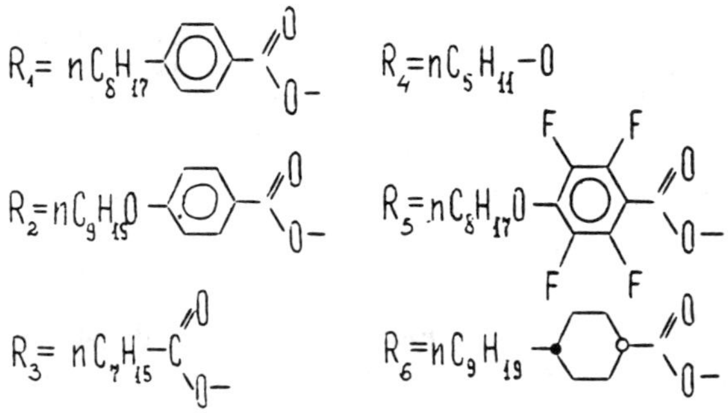

4.1. 2,3,6,7,10,11-hexa-substituted of triphenylene[9]

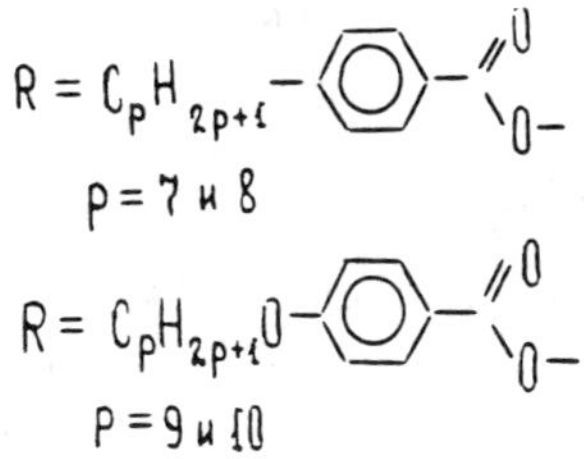

4.2. Triphenylene hexa-alkyl- and hexa-alkoxybenzoates[9,12]

$R = OC_5H_{11}$, OC_7H_{15}

4.3. Hexa-alkoxy-derivations of triphenylene[11,12]

$R = OCOC_{11}H_{23}$

4.4. Triphenylene hexa-*n*-dodecanoate[13]

$R = OC_nH_{2n+1}$ $(n = 1-13)$

$R = OCOC_nH_{2n+1}$ $(n = 1-15)$

4.5. Hexa-substituted of triphenylene[10]

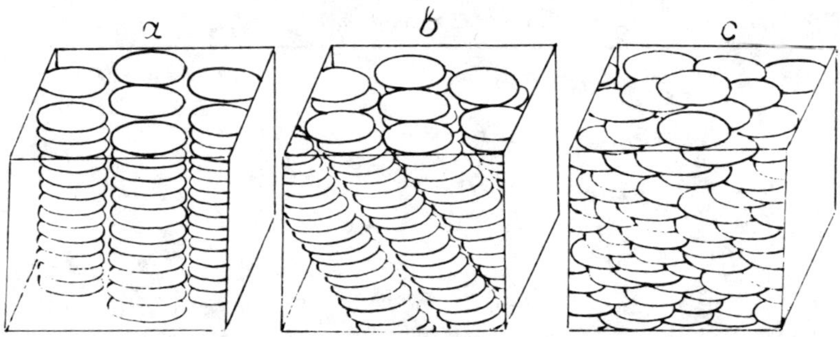

Figure 11.1. Structure of liquid crystals consisting of disc-like molecules: a—"column" structure with hexagonal packing of "columns" in which axes are perpendicular to molecular planes; b—"column" structure with hexagonal packing of inclined "columns"; c—absence of "columns"—"nematic" phase. According to classification of section 12.1 these liquid crystals are named SLC B–D, SLC H–D and NLC–D respectively. Represented structures are found directly from results of molecular dynamic "experiments."

parameter (equal, in this case, to the diameter of the disc-like molecule), then the Bragg parameters of the first orders of reflection from such a lattice will be

$$d_{100} = \sqrt{3}\, a/2,\ d_{110} = a/2,\ d_{200} = \sqrt{3}\, a/4,$$

in other words,

$$d_{100}:d_{110}:d_{200} = 1:0{,}5773:0{,}5.$$

Experimental data[4,5,15] for the three compounds representing the group of benzene-hexa-*n*-alkanoates, as well as for one solution, are presented in Table 11.2. An examination of the data in this table leads to the conclusion that the structure of the liquid-crystalline phase corresponds to a hexagonal lattice. As far as the lattice constant a is concerned, it increases from hexanoate to octanoate. It is remarkable that the experimental values of a agree reasonably well with the approximate molecular diameters, estimated on the basis of an idealized molecular model which takes into account the total length of the terminal "chains" (these diameters for hexanoate, heptanoate, and octanoate are equal to[4] 21.5, 24.0, and 26.5 Å, respectively[4]). Although the data obtained in Refs. 4, 5, and 15 differ, the ratio $d_{100}:d_{110}:d_{200}$ remains unchanged to a high degree of accuracy.

Apart from the structure formed by "columns" (or "filaments") and illustrated in Fig. 11.1a, the results obtained by x-ray scattering could also correspond to another structure.[4] In this last case, the

TABLE 11.2

Structure characteristics of benzene-hexa-*n*-alkanoates (BHA), which are determined on the basis of experimental data.[4,5,15*]

Characteristic	hexanoate $n = 6$	heptanoate $n = 7$	octanoate $n = 8$	heptanoate-octanoate mixture (1:1)
Average intermolecular period along hexagonal axe, Å	6[4]	6[4] 4.6[5] 4.6[15]	6[4] 4.6[5] 4.6[15]	4.6[15]
d_{100}, Å	18.1[4]	18.9[4] 15.0[5] 14.6[15]	20.1 15.9[5] 15.9[15]	15.6[15]
d_{110}, Å	10.6[4]	11.1[4] 8.7[5] 8.7[15]	11.7[4] 9.1[5] 9.1[15]	8.8[15]
d_{200}, Å	9.1[4]	9.7[4] 7.6[5] 7.6[15]	10.1[4] 8.0[5] 7.8[15]	7.7[15]
d_{100}:d_{110}:d_{200}	1:0.59:0.50[4]	1:0.59:0.51[4] 1:0.58:0.51[5] 1:0.59:6.52[15]	1:0.58:0.50[4] 1:0.57:0.50[5] 1:0.57:0.49[15]	1:0.56:0.49[15]
Lattice constant	21.0[4]	22.2[4] 17.4[5] 17.4[15]	23.3[4] 18.4[5] 18.2[15]	17.6[15]

* Early data[4] are given for completeness though they are probably less exact.

liquid crystals would consist of flat layers parallel to one another, regularly positioned and separated by a distance of 6 Å. A flat hexagonal lattice without ordering in the mutual arrangement of the molecules, belonging to different layers, would be observed in each layer. Such a structure could be expected if intermolecular attraction within each layer is much stronger than between different layers. However, since for systems of disc-like molecules, the reverse is true, the "column-like" structure is preferable.

Analogous conclusions were drawn in Ref. 16, which is concerned with the study of diffraction spectra of neutrons diffracted by a layer of octaphenylcyclotetrasiloxane. The existence of a mesophase in the temperature range 188–205°C was established. In a plane coinciding with the planes of the discs, a structure with two-dimensional long range order is observed, while in a direction perpendicular to these planes, a disordered structure occurs (more precisely, having only short range translation order).

Skipping ahead, we note that the results of the last investigations indicate that the axes of the "columns," consisting of the disc-like molecules forming the liquid crystals, are not perpendicular to the planes of these molecules, as proposed in the early works.[4,5,10,14] (Fig. 11.1a), but are tilted (Fig. 11.1b). This problem is examined in greater detail below.

Other early investigations[13] were concerned with studying hexa-substituted triphenylene (4 in Table 11.1). It is remarkable that in this work, concerning the study of hexa-*n*-dodecanoate triphenylene, three different liquid-crystalline mesophases, denoted by the authors as D_0, D_1, and D_2, were observed. Unfortunately, the data obtained concern primarily the description of the texture. The calorimetric measurements that were performed permitted concluding that the transition $D_0 \rightarrow D_1$ practically corresponds to a first-order transition (see Sec. 12.2).

Substances belonging to this class were also studied in Refs. 9–12. The textures were studied using an optical method. In addition, the methods of calorimetry and of nuclear magnetic resonance were used.[12] In this case, the objects of the investigation[10] were not chosen accidentally, but for the purpose of observing different, not "columnar" liquid-crystalline structures consisting of disc-like molecules. The authors of Ref. 10 pointed out two possible paths for destroying the "columns": a) introduction of an asymmetrical element into the central "core" of the molecule and b) introduction of an element into this core giving rise to strong steric obstacle to connection of molecules into "columns." It is the second path that led to success: the results of the study of the texture led to the conclusion that mesophases without "columns" (Fig. 11.1c), which can be called "nematic,"[10] are present. In such liquid crystals, the planes of the molecules are approximately parallel to one another, but there is no long range translation order in the positioning of the centers of mass of the molecules and there is no hexagonal ordering. The formation of this phase is characteristically observed at temperatures much higher than the region of existence of the liquid crystals with "columnar" structure[10] (see Secs. 11.2 and 12.1). The authors note that it is precisely the "nematic" phase of disc-like molecules that is the simplest model of carbon mesophases and is a characteristic example of a unique type of condensed state of matter, obtained deliberately.

Other representatives of this class of substances were investigated by Levelut.[11] Here we are talking about hexa-alkoxy-derivatives of triphenylene ($R = OC_5H_{11}$ and $R = OC_7H_{15}$), which were studied by x-ray diffraction analysis. The pattern obtained as a whole corresponded to the interpretation proposed by Chandrasekhar[4]: arrangement of molecules in "columns," forming a hexagonal lattice. However, a detailed analysis of the diffraction patterns obtained led to interesting conclusions. It turned out that in the case being studied order in the positioning of the aromatic sections of molecules forming the "columns" is quite distinct, while the liquid-crystalline disordering of these "columns" is due to the arrangement of the aliphatic "chains." This is apparently related to the presence of large solid "cores" (triphenylene and oxygen atoms) in the molecules forming these substances. The elastic properties of the molecular "columns" of liquid crystals of this type are also discussed in detail in Ref. 11.

An interesting type of liquid crystals of disc-like molecules was studied in Ref. 7. The substance investigated was hexa-*n*-octanoate of rufigallol (RHO) (See 2 in Table 11.1). Its uniqueness lies in the fact that the molecule has a "core" with low symmetry (a double screw axis *c*). The results obtained by x-ray diffraction analysis indicate the possibility of the formation of two types of mesophases (denoted as D_B and D_C) in this system.

The structure of the first mesophase is analogous to the case of hexa-*n*-octanoate of triphenylene: a two-dimensional lattice formed by parallel "columns" is present. This lattice is rectangular and represents a pseudohexagonal structure (Fig. 11.2). The degree of deviation from a regular hexagon can be characterized by the quantity $2\,(d_{20} - d_{11})/(d_{20} + d_{11})$. For the RHO examined, this deviation is 9%, while for hexa-esters of triphenylene, it is about 1%.[7] This relatively large distortion in the case of RHO is related with the shape of the molecular "core." A "liquid" translational short range order is observed along each "column" and, in addition, the average distance between the molecules is close to 5.6 Å. This comparatively large distance can be explained by the fact that the double bonds of oxygen in the carboxyl groups of molecule lie outside the planes containing its main elements.[7] The comparatively large distance between "cores" of molecules was also observed in the case of hexa-alkanoates of benzene (4.6 Å),[5] while esters of triphenylene exhibit the average value of this distance (4 Å).

Based on the experimental data, the following values were found for the rectangular lattice of RHO: a = 34.8 Å, b = 18.1 Å[7] (top of Fig. 11.2, Table 11.3). It is remarkable that the distance between the nearest lying "columns" is less than the average distance between molecules in the liquid phase. The area per "column," is about 314 Å,

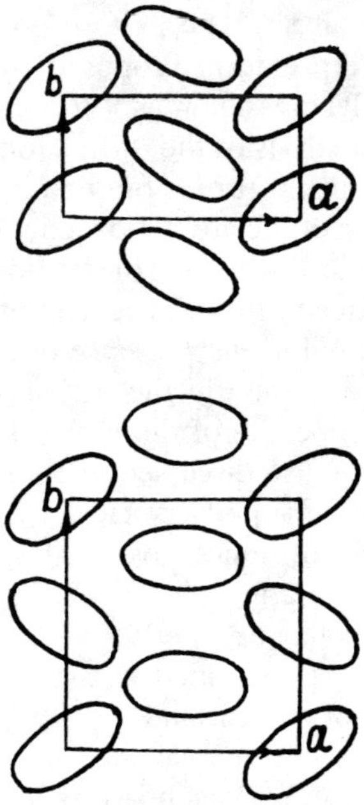

Figure 11.2. Schemes of lattices formed by "columns" phase D_B (upper) and phase D_C (lower).[17]

while the specific area per unit mass is $1.78 \cdot 10^7$ cm^2/g. In the case of hexa-*n*-octanoate of triphenylene, a specific area of $2.52 \cdot 10^7$ cm^2/g was obtained. The small specific areas are related, in the opinion of the authors, with the fact that the "chains" are not stetched out in the plane of the "core" in the mesophase, although in the case of

TABLE 11.3
Experimental and calculation values of characteristic distances $d_{i,j}$, Å for lattices mesophases D_B and D_C structures.[7]

	Mesophase D_B		Mesophase D_C			Mesophase D_C	
ij	Experiment	Calculation	Experiment	Calculation	ij	Experiment	Calculation
20	17.4		17.47	17.45	10		34.90
11	16.1		24.96	25.06	12	16.00	16.00
21		12.54		15.70	22		12.54
31	9.35	9.40		11.67	30		11.67
02		9.05		18.00	32	9.76	9.77
					04	8.98	9.00

the isotropic phase, the probability of an extended flat configuration becomes considerably higher.

It is remarkable that RHO in the state D_B is completely soluble in hexa-*n*-octanoate of triphenylene.[7] It is noted in Ref. 7 that at the present time this is a unique example of complete solubility in the state of the liquid crystal of two compounds consisting of disc-like molecules, belonging to different chemical classes.

The liquid-crystalline behavior in substances consisting of disc-like molecules was observed not only for monomolecular compounds, but also for polymers. Thus the texture of polymeric mesophases was investigated in Ref. 8 (3, Table 11.1).

Data, indicating the existence of tilted (or skewed) columnar structures of disc-like molecules are of great interest.[17,18] Thus, Frank and Chandrasekhar[17] established (on the basis of the results of optical observations) that the molecules in benzene-hexa-*n*-alkanoates are positioned in "columns," whose axes are not perpendicular to the planes of the molecules (as assumed previously). Different possible variants of such molecular packings were examined in Ref. 17. On the other hand, powder x-ray diffraction pictures of hexa-alkoxybenzoates of triphenylene showed[18] that "skewed" cybotactic groups, due to the existence of "columns" in the low temperature region with a tilted packing of molecules are observed in the pretransitional region. We note that these data correspond well with the results obtained independently based on an analysis of the results of molecular-dynamic modeling of liquid crystals formed by disc-like molecules (Fig. 11.1b, see also Sec. 12.1).

The possibility of forming "twisted" cholesteric mesophases in systems of disc-like molecules were discussed in Ref. 19. A "cholesteric" type mesophase was apparently observed for the first time experimentally in this work. In particular, it was established that when a chiral compound is added to the nematic phase formed by the disc-like molecules, a texture is formed that can be interpreted as a cholesteric texture. Mixtures including enantiometric esters with mesogenic cores and triphenylene heptyloxybenzoate, were studied. It turned out that optical textures typical of the cholesteric ordering are observed in such mixtures.[20] Destrade *et al.*[19] came to the conclusion that the structure of the observed mesophases corresponds to the scheme shown in Fig. 11.3. The unusual property here, compared with the ordinary CLC, is that the twisted structure arises with increasing temperature (after the nematic structure) and, in addition, the pitch of the helix decreases with increasing temperature. Destrade

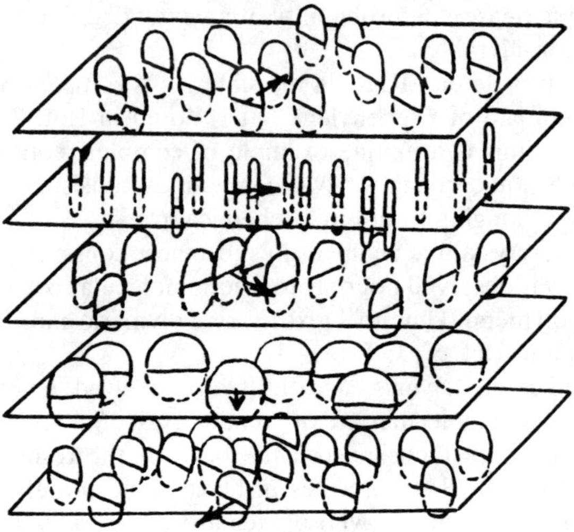

Figure 11.3. Structure of cholesteric phase (N_D^*) of a liquid crystal consisting of disc-like molecules.[19] The arrow shows direction of optical axis.

et al.[19] proposed a plausible model of the structure of the cholesteric phase consisting of the disc-like molecules, but there is still no reasonable theoretical explanation of this phenomenon, based on the properties of the intermolecular interaction. Experimental investigations of the diskotic mesophases of such "cholesteric" type will probably be expanded in the near future. As far as modeling of systems of disc-like molecules by the method of molecular dynamics is concerned, this approach permits establishing the stability of the core of "twisted" diskotic structures with a definite character of the pair interaction potential for the disc-like molecules. However, this twisting in the molecular-dynamic models has a somewhat different character than observed in Ref. 19 (see Sec. 12.1).

The "twisting" of the diskotic mesophases can arise in response to external actions, in particular, as a result of the interaction with surfaces bounding the diskotic liquid crystals (for example, if the surfaces have striations, obtained by rubbing) (Fig. 11.4).[9] It is much easier to create the "twisted" structure in the nematic mesophase, than in "columnar" structures. If the twisted diskotic phase is cooled, then when it transforms into the "columnar" structure, the twisting vanishes.

A thermotropic mesogenic substance consisting of diskotic parti-

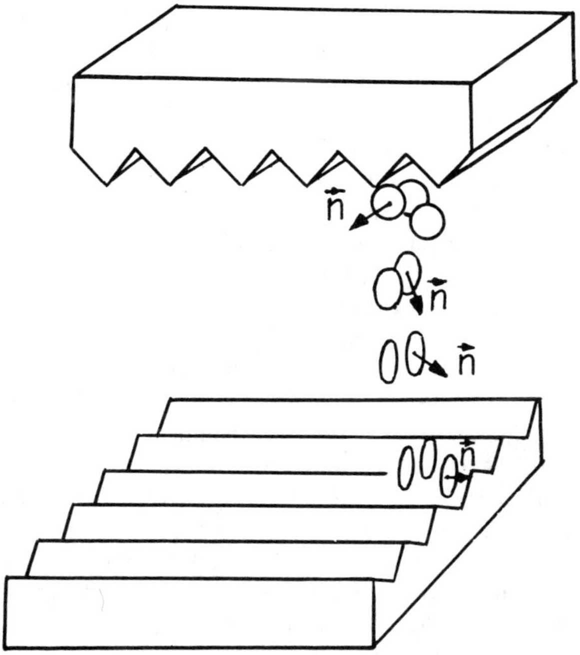

Figure 11.4. Orientations of molecules in twisted sample of phase D_F^*.[19] Molecules are shown as discs perpendicular to **n**.

cles, having metallo-organic complex compound, was synthesized for the first time[21]:

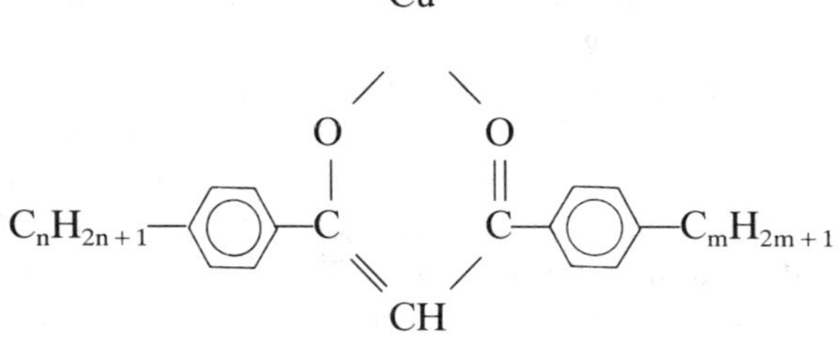

where $n = 10$ and $m = 10$. It was established that this compound is capable of forming a stable mesophase with binary or tetragonal symmetry. The transition temperatures and enthalpies were determined in this work.

The study of distortions, observed in a structure of different types of liquid-crystalline substances consisting of disc-like molecules, is of definite interest. Thus, using the method of polarization microscopy, Oswald[22] studied linear defects in thin layers of mesophases of hexapentoxytriphenylene. It turned out that $+\pi$ disclinations, bound to one another, are often encountered. Systems consisting of two rectilinear disclinations, situated at a right angle to one another, are also observed. In addition, associations with angles of 30° and 60° are present. This indicates that the δ-axes of $+\pi$ disclinations are not oriented randomly. The energy of attachment of the diskotic mesophase with glass was estimated. This energy is about 10^{-2} erg/cm^2. The magnetic susceptibility of mesophases of disc-like molecules was studied in Ref. 23.

A very instructive and interesting example is the structure of diisobutylsilandiole, whose identification remained an unsolved problem for 25 years.[24] The unusual textures observed between 88.4 and 98.7°C did not permit associating the phase with any of the known types. This problem was clarified by English researchers.[25]

Basing on x-ray diffraction data, and using the results obtained by the method of optical microscopy, thermal analysis, and the study of solubility, Bunning[25] came to the conclusion that this substance forms a mesophase, whose structural "units" are the dimer of molecules $(i\text{-}Bu)_2\,Si(OH)_2$, and in addition, in the dimer, these molecules are attached by hydrogen bonds. Such a dimer is shaped like a rectangle with smoothed angles (Fig. 11.5). In the crystalline phase, such dimers form tilted "columns," whose axes form a pseudohexagonal lattice (Fig. 11.6). The action of hydrogen bonds is also manifested along these "columns." In a transition between the solid crystal and the mesophase, the action of the hydrogen bonds along "columns" weakens, rotation of dimers around the axes of the "columns" becomes possible, and the "skewness" of these "columns" vanishes. In the course of the mesophase–isotropic liquid transition, the hydrogen bonds, connecting the molecules in the dimer, break down (Fig. 11.7).

The present status of the problem of the structure of the liquid crystals consisting of disc-like molecules was discussed by Chandrasekhar,[26] who summarized this field. The results of comprehensive investigations show that BHA molecules, consisting of a solid "core"

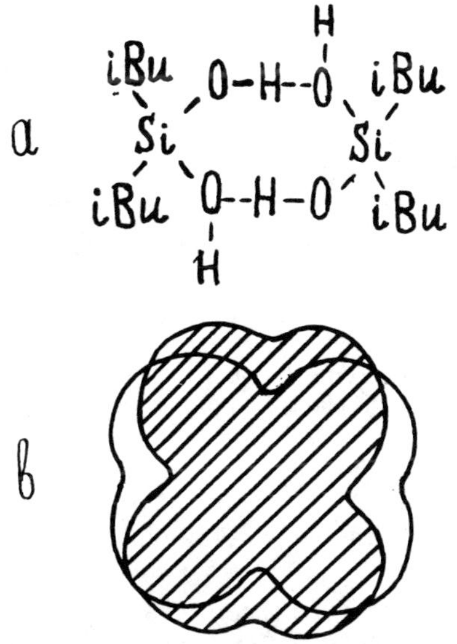

Figure 11.5. Structure of dimer of di-isobutylsilanediol formed by hydrogen bonds (a) and two possible orientations of dimer in mesophase (b).[25]

and flexible aliphatic "chains" (the diameter of the "core" does not exceed 1/3 of the diameter of the entire molecule), form a structure with tilted "columns," whose axes consist of pseudohexagonal lattice. It was established that di-isobutylsilandiole has an analogous lattice. High homologues of triphenylene esters also exhibit "columnar" structures, but a polymorphism is observed here: three mesophases, distinguished by the packing of molecules in "columns," can be observed. As far as triphenylene hexa-*n*-alkoxy- or alkyl-benzoates are concerned, they exhibit both "columnar" and nematic mesophases.[26]

The analysis of the experimental results of the structure investigations of LCD made it possible to theorize on the possibility of the existence of the spiral column-like structures[27,28] which were compared with spiral columns of St. Peter's high altar in Vatican City.[28] It is interesting to note that the same result was independently obtained on the basis of the molecular-dynamic simulation data[29–31] (see 12.1, Figures 12.21, 12.22).

Papers[17,32,33] were devoted to the study of the structure features of the inclined columns consisting of the disc-like molecules. The stability of columns of discotic liquid crystals during the longitudinal

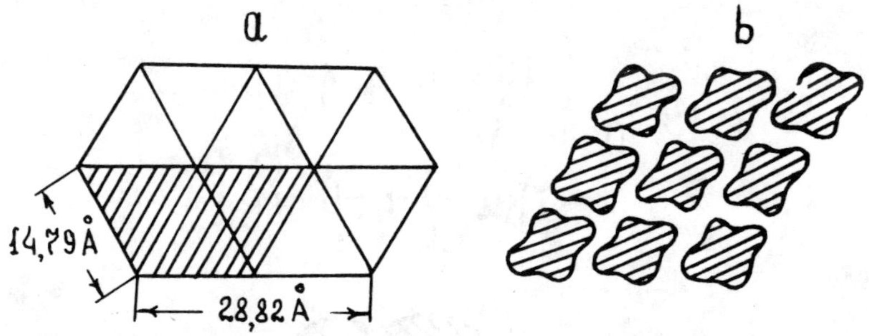

Figure 11.6. Single cell of crystal di-isobutylsilanediol (a) and packing of "columns" at quasi-hexagonal lattice in solid crystal of this substance (b).[25]

compression and bending[34,35] as well as structural and other properties of LCD being simultaneously columnar and lamellar[36] are of great interest.

It should be noted that in recent years a series of new methods of the experimental study of LCD was proposed; the investigation of LCD by the deuterium NMR method,[37] electron spin resonance study

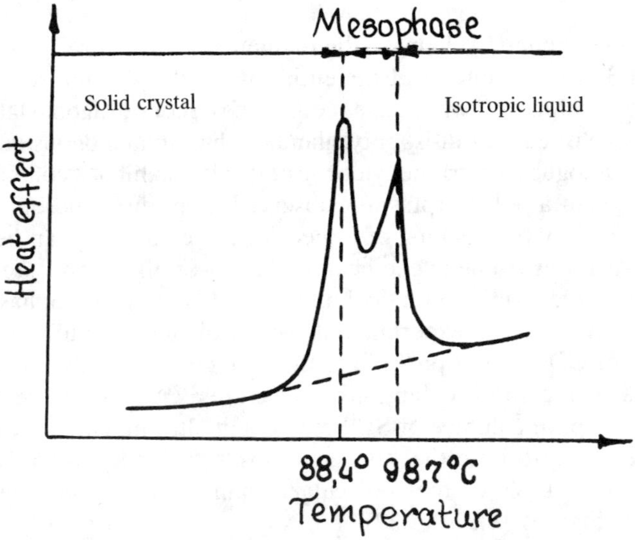

Figure 11.7. Curve of differential thermal analysis of di-isobutylsilanediol found by heating of sample (velocity of heating is 10°C min^{-1}).[25]

of nitroxide spin probes dissolved in the discotic mesophase,[38] syncrotron X-ray study of the orientation ordering structural phase transitions[39] are among them. The latter method made it possible to grow the stable threads from LCD with a length of nearly 1,5 mm and with diameter more than 50 μm.

The interesting connection between the molecular structure and liquid crystalline properties of a substance consisting of washer-like molecules (esters of myoinositol and mytilitol, for example) was found by German authors.[40] In the conclusion, we shall note that the reviews devoted to the structure of the LCD and some properties of these substances were published by Chandrasekhar.[41,42] In the review of A. Gasparous one can find the information on LCD and carbonaceous mesophase.[43] Carbonaceous mesophase and carbon fibers were also considered in the review of S. Otani.[44]

11.2. THEORETICAL MODEL OF A LIQUID CRYSTAL CONSISTING OF DISC-LIKE MOLECULES

One of the most interesting and important problems in the physics of the liquid-crystalline state in general and of liquid-crystalline systems consisting of disc-like molecules in particular is the problem of the nature of phase transitions. As E. I. Kats showed,[45] the characteristics of phase transitions from solid crystal to liquid crystals consisting of disc-like molecules, forming "liquid columns" or "filaments,"* and then to the nematic liquid crystals (in which such "columns" do not exist) and, finally, to the isotropic liquid can be studied theoretically using de Gennes' model of the nematic–smectic-A phase transition.[46,47]

In this section, we shall examine the results obtained with such a theoretical method, following Ref. 45.

According to the experimental data,[4] the "columns" of disc-like molecules form a hexagonal lattice. The density of the H_1 phase is modeled according to a two-dimensional (in the x and y directions) hexagonal law, and there is no periodicity in the perpendicular (z) direction. However, periodicity also arises in this direction in the case

*Such a mesophase, according to its elastic properties, is similar to an anisotropic "solid" mesophase of type H (although, possibly, it does not coincide with it). For this reason, in Ref. 45 and some other investigations, this phase is denoted as H_1.

of a transition into the solid crystalline phase. Then, for the density, we can write

$$\rho = \rho_\perp f(x,y) + \chi, \tag{11.1}$$

where

$$f(x,y) = \sum_i \cos \mathbf{P}_i \mathbf{r}_\perp; \qquad \sum_i \mathbf{P}_i = 0; \qquad r_\perp = (x,y);\ i = 1,2,3$$

$$x = \tfrac{1}{2}[\psi \exp(iq_0 z) + \psi^* \exp(-iq_0 z)],$$

$$q_0 = 2\pi/d,$$

while for H_1 phase, $\chi = 0$. Here $\rho_\perp$ is a coefficient with dimensions of density, and d is the period along the z axis. If the transition into the crystalline state is close to a second-order transition, then the free energy can be expanded in the vicinity of the transition, in the spirit of L. D. Landau's theory, in a series with respect to χ:

$$F = \int d^3r \left\{ a|\chi|^2 + b|\chi|^4 + C_\perp \left[\left| \frac{d\chi}{dx} \right|^2 + \left| \frac{d\chi}{dy} \right|^2 \right] + C_\parallel \left| \frac{d\chi}{dz} \right|^2 \right\}, \tag{11.2}$$

where b, $c_\perp$, and $c_\parallel$ are constants, and $a = \alpha(T - T^*)$. We can now introduce the orientational degree of freedom, using the general form of the orientational order parameter

$$Q_{\alpha\beta} = S_1\left(\mathbf{n}_\alpha \mathbf{n}_\beta - \frac{1}{3}\delta_{\alpha\beta}\right) + S_2\left(\mathbf{m}_\alpha \mathbf{m}_\beta - \frac{1}{3}\delta_{\alpha\beta}\right) + S_3\left(\mathbf{n}_\alpha \mathbf{m}_\beta + \mathbf{n}_\beta \mathbf{m}_\alpha\right), \tag{11.3}$$

where **n** and **m** are unit vectors, giving the orientation of the long axes of molecules (**n**) and planes, in which they are located (**m**), S_i are the corresponding order parameters. In the case of interest to us, taking into account the experimental data,[4] it may be concluded that $S_1 = 0$ (there is no ordering of the long axes) and $S_3 = 0$ (the molecules are nonpolar). As far as the vector **m** is concerned, it is analogous to the director of ordinary NLC.

The presence of a two-dimensional lattice in planes orthogonal to **m** leads to a fixed flux of the vector **m** through an arbitrary surface. The conditions of rigidity of the two-dimensional lattice can be written in the form

$$div\mathbf{m} = 0.$$

Physically, this corresponds to a much higher rigidity of the lattice compared with the bending and torsional deformations of "liquid columns." To satisfy this requirement, it is necessary to set $K_{11} = \infty$ in the expression for the elastic energy, coinciding with the Frank energy

$$F_\gamma = \int d^3r\left\{ \frac{1}{2} K_{ii}(div\mathbf{m})^2 + \frac{1}{2} K_{22}(\mathbf{m}rot\mathbf{m})^2 + \frac{1}{2} K_{33}(\mathbf{m}rot\mathbf{m})^2, \right. \quad (11.4)$$

where K_{ii} are the Frank moduli. The free energy then acquires the form

$$F_0 = \int d^3r\left\{ \tilde{\alpha}|\psi|^2 + \frac{1}{2}\beta|\psi|^4 + \frac{C_\perp}{2}[|(\nabla_\perp - iq_0\delta\mathbf{m})\psi|^2] + \frac{C_\parallel}{2}\left|\frac{d\psi}{dz}\right|^2 + \frac{1}{2}K_{22}(\mathbf{m}_0 rot\delta\mathbf{m})^2 + \frac{1}{2}K_{33}[\mathbf{m}_0 rot\delta\mathbf{m}]^2 \right\}, \quad (11.5)$$

where $\tilde{\alpha} = \frac{1}{2}\alpha + \frac{1}{2}C_\parallel q_0^2$, $\beta = \frac{1}{2}b$, $\mathbf{m}_0 = \mathbf{m} - \delta\mathbf{m}$, $\mathbf{m}_0\delta\mathbf{m} = 0$.

It is thought that the characteristic wave vectors of fluctuations of the order parameter are much smaller than q_0 and the transition to the complex order parameter ψ is made in the corresponding long-wavelength limit.

Further analysis can be performed following de Gennes' method[46] for the nematic–smectic-A transition. This leads to the fluctuation growth of K_{22} and K_{33} and other consequences, following from the analogy to superconductivity.

Usually, for the ordered phase, the correlation function $\langle[L(\mathbf{r}) -$

$L(0)]^2\rangle$ diverges logarithmically and the correlation function $\langle\psi^*(\mathbf{r})\psi(0)\rangle$: diverges in a power-law fashion:

$$\langle\psi^*(r)\psi(0)\rangle = \langle\phi^*(r)\phi(0)\rangle \exp\left\{-\frac{1}{2} q_0^2\langle[L(\mathbf{r}) - L(0)]^2\rangle\right\},$$

where the order parameter in the transformed form is expressed as

$$\phi = \psi \exp(-iq_0L),$$

and in addition, the correlation function of the longitudinal part can be found from the expansion (11.5). For the nematic–smectic-A transition (when the term proportional to K_{11} remains), we can write, using the Fourier components $\mathbf{q} = (q_\perp, q_z)$:

$$\langle L(q)L(-q)\rangle = \frac{T}{2C_\perp q_0^2|\phi|^2 q_z^2 + K_{11}q_\perp^4}.$$

The divergences related with the longitudinal part of $\delta\mathbf{m}$ follow from here. However, in the H_1 phase, $K_{11} \to \infty$ and $\langle|L(\mathbf{r}) - L(0)|^2\rangle \to 0$. For this reason, divergences do not arise in this case in accordance with the Landau–Peierls result.[48] As far as the transverse part of $\delta\mathbf{m}$ is concerned,[46]

$$\delta\mathbf{m} - \mathbf{A} = \nabla L,$$

its correlation function is finite:

$$\langle A_i(q)A_j(-q)\rangle = \frac{\tau(\delta_{ij} + q_iq_j/q^2)}{\frac{1}{2} C_\perp q_0^2|\phi|^2 q_\perp^2 + K_{33}q_z^4} \tag{11.6}$$

For a numerical analysis of fluctuations of ψ (which is virtually independent of the magnitude of the modulus K_{11}), it is convenient to use the Ginsburg–Landau parameter χ. Thus, for $K_{11} = \infty$ (H_1 phase), this parameter will be

$$\chi = \frac{0.39}{q_0}\frac{(\beta K_{33})^{1/2}}{(C_\parallel C_\perp^5)^{1/6}}\left(\frac{K_{22}}{K_{33}}\right)^{1/3}, \tag{11.7}$$

while for $K_{11} = 0$ (nematic–smectic-A transition)

$$\chi = \frac{0.89}{q_0}\frac{(\beta K_{33})^{1/2}}{(C_\parallel C_\perp^5)^{1/6}}\left(\frac{K_{33}}{K_{22}} + \frac{C_\parallel}{C_\perp}\right)^{-1/3}. \tag{11.8}$$

It is interesting to note that the parameter χ with $K_{11} = 0$ is much smaller than when $K = \infty$, if the anisotropy of "effective masses" is larger ($C_{\|}/C_{\perp} >> K_{33}/K_{22}$). For this reason, the transition into the ordered state (with $K_{11} = \infty$), could have been closer to a second-order phase transition. However, according to the experimental data,[4] the opposite is observed; therefore, $K_{33}/K_{22} \sim C_{\|}/C_{\perp}$. In addition to the fluctuation mechanism, various striction effects can also help determine whether or not a real phase transition is a first-order transition.

The analysis examined can be formally extended also to the case of phase transition from the nematic to the H_1 phase.[45] Instead of (11.1), the density is now given by the equation

$$\rho = \rho_0 + \psi_1 f(x, y), \tag{11.9}$$

and in addition, ψ_1 is the complex order parameter of the H_1 phase (for the nematic phase, $\psi_1 = 0$), while $f(x, y)$ is determined by the periodicity of the two-dimensional lattice (for hexagonal symmetry, it is given by the expression (11.1)). The free energy, corresponding to the transition from the nematic to the H_1 phase, can be written for the vicinity of this transition in the form

$$\begin{aligned} F = \int d^3r\Bigg\{ & a|\psi_1|^2 + \frac{1}{2} b|\psi_1|^4 + C_{\perp}\left[\left|\frac{d\psi_1}{dx}\right|^2 + \left|\frac{d\psi_1}{dy}\right|^2\right] \\ & + C_{\|}\left|\left[\frac{d}{dz} - i(\mathbf{p}_0\delta\mathbf{m})\right]\psi_1\right|^2 \\ & + \left[\frac{1}{2} K_{11}(div\delta\mathbf{m})^2 + \frac{1}{2} K_{22}(\mathbf{m}_0 rot\delta\mathbf{m})^2\right. \\ & \left.+ \frac{1}{2} K_{33}\,(\mathbf{m}_0 rot\delta\mathbf{m})^2\right]\Bigg\}, \end{aligned} \tag{11.10}$$

where $\mathbf{p}_0$ is a vector in the xy plane with components (p_0, k_0), where k_0 is the modulus of the reciprocal lattice vector.

If the phase transition nematic-phase–H_1 is close to a second-order transition, then fluctuation clusters of the H_1 phase already appear in the nematic region. Since such clusters cannot contribute to the deformation with $div\delta\mathbf{m} = 0$, they correspond to a growth of the modulus K_{11}. An estimate of this effect based on (11.10) leads to

$$\delta K_{11} \sim C_{11} P_0^2 T \xi, \tag{11.11}$$

where ξ is the coherence length. A more accurate expression for δK_{11} with the help of the response function gives the result

$$\delta K_{11} = \frac{T P_0^2 \xi_\parallel^2}{24\pi \xi_\perp}. \tag{11.12}$$

It is convenient to use the fluctuation growth of the modulus K_{11} accompanying the transition from the nematic to the H_1 phase to identifying this phase. The modulus K_{11} can be measured by a standard method using the Fredericks effect.[46] Analogous measurements in the H_1 phase permit determining the transition to the solid crystal according to the fluctuation growth of the K_{22} and K_{33} moduli. Measurements of the intensity of scattered light are also effective.

However, as follows from the experimental data,[4] the transition from H_1 phase can occur not to the nematic phase but directly to the isotropic liquid. One reason for this is the following. The fluctuations of **m** lead to a displacement of the transition temperature (renormalization of the coefficient a in the expansion (11.10), determining the characteristics of the transition):

$$a' = a + C_\parallel P_0^2 \delta m^2$$

and, correspondingly,

$$T_c = T^* + C_\parallel P_0^2 \delta m^2$$

In other words, the fluctuations **m** increase the temperature of the transition from the H_1 phase to the nematic phase. If the coefficients c_{11} and p_0 are large (while the Frank constants K_{22} and K_{33} are not too large), then T_c may already turn out to be in the region of stability of the isotropic phase.

It should be noted here that the heats of the transitions of the solid crystal–H_1-phase and the H_1-phase–isotropic liquid (40 kJ/kmole) exceeds by at least an order of magnitude the characteristic heats of transitions from the nematic phase to the isotropic liquid.[4] For this reason, the nematic phase may not be determined by a calorimetric method.[45]

To study the properties of the H_1 phase, following Ref. 45, we shall use the general expression for the total elastic energy (agreeing with the requirements of the Landau–Peierls theorem[48]):

$$W_0 = \frac{1}{2} C_{11}(U_{xx}^2 + U_{yy}^2) + C_{12}U_{xx}U_{yy} + (C_{11} - C_{12})U_{xy}^2$$
$$+ \frac{1}{2} C_{33}U_{zz}^2 + C_{13}U_{zz}(U_{xx} + U_{yy}) \tag{11.13}$$
$$+ \frac{1}{2} K_{33} \left\{ \frac{\partial^2 U_x}{\partial z^2} - \frac{\partial^2 U_y}{\partial z^2} \right\}^2 .$$

If dissipation is neglected, then three gapless acoustic modes follow from (11.13). The velocities of these modes depend on the orientations of the wave vector relative to the symmetry axes of the system. If the polar system

$$\omega_{1,2,3} = v_{1,2,3}\,(\theta,\vartheta)q,$$

is introduced, then when $C_{11} << C_{33}$, in the first approximation,

$$v_1 = (C_{33}/\rho)^{1/2}\,(1 + \delta\cos^2\theta), \tag{11.14}$$
$$v_{2,3} = (C_{33}\delta/\rho)^{1/2}\,\sin\theta\cos\theta f_{2,3}\,(\vartheta),$$

and, in addition, $f_{2,3}\,(\vartheta)$ is determined by the two-dimensional symmetry of the lattice and for a hexagonal lattice, is nearly independent of ϑ.

It thus turns out that with respect to its elastic properties, the H_1 phase is analogous to the usual three-dimensional crystal (in the general direction of **q**) and has three acoustic modes. The first mode corresponds to the usual sound in the liquid (with weakly anisotropic velocity), while the two other modes correspond to deformation oscillations of the planar lattice (in these oscillations, the volume remains practically constant). The velocity of these modes depends considerably on the polar angle θ and vanishes when $q_z = 0$ and $q_\perp = 0$. Thus, here, a difference is observed from the SLCA, when, in general, only two acoustical modes are observed.

The features of the acoustical spectrum noted above are manifested in the nature of the scattering of light. The cross section for scattering by fluctuations $\langle u_x^2 \rangle$ and $\langle u_y^2 \rangle$ into a unit solid angle can be defined by the expression

$$\frac{d\delta}{d\Omega} = \frac{\omega^4}{16\pi^2} \langle \delta\epsilon_{ik} \delta\epsilon_{ml} \rangle \, \mathbf{p}_i \mathbf{p}'_m \, \mathbf{p}_k \, \mathbf{p}'_l, \tag{11.15}$$

where **p** and **p**′ are the polarizations of the incident and scattered beams. As far as fluctuations of the dielectric permittivity $\delta\epsilon_{ik}$ are concerned, they are related with fluctuations of the displacement

$$\delta\epsilon_{ik} = M u_{ik}. \tag{11.16}$$

The scattering intensity for polarizations **p** and **p**′ in a plane perpendicular to **m**, according to (11.13), (11.14), and (11.16), is expressed in the form

$$I = \frac{M^2\omega^4}{16\pi^2} \frac{Tq_\perp^2}{C_{11}q_\perp^2 + K_{33}q_z^4}.$$

In the case examined, the scattering is not of the nature of critical opalescence. The scattering intensity is small in the entire region of existence of the H_1 phase; there is no singularity in the longitudinal scattering. This distinguishes the case of the H_1 phase from degenerate systems such as NLC and SLC.[49]

An example of a simple theoretical model, for which a phase diagram of the solid crystal—H_1-phase–isotropic liquid is observed, is given in Ref. 45. It is assumed that the anisotropic pair interaction and the expansion for the two-particle correlation function have the usual form:

$$W_{12} = W_0(r_{12}) + W_2(r_{12})P_2(\cos\theta_{12}),$$

$$g = g_0(r_{12}) + g_2(r_{12})P_2(\cos\theta_{12}),$$

where r_{12} is the distance between the centers of mass of the molcules, θ_{12} is the angle between the normals to the plane of the molecules, P_2 is the Legendre polynomial. The equation for the distribution function ρ_1 in the case when only the smallest Fourier components of the interaction are included acquires the usual form[50]:

$$ln\Lambda\rho_1 = \alpha_0 + 2\alpha_1 f(x,y)\sigma + \beta_0 P_2(\cos\theta)S + 2\beta_1 P_2(\cos\theta)f(x,y)\eta,$$

where σ is the translational order parameter (i.e., $\langle f(x,y)\rangle$), S is the orientational parameter (i.e., $\langle P_2(\cos\theta)\rangle$), η is the mixed parameter (i.e., $\langle f(x,y)P_2(\cos\theta)\rangle$), Λ is a normalization constant, while α_0, β_0,

α_1, and β_1 are easily found for fixed T and $v \simeq \alpha^3$ (volume of the elementary cell).

In the case that $\sigma \neq 0$, we have a solid two-dimensional crystal; if $S \neq 0$, then a nematic phase is observed, while the case $\eta \neq 0$ corresponds to the H_1 phase. It follows from here that for the existence of a phase diagram without the nematic region it is necessary that $\beta_0 = 0$ and $\alpha_1 = 0$. Analysi of this case leads to the conclusion that for such a model, the phase transition H_1-phase–isotropic liquid is a first-order transition, while the critical value of the parameter $\beta_{1c} = 0.147$.

In recent years, intensive investigations were made in the field of searching for new ways of theoretical prediction of discotic liquid crystal properties. The generalization of the molecular-statistical MacMillan's theory of SLC was performed for the case of liquid crystals consisting of the disc-like molecules.[51] Phase transitions in discotic liquid crystals are analysed on the basis of the Landau theory and possible types of order in such systems are studied.[52,53] A simple model of the columnar phase of disc-like molecules liquid crystals was proposed.[54]

The analysis of the possible structural phase transitions due to the change of the symmetry of the two-dimensional lattice of liquid columns and the tensor orientational order parameter in the discotic column-like liquid crystals was carried out by the Landau theory of the structure phase transition.[55,56]

The de Skoulious-de Gennes model was used for the investigation of the periodical modulation of a liquid column lattice.[57] Dynamic properties of discotic liquid crystals were studied by Soviet authors.[58] The theory of macroscopic dynamics near the nematic-columnar transition in discotic liquid crystals was developed by Helmut R. Brand.[59]

CHAPTER XII.

The Results of Computational-Theoretical and Experimental Study of Liquid Crystals Consisting of Disc-Like Molecules

'Tis pity if the case require
(Or so we say) that in the end
We speak the literal to inspire
The understanding of a friend.
But so with all, from babes that play
At hide-and-seek to God afar,
So all who hide too well away
Must speak and tell us where they are.

Robert Frost, "Revelation"

12.1. THE RESULTS OF MOLECULAR-DYNAMICS STUDIES OF THE PARTICLE DYNAMICS, STRUCTURES AND PROPERTIES OF LIQUID CRYSTALS CONSISTING OF DISC-LIKE MOLECULES

The purpose of the investigations in Refs. 60–63 was to study the models of systems of disc-like molecules by the method of molecular dynamics taking into account the most important features of the molecular structure and the results of molecular-dynamic computer "experiments," performed previously for systems of elongated particles (NLS, SLC, CLC).[64–66] As a result, it was possible to determine the details of the structure that characterized the tranlational and orientational ordering of particles, to calculate some thermophysical properties, and to determine the characteristics of the motion of the molecules. Analysis of the data obtained permitted explaining on a molecular level not only the structure, mechanism of the motion of molecules, and other properties of systems previously studied experimentally,[4–7] but also to predict some new liquid crystalline structures in substances formed by disc-like molecules.[65,66]

Systems of 168 particles, shaped like flattened ellipsoids of revolution, were used as molecular dynamics model. The particles were placed in the main specimen, shaped like a rectangular parallelepiped, whose dimensions L_x, L_y, and L_z form the ratio 4.70:3.37:3.89. The ratio of the lengths of the short and long axes γ was chosen equal to 0.3 based on x-ray diffraction studies.[4,26] To decrease the influence of surface effects, periodic boundary conditions were imposed on the faces of the main specimen.

To model and to determine the properties of different types of systems of disc-like molecules, a series of molecular dynamics computer "experiments" was performed for a number of intermolecular pair interaction potentials. In so doing, the interaction of homogeneous flattened ellipsoids of revolution, interaction of flattened ellipsoids with centers of attraction and repulsion, interaction of flattened ellipsoids including the dipole-quadrupole interactions were taken into account. These interaction potentials can be represented in the following general form:

$$\phi(\mathbf{u}_1,\mathbf{u}_2,\mathbf{r}) = 4\epsilon_0\epsilon(\mathbf{u}_1,\mathbf{u}_2)$$

$$\left\{\left[\frac{\sigma_0\sigma(\mathbf{u}_1,\mathbf{u}_2,\hat{\mathbf{r}})}{r}\right]^{12} - \left[\frac{\sigma_0\sigma(\mathbf{u}_1,\mathbf{u}_2,\hat{\mathbf{r}})}{r}\right]^{6}\right\}$$

$$+ \sum_i 4\epsilon_0\epsilon_i' \left(\frac{\sigma_0\sigma_i'}{R_i}\right)^{12} - \sum_j 4\epsilon_0\epsilon_j'' \left(\frac{\sigma_0\sigma_j''}{R_j}\right)^{6}$$

$$- 4\epsilon_0\mu(\mathbf{W}_1\times\mathbf{W}_2\cdot\hat{\mathbf{r}})(\mathbf{W}_1\cdot\mathbf{W}_2)\left(\frac{\sigma_0}{r}\right)^{7}, \quad (12.1)$$

where ϵ and σ are functions of relative orientation of interacting molecules and the anisotropy parameter γ, determined according to Ref. 67; the anisotropy parameter $\gamma = \sigma_{\|}/\sigma_{\perp}$; $\mathbf{u}_i$ and $\mathbf{u}_2$ are unit vectors, oriented along the axis of rotation of the ellipsoids (short axis); $\mathbf{r}$ is a vector connecting the centers of mass of the particles; $\hat{\mathbf{r}} = \mathbf{r}/r$; R_i is the distance between the centers of repulsion of the interacting molecules; R_j is the distance between the centers of attraction; and, ϵ_0 and σ_0 are scale parameters for the energy and length, respectively. The first term on the right side of (12.1) describes the interaction of two homogeneous ellipsoids of revolution. The second term takes into account the role of the centers of repulsion, reflecting the presence of atomic groups in the molecules preventing two ellipsoids from coming close to one another. The third term on the right side of (12.1) takes into account the centers of attraction. Finally, the last term takes into account the dipole-quadrupole interaction of molecules, which according to the Goossens' theory,[68] which has been widely used in recent years, determines the twisted CLC structure. Here, the parameter μ characterizes the relative magnitude of this interaction and is related with the electron density distribution of the molecules.

The starting configurations were given so that the axes of rotation of the ellipsoids were oriented along one of the edges of the main specimen (parallel to the x axis), while the disc-like molecules themselves constituted columns, forming a hexagonal lattice. In separate cases, the coordinates of the particles were chosen from data generated by preliminary computer experiments.

The equations of motion were integrated in the same manner as in the investigation of systems of elongated ellipsoidal particles.[64–66] However, in the case of systems of flattened ellipsoids, it was necessary to use a somewhat smaller time step ($\Delta t^* = 0.004$) in order to assure that the equations of motion are integrated accurately and the results are reliable. All calculations were performed in a reduced form and, in addition, the parameters ϵ_0, σ_0, and the mass of the molecule m were used as the scaling units. The functions of dynamic variables were averaged over 1000 steps after the system reached an equilibrium state. It was assumed that an equilibrium state is reached when the orientational order parameter attains the steady-state constant values (corresponding to a Maxwell distribution, equipartition of the kinetic energy over the degrees of freedom, and constant average values of the usual functions of dynamic variables were established much earlier).

We shall first examine the basic results of the molecular dynamics investigation of systems consisting of homogeneous flattened ellipsoids of revolution, whose interaction is described by a potential (12.1) with $\epsilon_i' = \epsilon_j'' = \mu = 0$ (the dependence of the interaction energy on the distance for characteristic relative orientations of interacting molecules are shown in Fig. 12.1). In this case, characteristically, in the course of the transition to the equilibrium state, the initial packing of particles in columns, forming a hexagonal lattice, is rapidly

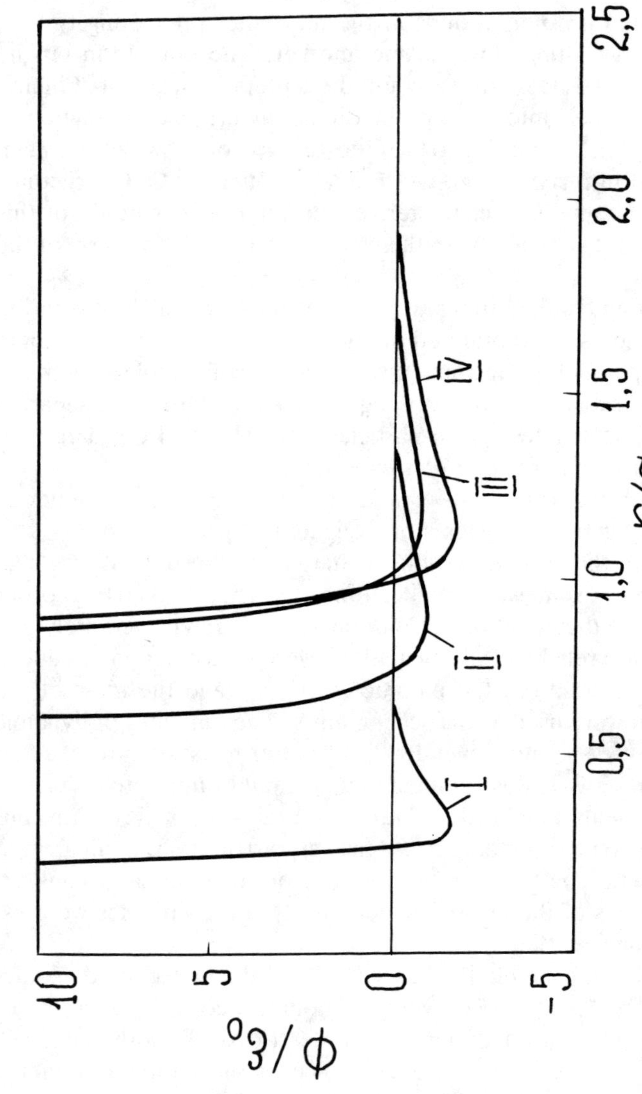

Figure 12.1. Potential of interaction (12.1) at $\varepsilon_i' = \varepsilon_j'' = \mu = 0$ of oblate ellipsoids of revolution ($\gamma = 0.3$): I—orientation ∥; II - | ○; III - —○; IV - — —.

"smeared," and as a result a structure characterized by the absence of long range translational ordering in the positions of the centers of mass of particles and absence of distinct "filaments" or "columns," formed by the molecules, is established. This is clearly indicated by the data in Fig. 12.2, which shows the *z-y* projections of the trajectories of the centers of mass of particles located in layers with thickness 0.34 σ_0. Analogous projections of trajectories found for other layers indicate the breakdown of the initial structure of the columns.

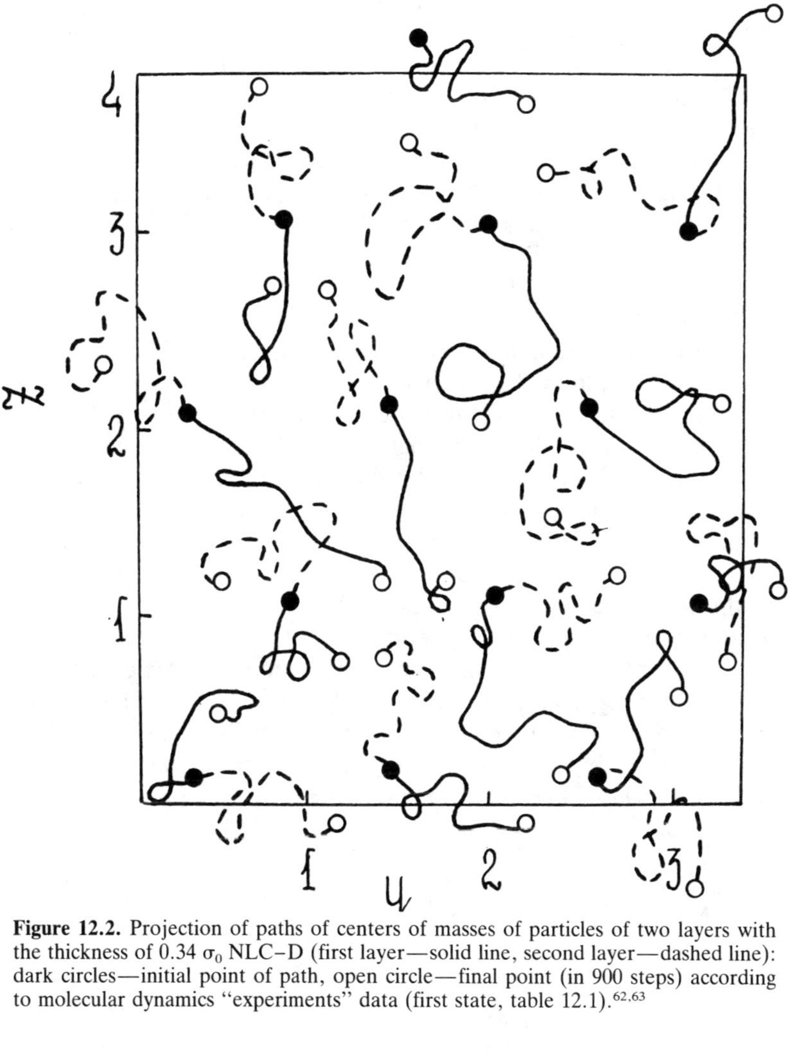

Figure 12.2. Projection of paths of centers of masses of particles of two layers with the thickness of 0.34 σ_0 NLC–D (first layer—solid line, second layer—dashed line): dark circles—initial point of path, open circle—final point (in 900 steps) according to molecular dynamics "experiments" data (first state, table 12.1).[62,63]

Here, the short axes of he ellipsoidal molecules (axis of revolution of flattened ellipsoids) remain approximately parallel to one another (Fig. 11.1c). The projection of molecules onto the plane parallel to the predominant orientation of the axes of revolution (director) is reminiscent of the analogous projection of ordinary NLC of elongated molecules.

Orientational ordering in systems of this type* turns out to be higher than in ordinary NLC.[46] Thus, the parameters $\langle P_2 \rangle$ for the liquid crystalline states examined varied from 0.60 to 0.75, the parameter $\langle P_4 \rangle$-varied from 0.25 to 0.44. This is related with the fact that it is more difficult for diskotic particles to reorient than for elongated particles to reorient. The orientational distribution functions $f(\cos\theta)$ for systems of homogeneous flattened ellipsoids were steeply dropping curves (Fig. 12.3), qualitatively similar to analogous dependences that we obtained earlier for nematic systems of elongated ellipsoidal particles.[64,65] The features noted permitted establishing a definite analogy between the properties of systems of this type and ordinary NLC formed by elongated molecules. For this reason, we termed such systems nematic liquid crystals of disc-like molecules (NLC-D).**

We shall now examine the case of sytems of disc-like molecules, whose interaction is described by the potential (12.1) taking into account the first and third terms on the right side (i.e., $\epsilon_i' = \mu = 0$). In this case, a molecular model with a center of attraction located at the center of mass of the molecules was used; therefore, the third term in (12.1) can be written as $-4\epsilon_0\epsilon''(\sigma_0\sigma''/r)^6$. Thus the interaction of benzene rings, located at the centers of two interacting disc-like molecules, was modeled. As shown in Fig. 12.4, this model is similar to the model that we used for modeling SLC (Fig. 7.8). The dependences of the energy of interaction on the distance between the centers of mass of interacting particles for some characteristic configurations are presented in Fig. 12.5. We note that the values of ϵ'' and σ'' were set equal to 3.0 and 0.3, respectively. In the future, when systems of this kind will be better studied experimentally, it will be possible to chose these parameters from data obtained by special

*With respect to systems of disc-like molecules, it is convenient to define the director as the unit vector parallel to this axis of predominant orientation of the short axes (axes of rotation) of flattened ellipsoids.

**An analogous term was used independently in Ref. 10.

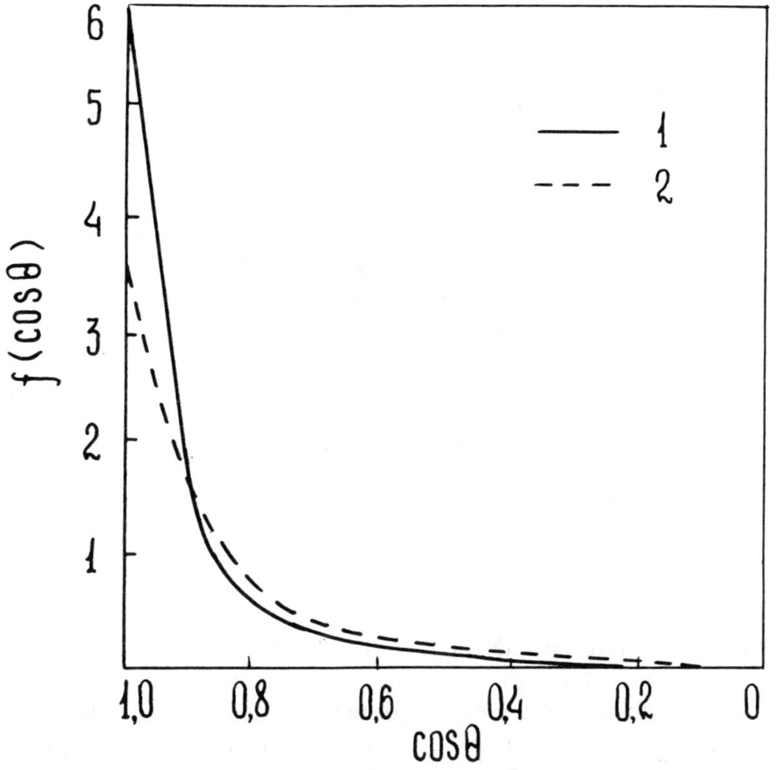

Figure 12.3. Orientational distribution functions f (cos θ) of particles NLC–D: 1—$T^* = 0.68$; $P_2 = 0.734$; 2—$T^* = 0.78$; $P_2 = 0.631$.[62,63]

physical experiments or from conditions guaranteeing quantitative agreement with experimental data.

In the case examined, in contrast to the preceding case, the initial columnar structure remains during the course of the transition of the system into the equilibrium state and after this transition (although, of course, there is some "smearing" of the initial "columns" by the thermal motion of particles) up to a quite high temperature $T^* = 1.32$. The trajectories of the centers of mass of particles in this case have a specific character (Fig. 12.6), indicating the oscillations of the centers of mass around points lying on the axes of the "columns" (or "filaments"). These trajectories differ considerably from the preced-

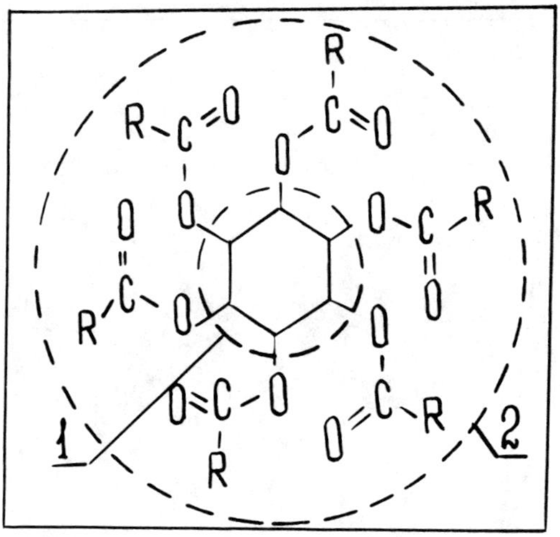

Figure 12.4. Model of disc-like molecule with center of attraction (R from $n\text{-}C_5H_{11}$ till $n\text{-}C_8H_{17}$): 1—center of attraction; 2—external ellipsoid.[62,63]

ing case and indicate the stability of the "columns" or "filaments," constituting the liquid crystals (Figs. 11.1a and b).

The distribution of the centers of mass of the particles making up a "column" was analyzed with the help of a binary distribution function $q(x)$, representing a one-dimensional variant of the usual binary distribution function.[67] The rapid damping of the peaks in $g(x)$ indicates a distribution that has the typical "liquid" character (presence of short range order in the absence of long range order (Fig. 12.7)). Thus long range translation order in the positioning of the centers of mass of molecules of each "column" in absent. The projection of each "column" on a plane parallel to the predominant orientation of the axes of rotation of molecules, is similar to the corresponding projection of molecules in a layer of the usual SLC. We have termed systems of this type smectic liquid crystals of disc-like molecules (SLC-D).

The degree of orientational order of SLC-D is even higher than in the case of NLC-D: the parameter of orientational order $\langle P_2 \rangle$ fell between 0.75 and 0.97. The orientational distribution functions $f(\cos\,\theta)$ had a character analogous to the case of NLC-D and the usual NLC of elongated molecules.[65,66] but were steeper and had a higher peak (Fig. 12.8).

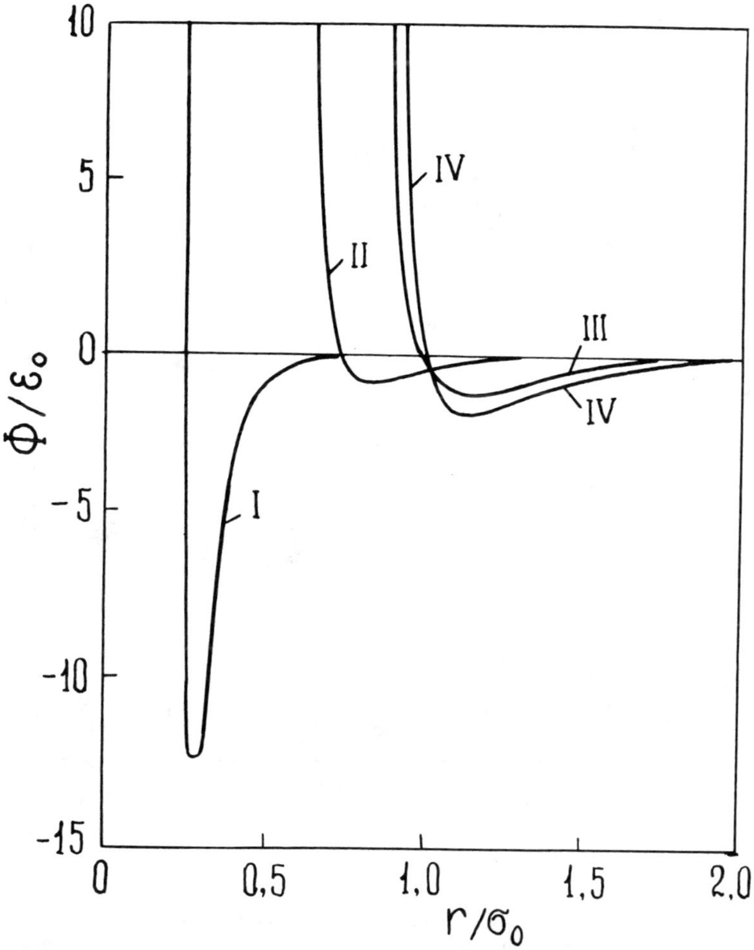

Figure 12.5. Potential of interaction (12.1), $\varepsilon_i^1 = \mu = 0$. Orientations correspond to designations of Figure 12.1: $\gamma_1 = 0.3$; $\gamma_2 = 1.0$; $\varepsilon_2 = 3.0$; $\sigma_2 = 0.3$.

We note that the arrangement of the projection of axes of the "columns" on a plane perpendicular to their axes, in many cases, correspond to a two-dimensional liquid (Fig. 12.9a), and in other cases (at lower temperatures) to a two-dimensional crystal (hexagonal lattice, Fig. 11.1a). It is natural to denote systems of the first type by SLCA-D and those of the second type SLC B-D. Structures of precisely the last type were observed experimentally[4] and studied relatively well.

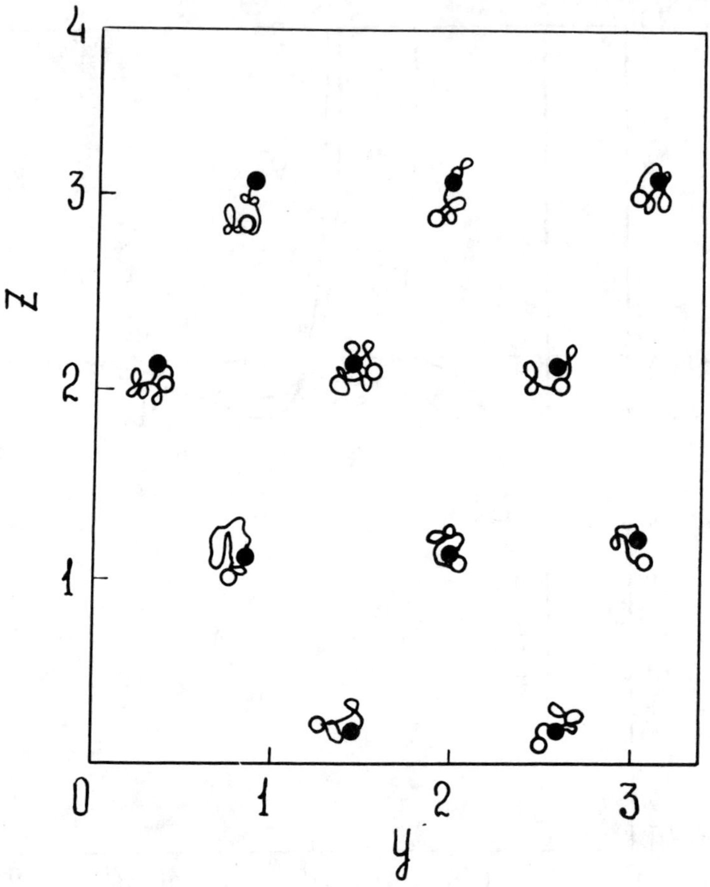

Figure 12.6. Projection $z - y$ of centers paths of particle masses of SLC–D for the layer the thickness of which is 0.34 σ_0 according to molecular dynamic experiment data (state 4, Table 12.1). Black circles are the initial point, open circles—the final point (in 1000 steps).[62,63]

We shall now proceed to an examination of the features of the dynamics of disc-like particles in NLC-D and SLC-D. We studied the translational motion with the help of the time correlation functions (TCF) of the translational velocities

$$\psi_v^{\alpha}(t) = \langle v^{\alpha}(0)v^{\alpha}(t)\rangle / \langle v^{\alpha}(0)v^{\alpha}(0)\rangle, \ (\alpha = x,y,z), \qquad (12.2)$$

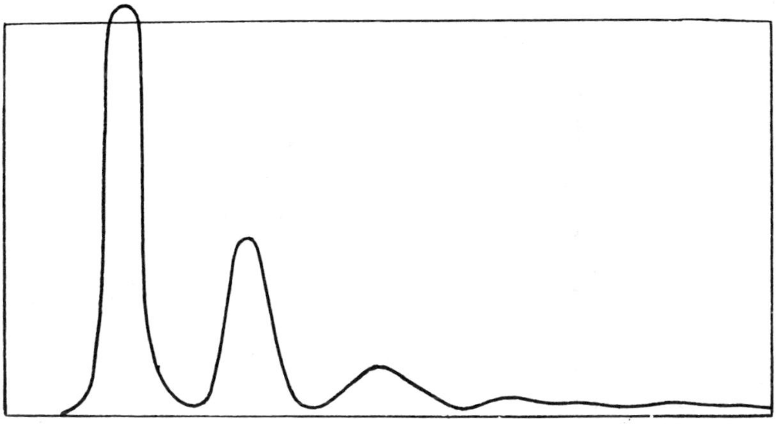

Figure 12.7. One-dimension distribution function of mass centers of molecules in "column" of SLC–D (state 5, Table 12.1). Calculation is made for cylindrical volume with radius $R/\sigma_0 \leqslant 0.25$.[62]

where $v^\alpha(t)$ is the projection of the velocity of the center of mass of a particle on one of the coordinate axes at the time t, while the angular brackets indicate averaging over particles and initial time. Taking into account the characteristics of the shape of molecules and the structure of the system as a whole, it is expedient to study the nature of the motion of the particles in directions parallel to the director and perpendicular to it (TCF $\psi_v^{\parallel}$ and $\psi_v^{\perp}$, respectively).

The general form of the TCF ψ_v for the liquid crystalline systems consisting of disc-like molecules NLC-D studied is typical for liquids in the sense that there is a distinct region of negative values and then a comparatively rapid damping of TCF. The most distinct difference between $\psi_v^{\parallel}$ and $\psi_v^{\perp}$ lies in the fact that $\psi_v^{\parallel}$ decreases much more rapidly than $\psi_v^{\perp}$ (Fig. 12.10 and 12.11), i.e., a situation opposite to that occurring in liquid crystals of elongated particles is observed.[64,65] This nature of TCF reflects the difference between the translational motion of disc-like particles relative to the director **n**, indicating that they move more freely in a direction perpendicular to **n**. This result is physically reasonable. It is interesting to compare TCF of translational velocities for NLC-D and SLC-D. In the case of NLC-D, the TCF $\psi_v^{\parallel}$ damps out monotonously and relatively smoothly, while for SLC-D (Fig. 12.12), this function has a distinct oscillatory character, reflecting the presence of oscillations of particles in a direction per-

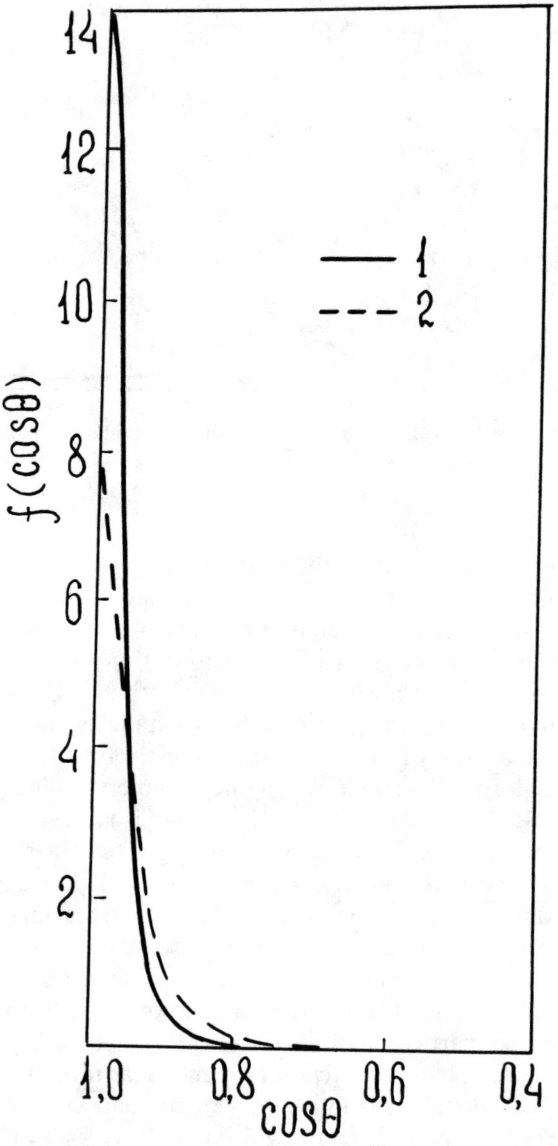

Figure 12.8. Orientational distribution functions $f(\cos\theta)$ of particles of SLC–D: 1—state 3, Table 12.1; 2—state 4, Table 12.1.[62]

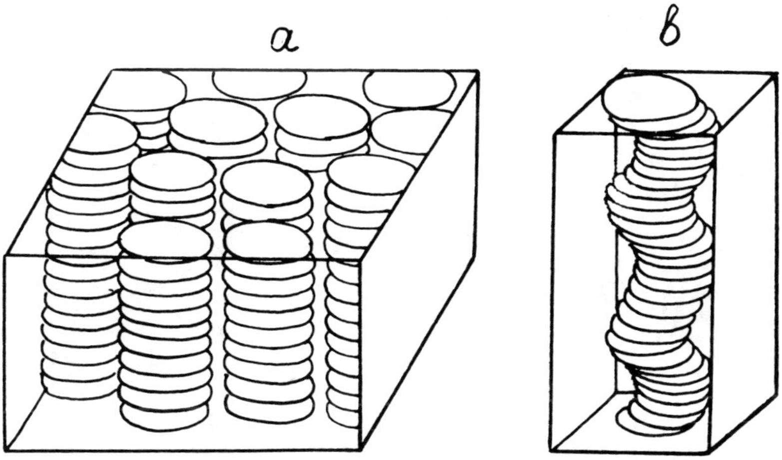

Figure 12.9. Structure of "column-like" liquid crystal, consisting of disc-like molecules with the absence of ordering of axes of "columns" (SLC–D) (a) and with spiral "columns" of SLC–D (b).[62]

pendicular to the director (i.e., perpendicular to the axes of "columns"). For NLC-D, the following feature of $\psi_v^{\parallel}$ ($T^* = 0.67$) was noted: a minimum and a small peak in the region of positive values, which disappeared with increasing temperature ($T^* = 0.78$).

Integration of TCF $\psi_v^{\parallel}$ and $\psi_v^{\perp}$ with respect to time permitted determining the components D_i of the self-diffusion tensor:

$$D_i = \frac{kT}{m} \int_0^{\infty} \psi_v^i\,(t)dt, \tag{12.3}$$

whose values are represented in Table 12.1. We note that a feature of NLC-D and SLC-D is the higher values of $D_{\perp}$ with respect to $D_{\parallel}$, which, apparently is related with the diskotic shape of the particles and their higher orientational order. Thus the components of the translational self-diffusion tensor of NLC-D $D_{\parallel}^*$ at temperatures $T^* = 0.68–0.78$ varied from 0.009 to 0.013, while $D_{\perp}^*$ varied from 0.046 to 0.047; the effective coefficient of self-diffusion D_0^* was close to 0.03. A smaller change with temperature is characteristic of SLC-D (the potential (12.1) including the first and third terms on the right side): at $T^* = 0.79–1.32$, $D_{\parallel}^*$ varies from 0.005 to 0.006. In this case, the components of the self-diffusion tensor are much lower than

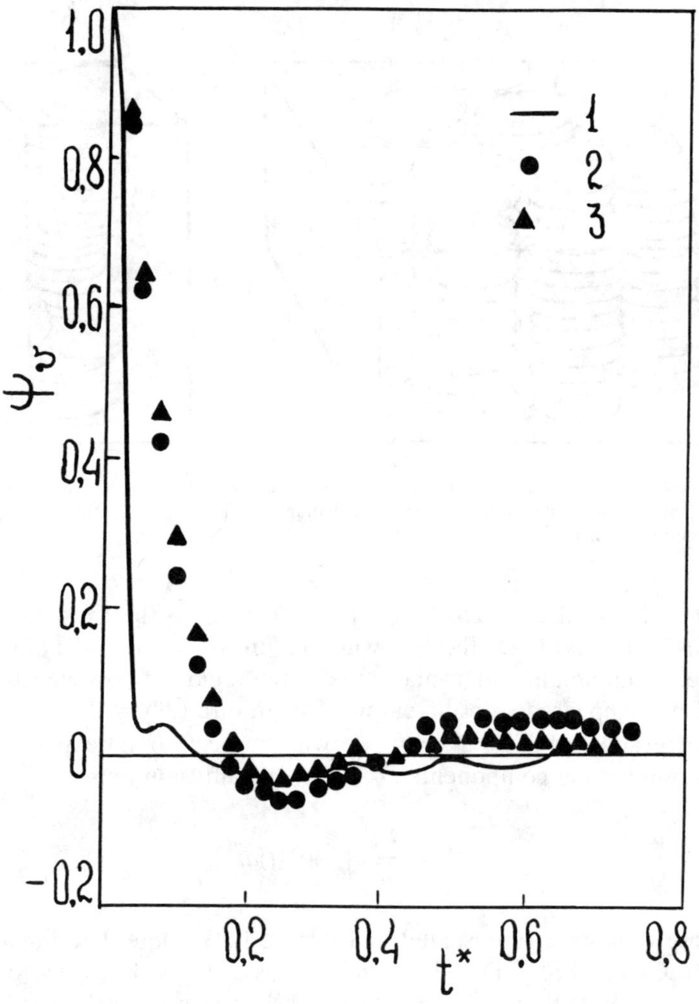

Figure 12.10. Time correlation functions of linear velocities of particles of NLC–D (state 1, Table 12.1): 1—along the x-axis (in parallel to director); 2—along the y-axis; 3—along the z-axis.[63]

for NLC, while the effective coefficient of self-diffusion D_0^* is close to 0.010. However, with increasing temperature, these quantities increase to values characteristic of NLC-D (Table 12.1).

The rotational motion of disc-like molecules was studied with the

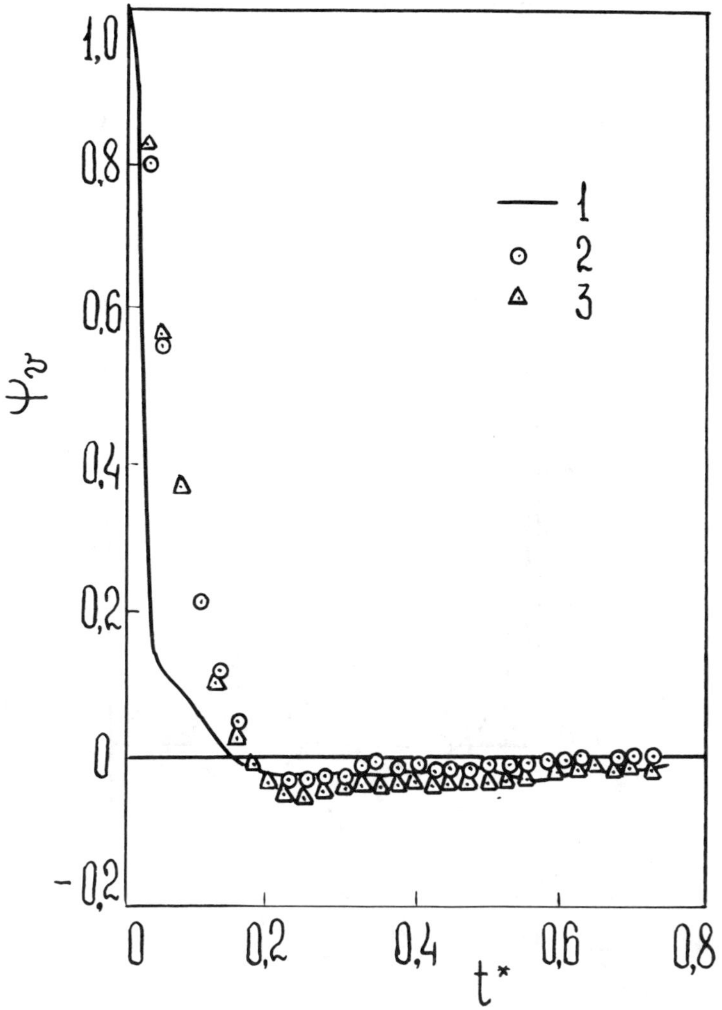

Figure 12.11. Time correlation function of linear velocities of particles of NLC–D (state 2, Table 12.1); see designations of Figure 12.10.[63]

help of TCF of angular velocitics

$$\psi_\omega = \langle \boldsymbol{\omega}(0)\boldsymbol{\omega}(t)\rangle / \langle \boldsymbol{\omega}(0)\boldsymbol{\omega}(0)\rangle. \tag{12.4}$$

The results of calculations for NLC-D are presented in fig. 12.13. For SLC-D ($T^* = 0.79–1.32$), these functions had a distinctly os-

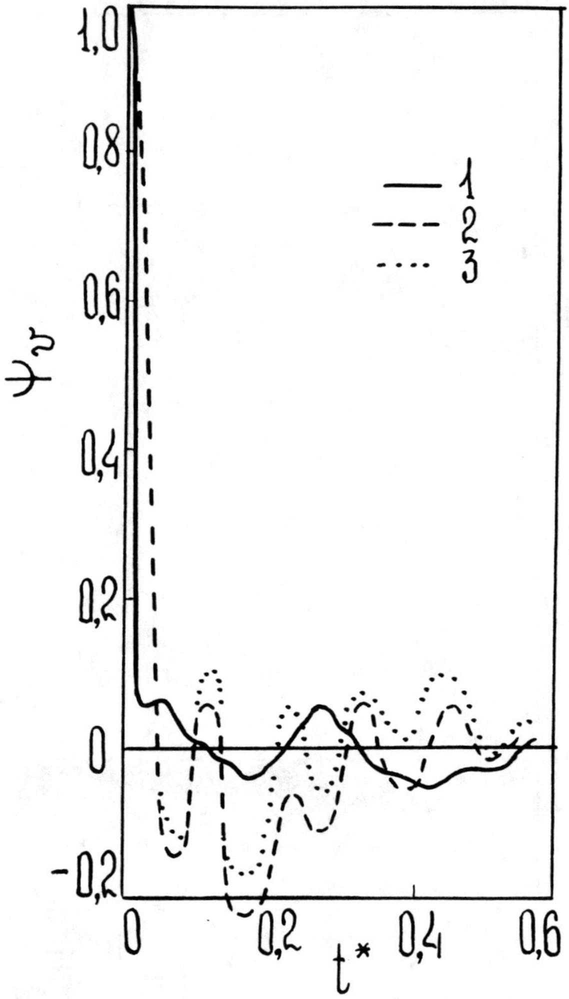

Figure 12.12. Time correlation functions of linear velocities of particles of SLC–D (state 3, Table 12.1): 1—along the *x*-axis (in parallel to director); 2—along the *y*-axis; 3—along the *z*-axis.[63]

cillatory character (Fig. 12.14), typical for analogous TCF of solids and, in addition, the correlation of angular velocities of the particles remained for relatively long times. This form of TCF ψ_ω indicates the presence of collective oscillatory motions of the axes of rotation of

Table 12.1.
Some results of molecular dynamic modelling of liquid crystal systems (NLC-D, SLC-D) of disc-shaped molecules (packing coefficient $\eta = 0.427$)[62,63]

Retained terms of the right part (12.1)	1	1	1 and 3	1 and 3	1 and 3
Number of state	1	2	3	4	5
T^*	0.68	0.78	0.79	1.32	1.75
U^*/N	− 9.767	− 9.536	−19.98	−17.20	−14.15
$\langle P_2\rangle$	0.743	0.631	0.972	0.909	0.791
$D^*_\parallel$	0.009	0.013	0.005	0.006	0.018
$D^*_\perp$	0.047	0.046	0.011	0.013	0.028
$D_\parallel/D_\perp$	0.19	0.28	0.45	0.46	0.64
D_O	0.034	0.035	0.009	0.011	0.025
D^*_R	—	—	0.000	0.048	0.103
τ^*_1	16.8	9.53	55.8	38.5	20.6
τ^*_2	5.8	3.12	19.4	13.8	7.6
τ_1/τ_2	2.9	3.1	2.9	2.9	2.7
Type of mesophase	NLC-D	NLC-D	SLC-D	SLC-D	NLC-D

the "disks." However, with increasing temperature ($T^* = 1.75$ and higher), the oscillatory character of the TCF ψ_ω vanished, and it acquired a form typical of ordinary liquids and liquid crystals.

The coefficient of rotational diffusion D_R of SLC-D, found by integrating the TCF ψ_ω:

$$D_R = \frac{kT}{m}\int_0^\infty \psi_\omega\,(t)dt, \qquad (12.5)$$

was close to zero for $T^* = 0.79$, but increasing rapidly, at $T^* = 1.32$ and 1.75, it constituted 0.048 and 0.103, respectively.

Figures 12.15 and 12.16 show the vectorial and tensorial reorientational TCF

$$\psi_l(t) = \langle P_l[\mathbf{u}(0)\cdot\mathbf{u}(t)]\rangle, \qquad (l = 1, 2) \qquad (12.6)$$

where P_l is a Legendre polynomial of degree l. The functions $\psi_l\,(t)$ characterize the rate of loss of correlation of orientations of disc-like molecules. A very slow decrease of ψ_l with time (especially for SLC-D) indicates the difficulty of reorientation of disks, which is typical

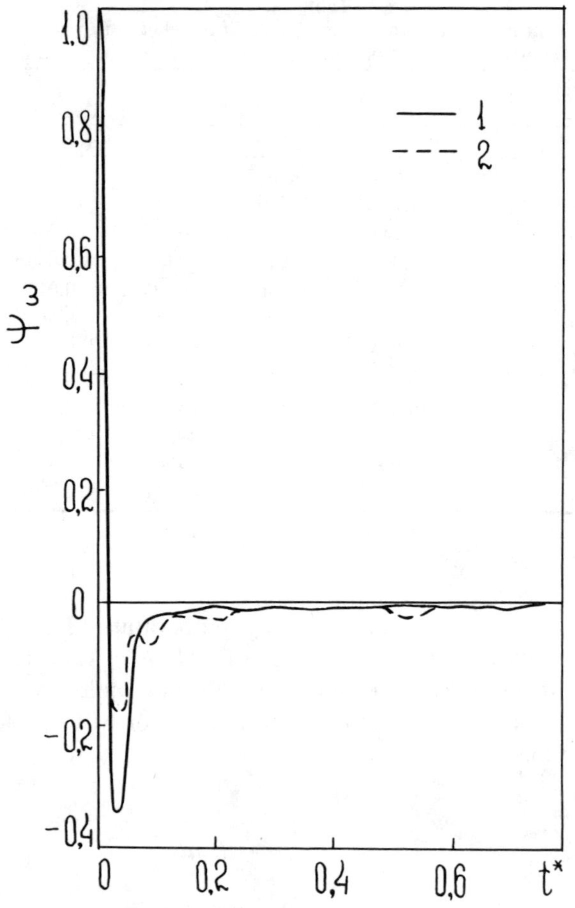

Figure 12.13. Time correlation functions of angular velocities of particles of NLC–D: 1—state 2, Table 12.1; 2—state 5, Table 12.1.[63]

for dense systems, consisting of very nonspherical particles. The correlation times

$$\tau_l = \int_0^\infty \psi_l(t)dt, \qquad l = 1, 2 \tag{12.7}$$

for SLC-D greatly exceed the values of τ_l for NLC-D (see Table 12.1). This is related with the fact that the presence of "columns" or

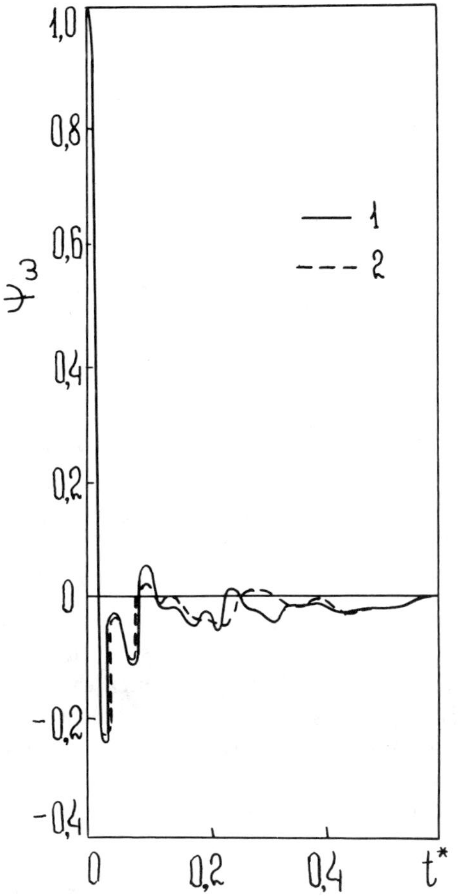

Figure 12.14. Time correlation functions of angular velocities of particles of SLC–D: 1—state 3, Table 12.1; 2—state 4, Table 12.1.[63]

"filaments" apparently inhibits reorientation of disc-like particles. The ratio of the correlation times τ_1/τ_2 in all "experiments" was close to 3, which indicates the diffusion nature of the rotational motion of particles in the system.

The properties examined above refer to liquid crystalline state. When the molecular dynamics "experiment" was performed at a lower temperature, the usual crystalline state was observed (in particular, long range translational ordering of the centers of mass of

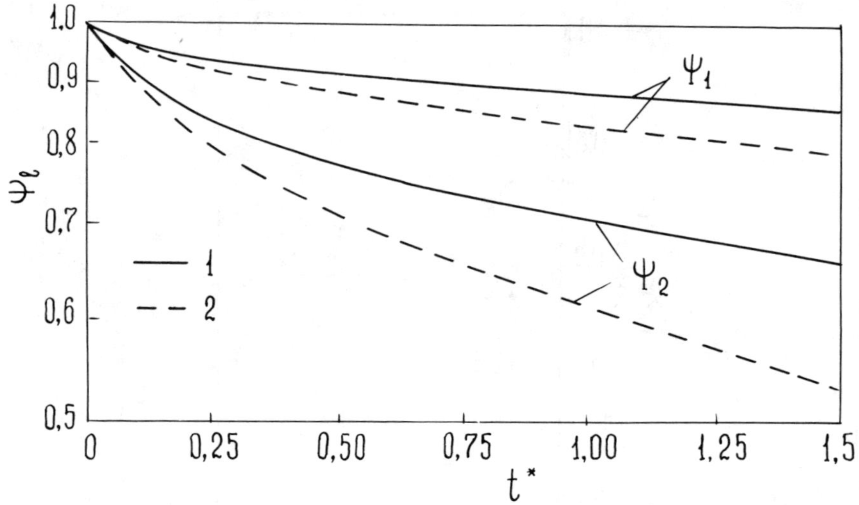

Figure 12.15. Reorientational time correlation functions of NLC–D: 1—state 1, Table 12.1; 2—state 2, Table 12.1.[63]

particles was present). On the other hand, if the temperature is high, then the system in the course of the transition into the equilibrium state relaxes to an isotropic liquid, for which the absence of orientational order is characteristic (values of $\langle P_2 \rangle$ close to zero and other properties typical of an ordinary isotropic liquid). It should be emphasized, however, that according to the results of our molecular dynamics calculations, the region of the liquid crystalline state in

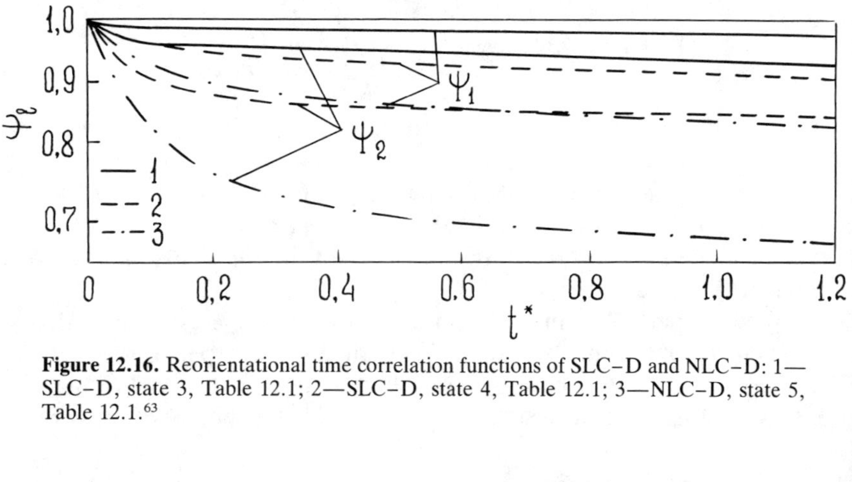

Figure 12.16. Reorientational time correlation functions of SLC–D and NLC–D: 1—SLC–D, state 3, Table 12.1; 2—SLC–D, state 4, Table 12.1; 3—NLC–D, state 5, Table 12.1.[63]

diskotic systems is quite broad and larger than in typical ordinary liquid crystalline systems consisting of elongated particles and of their molecular dynamics models. The result as a whole agrees with the results of experimental works. Thus, according to Ref. 13, the region of existence of the liquid crystalline phase of diskotic molecules triphenylene hexa-*n*-dodecanoate lies between 353.2 and 395.5 K, while for the substances studied in Ref. 10, the region of existence constitutes up to 90 K.

The observation of a transition into NLC-D with increasing temperature (Table 12.1) in our model of SLC-D (interaction potential (12.1) including the first and third terms on the right side) is of special interest. This transition is indicated by qualitative changes in the trajectories of the particles, indicating breakdown of the "columns" as well as considerable changes in other characteristics such as the coefficients of diffusion, correlation times τ_1 and τ_2, etc.*

Few of the liquid crystalline systems, whose properties we have studied by the method of molecular dynamics and which are examined in the present section, have been studied experimentally. However, data obtained permit referring the results of separate observations to NLC-D and to other previously unknown types of structures.[8,10,13] Thus, in Ref. 13, by way of calorimetric measurements and optical study of textures, it was shown that in some substances consisting of disc-like molecules, not one but three different mesophases are observed, analogously to ordinary liquid crystals of elongated molecules. Later, a substance which at definite temperatures behaves similarly to a nematic liquid crystal of disc-like molecules, was synthesized.[10] The very limited information on the properties of polymeric mesophases, analogous to the nematic phase of disc-like molecules, was given in Ref. 8. Unfortunately, primarily only photographs of textures of such systems are published.

We shall now examine other types of liquid crystalline structures of disc-like molecules. Many of these structures have not yet been observed experimentally, but as the results of the molecular dynamics experiments performed based on the models used indicate, they are stable and it may be expected that they will be observed experimentally in the near future.

The molecular model (12.1) including the first, second, and third terms on the right side, describing the interaction of flattened ellipsoids of revolution with a center of attraction and repulsion (flattened

*A theoretical analysis of a transition of this type is given in Ref. 45.

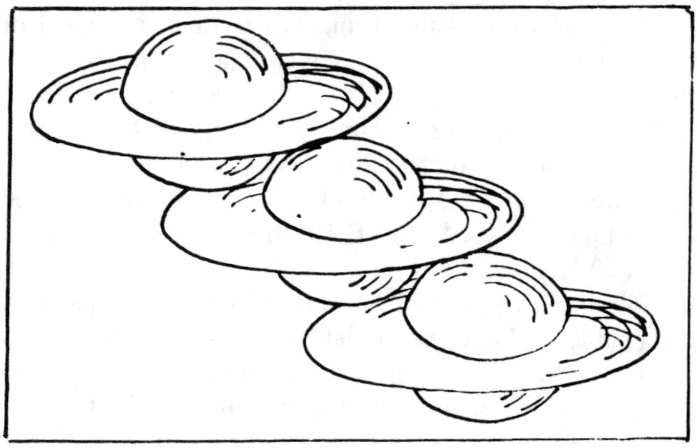

Figure 12.17. Simple molecular model (oblate ellipsoid of revolution with central convex spherical part) which explains the arising of column-like inclined structures.[60–63]

ellipsoid of revolution with a convex central part, Fig. 12.17), permits predicting the tilted filamentary structure (Fig. 11.1b). The molecular dynamics experiment in which such a structure was used as the starting structure indicated that at temperatures T approximately equal to 0.7–1.2, this structure is retained (although, of course, there is some "smearing" by thermal motion) for a long period of time (up to $5 \cdot 10^3$ steps), which indicates its stability. This structure turns out to be energetically more favorable (Fig. 12.17) and corresponds to contemporary experimental data.[26] As in the in case of SLCA-D and SLCB-D, long range translational ordering of the centers of mass of particles within each "column," is absent, but short range order is present. Depending on the presence of ordering of the projections of axes of columns on the plane perpendicular to them, we can talk about smectic liquid crystals C of disc-like molecules (SLCC-D) or smectic H disc-like molecules (SLCH-D).

It is understandable that the relative displacement of two molecules with the distance between them and the angle α remaining constant (Fig. 12.17) does not change their interaction energy. For this reason, the model also predicts the structure of twisted "columns," when each diskotic molecule in a column is displaced to such an extent relative to its neighbor that a twisted structure forms (Fig. 12.9b). This displacement can be caused by weak long range forces or the properties of the molecular structure. The geometric location of the centers of the molecules in this case forms a twisted curve. Within the limits of each twisted "column", there is no long range ordering of the centers of mass of the molcules. Depending on the presence or absence of relative ordering of the twisted columns and the tilting of planes of molecules to the dominant direcion of the axis of the "column" (such tilting can arise, for example, if one of the relative positions of neighboring molecules of the "column" is observed more often than others), we can talk about liquid crystals SLCA-D, SLCC-D, and SLCH-D.

The model of a flattened ellipsoid, which has a feature along the **W** axis parallel to its large median plane, predicts interesting new structures. This feature can be related to the shape of the particle (triaxial ellipsoid of revolution), singularity in the distribution of the electron density (dipole or quadrupole moment), and an insignificant singularity of the shape. Polymer molecules are an example.[8] Some angle, characteristic for ordinary CLC, can arise between the directions $\mathbf{W}_1$ and $\mathbf{W}_2$ of two neighboring molecules, whose large median planes are approximately parallel, for example, as a result of the dipole-quadrupole interactions (Fig. 12.18a). The interaction of such particles can be described by potential (12.1) including the first and last terms on the right side.* The plane parallel to the planes of the molecules, is perpendicular to the axis of the cholesteric twisting, but no "columns" appear, and there is no long range translational ordering of the centers of mass. Such liquid crystals can be called cholesteric liquid crystals of disc-like molecules (CLC-D).

If the molecular model is additionally characterized by the presence of a center of attraction (potential (12.1), including the first, third, and fourth terms on the right side), then a combination of cholesteric twisting and "columns" is observed (Fig. 12.18b). In other words, the molecules comprise the "columns" for a typical cholesteric helix.

In connection with the difficulty of introducing periodic boundary conditions in studying the spiral structures and the practical role of films, in this work, thin films were studied and, in addition, the technique was analogous to that described in Ref. 66. The molecular dynamics modeling demonstrates the stability of spiral structures. One of the results of such modeling is the derivation of a considerably shorter pitch of the cholesteric helix in liquid-crystalline substance consisting of "columns" of disc-like molecules compared with the ordinary CLC of elongated particles (the characteristic pitch here is

---

*In writing down the potential (12.1), for simplicity, the triaxial nature of the ellipsoid is neglected.

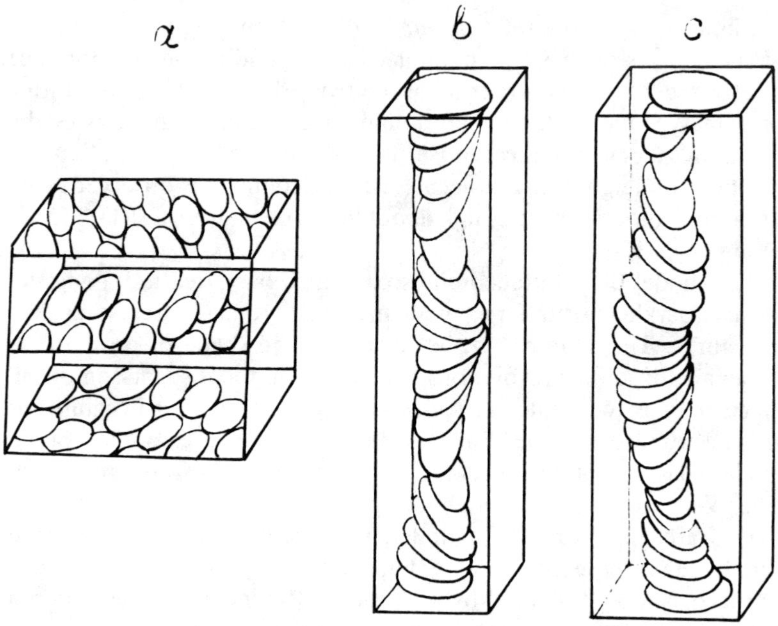

Figure 12.18. Configurations of disc-like molecules which compose CLC–D (a) and form the "columns" of CLC (A or B)–D (b) and CLC S (A or B)–D (c).

200 Å compared with 700 Å for ordinary CLC). The reason for this difference lies in the fact that the effect of factors giving rise to the appearance of a helix has a much stronger effect for "columns." In ordinary CLC, the thermal distribution leads to the fact that the real angle of relative rotation of molecules turns out to be much smaller than the angle corresponding to minimum of the interaction potential with particles positioned so that the line joining the centers of mass is perpendicular to the long molecular axes.

Comparing "twisted" structures, observed experimentally[19] (Fig. 11.3) with the structures obtained by modeling,[62,63] the definite difference should be noted. In the first case, the appearance of the helix is related with the rotation of the "director," corresponding to the predominant orientation of the short axes of molecules and coinciding with the orientation of the optic axes, while in the second case, we can talk about two "directors." One of them is determined by the dominant orientation of the short axes, and the second is determined by the dominant direction of the long axes, lying in the planes of the molecules (here, the latter differ from a regular circle). It is the second "director" that "rotates" (Fig. 12.18). It should be said that the structure corresponding with Fig. 12.18 is energetically more favorable than the structure shown schematically in Fig. 11.3.

If the axes of the "columns" are perpendicular to the planes containing the two long axes of the molecules, while the projections of the axes on the plane perpendicular to them are situated analogously to the centers of particles of a "two-dimensional" liquid, we can talk about cholesteric liquid crystals A of disc-like molecules (CLCA-D); if this arrangement is ordered ("two-dimensional" crystal), then we have a cholesteric liquid crystal B of disc-like molecules (CLCB-D). In some cases, it appears that the tilted structures CLCC-D and CLCH-D form: in the first case, the axes of the tilted columns form a two-dimensional liquid, while in the second, they form a two-dimensional crystal. Such structures are predicted by a model corresponding to the interaction potential (12.1) including all terms on the right side. The tilting of the columns could be due to the presence of centers of repulsion in molecules ("broadening" of the center of the molecules, Fig. 12.17).

Somewhat more complicated structures correspond to the case when the cholesteric twisting is combined with a screw-like form of the line connecting the centers of the molecules comprising the column (Fig. 12.18c). The pitch of this screw line of centers of molecules is determined by weak interaction forces with distant neighbours or by the details of the molecular structure. Here, we can talk about liquid crystals CLCSA-D, CLCSB-D, CLCSC-D, and CLCSH-D.

The basic features of the structure of liquid-crystalline mesophases consisting of disc-like molecules, examined in this section, are presented in Table 12.2. Most of the liquid-crystalline structures examined have not yet been studied experimentally. For this reason, we can hope that the information obtained by the method of molecular dynamics on the properties of liquid crystals of this type as well as the indications of the stability of other systems of disc-like molecules (Table 12.2) will facilitate further progress in experimental and computational-theoretical investigations of these interesting varieties of liquid crystals. We can expect that not only will the important characteristics of the mesophases mentioned be studied and the sphere of their practical application established, but new structures formed by disc-like molecules will also be discovered.

Numerical simulation of the system of the infinitely thin rigid discs of the same diameters was performed by the Monte-Carlo method,[69] and a weak phase transition of the first kind was found, the equation

Table 12.2.
Structure features of liquid crystal mesophase consisting of disc-shaped and elongated molecules based on data of molecular dynamics "experiments"[62,63]

Type of liquid crystal mesophase	Preferable orientation of short (or long) molecular axes	"Columns" (smectic layers)	Direction of director coincides with direction of "column" (perpendicular to plane of smectic layer)	Two-dimensional lattice ("two-dimensional solid")	Axis of columns are right (smectic layers are flat)	Angle between neighbouring molecules leading to formation of cholesteric helix
Systems of disc-shaped molecules						
IL	−	−		−		−
NLC-D	+	−		−		−
SLCA-D	+	+	+	−	+	−
SLCB-D	+	+	+	+	+	−
SLCC-D	+	+	−	−	+	−
SLCH-D	+	+	−	+	+	−
SLCSA-D	+	+	+	−	−	−
SLCSB-D	+	+	+	+	−	−
SLCSC-D	+	+	−	−	−	−
SLCSH-D	+	+	−	+	−	−
CLC-D	+	−		−		+
CLCA-D	+	+	+	−	+	+
CLCB-D	+	+	+	+	+	+
CLCC-D	+	+	−	−	+	+
CLCH-D	+	+	−	+	+	+
CLCSA-D	+	+	+	−	−	+
CLCSB-D	+	+	+	+	−	+
CLCSC-D	+	+	−	−	−	+
CLCSH-D	+	+	−	+	−	+
Systems of elongated molecules						
IL	−	−		−		−
NLC	+	−		−		−
SLCA	+	+	+	−	+	−
SLCB	+	+	+	+	+	−
SLCC	+	+	−	−	+	−
SLCH	+	+	−	+	+	−
CLC	+	−		−		+

of state for the isotropic phase was obtained, and the dependence of order parameters on density and their fluctuations was determined.

12.2. THE RESULTS OF EXPERIMENTAL INVESTIGATIONS

The thermodynamic properties (temperatures and heats of transitions) of benzene-hexa-*n*-alkanoates (BHA) were investigated by Chandrasekhar and his coworkers.[4,5] The technique based on the use of a Perkin–Elmer model DSC-2 differential scanning calorimeter, which provides high accuracy, was used in Ref. 5. The corresponding data are presented in Table 12.3. In addition, some results of an experimental study of the phase diagrams of the substances BHA 6 and BHA 8 up to high pressures (Fig. 12.19 and 12.20) are presented in Ref. 5. It is interesting that in BHA 6 an additional phase appears at pressures above 160 bar, while in BHA 8, the mesophase is observed only at pressures below 140 bar.[5] The experimental values of dT/dP and ΔH for BHA 8 at atmospheric pressure correspond to ΔV equal to 12.13 and 3.5 cm^3/mole for solid–mesophase and mesophase–isotropic-liquid transitions, respectively.

The data on Table 12.3 supplement the results presented for ben-

Table 12.3.
Temperatures and heats of phase transitions of benzene-hexa-n-alkanoates (BHA)[4,5*]

Substance	Type of phase transition**	Temperature of transition, K	Heat of transition kJ/mole
BHA5	SC → IL	379	26
BHA6	SCII → CI	348.9	16
	SCI → IL	367.7	33
BHA7	SC → LC-D	354.4	32
	LC-D → IL	360.2	22
	IL → LC-D	356.7	
BHA8	SC → LC-D	353.0	43.8
	LC-D → IL	356.6	19
	IL → LC-D	355.0	
BHA9	SC → IL	353.6	70.0
	IL → LC-D	349.8	14
BHA10	SCII → CI	323.7	43.6
	SCI → IL	358.7	86.7

*For BHA5 are given data in reference 4, for rest substances are given data in Reference 5.

**See footnotes for Table 9.2, LC-D-liquid crystal of disc-like molecules.

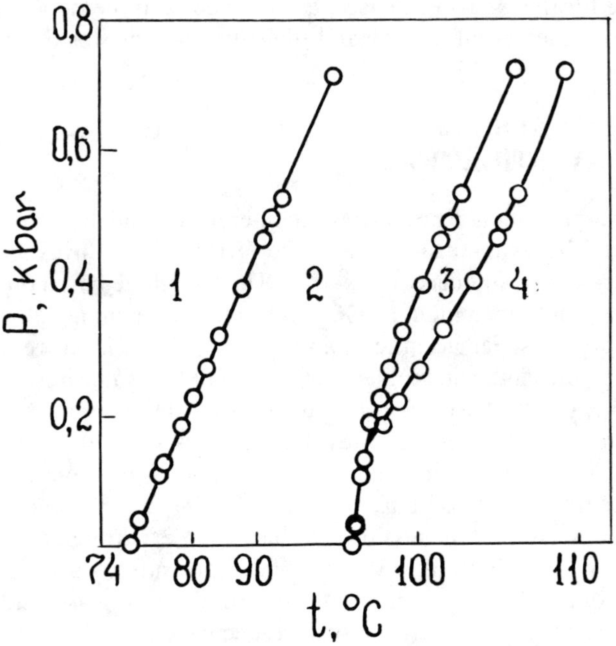

Figure 12.19. Phase diagram of BHA6 which was found in heating:[5] 1,2—solid crystal; 3—liquid crystal; 4—isotropic liquid.

zene-hexa-*n*-alkanoates in Ref. 6 and in Table 12.4. Unfortunately, in this work, the method used to obtain the data presented, which do not agree completely with the results in Refs. 4 and 5, was not indicated.

One of the first observations of mesophases in a system of disc-like molecules was performed in Ref. 16. Analysis of the results of measurements of diffraction neutron spectra ($\lambda = 2.515$ Å) on a layer of octaphenylcyclotetrasiloxane led to the conclusion that a mesophase exists in this substance in the temperature range 461–478 K.

Hexa-aloxy-derivatives of triphenylene ($n = 5$ and 7) have been studied by French investigators.[12] The data they obtained are presented in Table 12.5. The authors note that in the substance $n = 5$ polymorphism is observed in the solid-crystalline state (with transitions at 336 and 338 K), and in addition, the change in enthalpy ΔH_1 corresponds to melting of the high-temperature phase.

Triphenlylene hexa-*n*-dodecanoate was investigated in Ref. 13. The transition enthalpies were determined by the method of differential

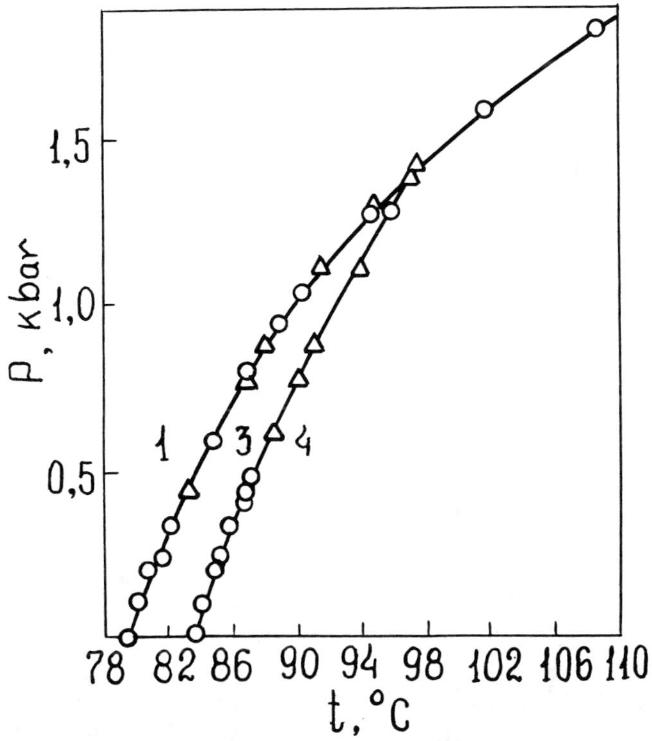

Figure 12.20. Phase diagram of BHA8 which was found in heating[5]; see designations in Figure 12.19.

scanning calorimetry using Dupont 990 thermal analyzer. It turned out that in this substance three mesophases (D_0, D_1, and D_2) are observed, corresponding to the following scheme[13]

$$\text{SC} \xrightarrow{353} \text{LC-D}_0 \underset{\underset{(\sim 0)}{365}}{\overset{366}{\rightleftarrows}} \text{LC-D}_1 \underset{\underset{(150)}{383}}{\overset{384}{\rightleftarrows}} \text{LC-D}_2 \underset{\underset{(2420)}{394.1}}{\overset{395.5}{\rightleftarrows}} \text{IL}.$$

Here, the transition temperatures (in K) are indicated near the corresponding arrows, characterizing heating or cooling; the heats of the transitions (in J/mole) are indicated below in the parentheses. Here, not only is the observation of three mesophases in this substance, but also the practical correspondence of the transition between mesophases D_0 and D_1 to a second-order transition is of interest.

Triphenylene hexa-akyl- and alkoxy-benzoates were studied by French investigators,[10] who discovered together with the ordinary

Table 12.4.
Temperatures (K, upper lines) and heats (kJ/mole, lower lines) of phase transitions in benzene-hexa-n-alkanoates (BHA)[6]

Substance	Phase state of substance* SC II		SC I (or C)		LC-D		IL
BHA6	+	348.9	+			367.7	+
		16				33	
					+	362**	
BHA7			+	353.7	+	359.3	+
				32		22	
BHA8			+	353.0	+	356.9	+
				44.8		19	
BHA9			+			353.7	+
						70	
					+	351.9***	+
						14	

*Phase observed in given substance are marked by " + " in corresponding columns.
**Virtual temperature of transition is determined by extrapolation of lines of transition at phase diagram.
***Temperature of transition is determined by repeated heat of monotropic mesophase.

phase a "nematic" mesophase of disc-like molecules, characterized by the absence of "columns." The transition temperatures for substances of this type are presented in Table 12.6.

As noted above, new types of mesophases of disc-like molecules were discovered by Billard and coworkers,[7,70] and in addition, the structure of these mesophases was studied in detail in Ref. 7 (see Sec. 11.1). The temperatures of the corresponding transitions (in K) are presented in the following scheme (the notations of the mesophases used in Ref. 7 are retained):

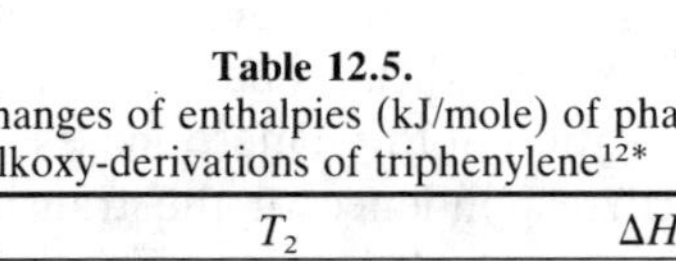

In one of the preceding sections, we examined the interesting ex-

Table 12.5.
Temperatures (K) and changes of enthalpies (kJ/mole) of phase transitions in hexa-alkoxy-derivations of triphenylene[12]*

n	T_1	T_2	ΔH_1	ΔH_2
5	342	395	32.5	8.13
7	341	366	57.1	4.69

*Index "1" refers to transition SC-LC-D, index "2" refers to transition LC-D-IL.

Table 12.6.
Temperatures of transitions (K) in hexa-alkyl and alkoxybenzoates of triphenylene[10]

	Phase state of substance*						
R (see Table 12.1)	SC		D		N_D		IL
C_7H_{15} –	+	403	+	483			+
C_8H_{17} –	+	452			+	465	+
C_9H_{19} –	+	453			+	495	+
$C_{10}H_{21}$ –	+	428			+	450	+

*Here are used the names of mesophases (D, N_D) given in Reference 10. Phases observed in given substance are indicated by "+" in corresponding columns.

ample of di-isobutylsilanediole, in which the "only" diskotic element, constituting the mesophase, is a dimer.[24,25] The transition temperatures (in K) in this substance are as follows[25]:

$$\text{SC} \underset{350.2}{\overset{361.6}{\rightleftarrows}} \text{LC-D} \underset{368.8}{\overset{371.9}{\rightleftarrows}} \text{IL},$$

while the transition enthalpies[23] are:

$$\Delta H_1\ (\text{SC} \longrightarrow \text{LC-D}) = 7.6\ \text{kJ/mole},$$

$$\Delta H_2\ (\text{LC-D} \longrightarrow \text{IL}) = 7.2\ \text{kJ/mole}.$$

It is noted in Ref. 25 that the value of the ratio $\Delta H_2/\Delta H_1$ (0.94) is high compared to the characteristic values of this ratio for other substances, capable of forming LC-D (0.2–0.7). It appears that this is related with the breaking of intermolecular bonds, accompanying the transition LC-D–IL in di-isobutylsilanediole.

The mesophases discovered by Kelker and Wirzing[8] in polymer systems are of great interest. The temperatures and heats of transitions in these substances are presented in Table 12.7. It should be

Table 12.7.
Temperatures and enthalpies of transitions in polymeric systems[8]*

Number of substance	n	R	Temperatures of phase transitions, K	Enthalpies of transitions, J/g
1	6	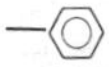	SLC 437 NLC 457 IL	
2	2	$—OCH_3$	VS 373 NLC 394 IL	2.3
3	6	$—O—CH_3$	VS 368 NLC 378 IL	2.1
4		Copolymer with (2) and chyral fragment	VS 453 CLC 503 IL	
5	6	$—OC_6H_{13}$	VS 333 SLC 388 IL	15.5
6	3	$—C_6H_{13}$	VS 373 SLC 393 IL	6.8
7	6	$—CH_3$	VS 343 NLC 357 IL	1.3

*See footnotes for Table 9.2. vs-vitreous solid

noted that the mesophases are identified based on optical observations and the nature of these mesophases cannot yet be viewed as completely clear.

The magnetic susceptibility of mesophases formed in systems of disc-like molecules was studied in Ref. 23. The experimental data obtained as a result of studying the state diagrams of solutions of liquid crystals formed by disc-like molecules are examined in Refs. 6, 9, 15, and 25 (see Sec. 14.1).

The peculiarity of properties of NLC and prospects of the practical application of these substances promoted the continuous extension of investigations into their properties. Among these works are the study of new classes of compounds (the investigation of the metalloorganic discotic liquid crystal,[71] for example), the observation of the phase behaviour of the homologous series of the discotic compounds demonstrating the reentrant phases,[72] and a new method of the investigation of structural features of LCD (magnetic susceptibility as a probe of the solid-discotic phase transition,[73] for example).

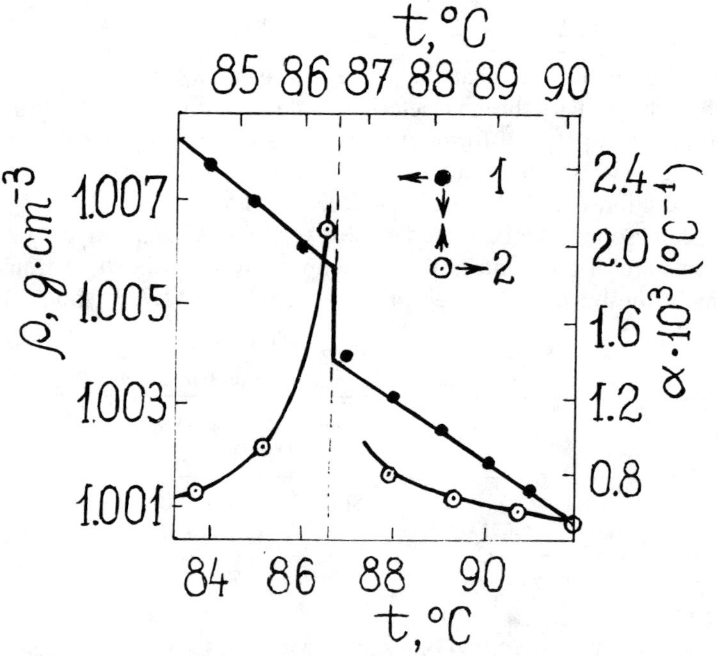

Figure 12.21. Density versus temperature for C_8HATX near the N_D—D_{rd} transition (1) and thermal expansion coefficient versus temperature near the N_D—D_{rd} transition (2).[78]

The results of the study of some thermodynamic properties of the discotic liquid crystals were generalized in paper 74.

The results regarding transitional diffusion in discotic mesophases studied by the NMP pulsed field gradient method[75] are of special interest. The investigations were performed for hexa (hexyloxy) triphenylene ($n = 6$), hexa (octyloxy) triphenylene ($n = 8$), triphenylene hexaundecanoate and truxene hexadodecanoate. The self-diffusion coefficients were found to range between $2 \cdot 10^{-11}$ and $6 \cdot 10^{-11}$ m^2/s and in the mesophase region $D_\perp / D_\parallel > 3.5$. This result agrees with the molecular-dynamic data according to which $D_\perp / D_\parallel = 3.5$–$5.0$[76,77] (see 12.1).

A wide range of experimental data was obtained by dilato-metric and thermobarometric studying members of the hexa-alkanoyloxy truxene homologous series (CnHATX, $n = 8$–13).[78] Density, volume changes for transitions, thermal expansion, isothermal compressibility coefficients and pressure-temperature phase diagrams were determined for the discoid nematic (N_D), rectangular columnar ($D_{\rm rd}$) and hexagonal columnar ($D_{\rm hd}$ mesophases. The analysis of these results makes it possible to draw many important and interesting conclusions. Figures 12.21 and 12.22 give results for density measurements versus

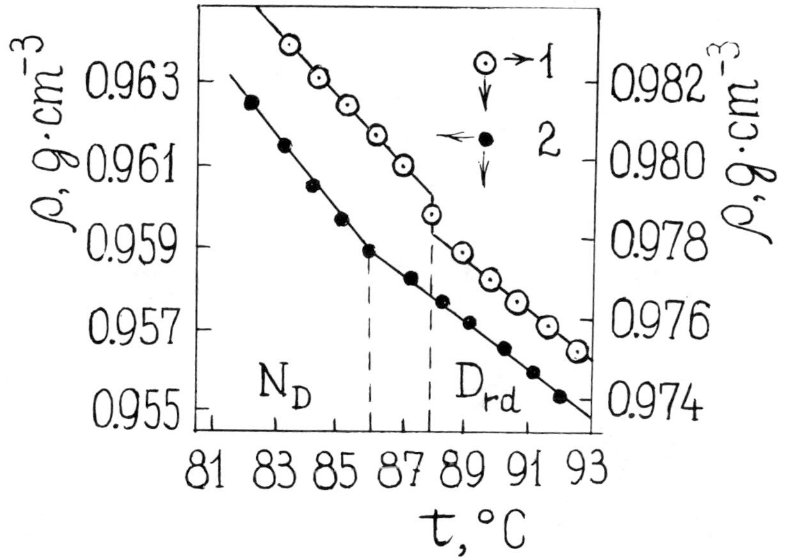

Figure 12.22. Density versus temperature for C_nHATX near N_D—D_{rd} transitions: 1 – $n = 10$; 2 – $n = 12$.[78]

temperatures for $n = 8$, 10 and 12 at the N_D–D_{rd} transition. Substantial changes for the density for $N = 8$ and 10 were observed. However for $n = 12$ no volume change can be determined although anomalies appear for the density near transition temperature (detected by optical microscopy). It seems the pretransition effect conceals detection of the transition. At the D_{rd}–D_{hd} transition ($n = 12$, 109°C) only a change of a slope appears (transition with no apparent volume change)[78]. Figure 12.21 also gives experimental data for the thermal expansion coefficient ($n = 8$).

Unfortunately, at the present time, experimental data on the thermodynamic and, especially, transport properties of liquid crystals consisting of disc-like molecules are still sparse. Investigators note the high viscosity of "columnar" diskotic structures.[6] On the other hand, the "nematic" mesophase of disc-like molecules is apparently much more fluid.[10] The difference between the elastic, viscous, and other properties of diskotic phases and the properties of ordinary liquid crystals was noted by Chandrasekhar.[26] To a certain extent, the results of molecular dynamics modeling of liquid crystals of disc-like molecules fill this omission (see Sec. 12.1).

References for Part IV

CHAPTER XI

1. V. N. Velyakov, Ot perstnya nastroeniya do stereo-televizora, *Tekhnika i nauka*, No. 4, pp. 14–15 (1982).
2. S. Otani, Proceedings of the 8th International Liquid Crystal Conference, *Mol. Cryst. Liq. Cryst.*, **63**, 249 (1981).
3. H. Gasparoux, *Liq. Cryst. One- and Two-Dimens.*, Order. Proc. Conf., Berlin (1980), pp. 373–382.
4. S. Chandrasekhar, B. K. Sadashiva, and K. A. Suresh, *Pramana*, **9**, 471 (1977).
5. S. Chandrasekhar, B. K. Sadashiva, K. A. Suresh, N. V. Madhusudana, S. Kumar, R. Shashidhar, and G. Venkatesh, *J. Phys. (Fr.)*, **40**, C3120 (1979).
6. J. Billard and B. K. Sadashiva, *Pramana*, **13**, 309 (1979).
7. J. Billard, J. C. Dubois, C. Vaucher, and A. M. Levelut, *Mol. Cryst. Liq. Cryst.*, **66**, 115 (1981).
8. H. Kelker and U. G. Wirzing, *Mol. Cryst. Liq. Cryst. Lett.*, **49**, 175 (1979).
9. C. Vauchier, A. Zann, P. Barny, J. C. Dubois, and J. Billard, *Mol. Cryst. Liq. Cryst.*, **66**, 103 (1981).
10. H. T. Nguyen, C. Destrade, and H. Gasparoux, *Phys. Lett. A*, **72**, 251 (1979).
11. A. M. Levelut, *J. Phys. Lett. (France)*, **40**, L81 (1979).
12. J. Billard, J. C. Dubois, H. T. Nguyen, and A. Zann, *Nouv. J. de Chem. (France)*, **2**, 535 (1978).
13. C. Destrade, M. C. Mondon-Bernaud, and H. T. Nguyen, *Mol. Cryst. Liq. Cryst. (Lett.)*, **49**, 169 (1979).
14. S. Chandrasekhar, *Curr. Sci.*, **47**, 523 (1978).
15. R. E. Goozner and M. M. Labes, *Mol. Cryst. (Letters)*, **56**, 75 (1979).
16. F. Volino and A.-J. Dianoux, *Ann. Phys.*, **3**, 151 (1978).
17. F. C. Frank and S. Chandrasekhar, *J. Phys. (France)*, **41**, 1285 (1980).
18. A. M. Levelut, F. Hardoum, H. Gasparoux, C. Destrade, and H. T. Nguyen, *J. Phys. (France)*, **42**, 147 (1981).
19. C. Destrade, H. T. Nguyen, J. Malthête, and J. Jacques, *Phys. Lett. A*, **79**, 189 (1980).
20. D. Demus and L. Richter, *Textures of Liquid Crystals*, Verlag Chemie, Weinheim, New York (1978).
21. A. M. Giroud-Godquin and J. Billard, *Mol. Cryst. Liq. Cryst.*, **66**, 147 (1981).
22. P. Oswald, *J. Phys. Lett. (France)*, **42**, L171 (1981).
23. G. Sigaud, M. F. Achard, C. Destrade, and H. T. Nguyen, *Liq. Cryst. One and Two Dimens.*, Order. Proc. Conf., Berlin (1980), pp. 403–408.
24. G. W. Gray, *Mol. Cryst. Liq. Cryst.*, **63**, 3 (1981).
25. J. D. Bunning, J. E. Lydar, C. Eaborn, P. M. Jackson, J. W. Goodby, and G. W. Gray, *J. Chem. Soc. Farad. Trans.*, **78**, 713 (1982).
26. S. Chandrasekhar, *Mol. Cryst. Liq. Cryst.*, **63**, 171 (1981).
27. C. Destrade, H. T. Nguyen, J. Malthête and J. Jacques, *Phys. Lett.*, **A79**, 189 (1980).
28. J. Malthête, J. Jacques, H. T. Nguyen and C. Destrade, *Nature*, **298**, 46 (1982).
29. A. L. Tsykalo, *Doklady AN SSSR*, **256**, 1367 (1981).

30. A. L. Tsykalo and A. D. Bagmet, *Krystallografiya*, **27**, 941 (1982).
31. A. I. Tsykalo, *Mol. Cryst. Liq. Cryst.*, **128**, 99 (1985).
32. L. Mamlok and J. Malthête, *J. Phys. Lett. (Fr.)*, **43**, 641 (1982).
33. Minoru Takabatake and Shigeo Iwayanagi, *Jap. J. Appl. Phys.*, **21**, 685 (1982).
34. M. Cagnon, M. Charbia and G. Durand, *Phys. Rev. Lett.*, **53**, 938 (1984).
35. M. Charbia, M. Cagnon and G. Durand, *Phys. Lett. (Fr.)*, **46**, 683 (1985).
36. J. Billard, *C. r. Acad. Sci. Ser.* 2, **299**, 905 (1984).
37. Z. Luz, D. Goldfarb and H. Zimmermann, *Nucl. Magn. Res. Liq. Cryst.*, Proc. NATO Adv. Study Inst., San Miniato, July 26–Aug. 7, 1983. Dordrecht e.a. 343 (1985).
38. E. Meirovitch, Z. Luz and H. Zimmermann, *J. Phys. Chem.*, **88**, 2870 (1984).
39. C. R. Safinya, K. S. Liang, William A. Varady, Noel A. Clark and G. Anderson, *Phys. Rev. Lett.*, **53**, 1172 (1984).
40. Bernd Kohne, Klaus Praefcke, Werner Stephan and Peter Nürnberg, *Z. Naturforsch.*, **B40B**, 981 (1985).
41. S. Chandrasekhar, *Mol. Cryst. Liq. Cryst.*, **63**, 171 (1981).
42. S. Chandrasekhar, *Adv. Liq. Cryst.*, **5**, 47 (1982).
43. H. Gasparoux, *J. chim. phys. et phys.-chim. biol.*, **81**, 11 (1984).
44. S. Otani, *Mol. Cryst. Liq. Cryst.*, **63**, 249 (1981).
45. E. I. Kats, *Zh. Eksp. Teor. Fiz.*, **75**, 1819 (1978).
46. P. de Gennes, *Fizika zhidkikh kristallov* (Physics of Liquid Crystals), Mir, Moscow (1977).
47. S. A. Pikin, *Strukturnye prevrashcheniya v zhidkikh kristallakh* (Structural Transformations in Liquid crystals), Nauka, Moscow (1981).
48. L. D. Landau and E. M. Lifshitz, *Statisticheskaya fizika* (Statistical Physics), Nauka (1976).
49. V. L. Pokrovskii and E. I. Kats, *Zh. Eksp. Teor. Fiz.*, **73**, 774 (1977).
50. K. K. Kobayashi, *Mol. Cryst. Liq. Cryst.*, **13**, 137 (1971).
51. S. Chandrasekhar, *Phil. Trans. Roy. Soc. London*, **A309**, 93 (1983).
52. E. I. Katz, *Usp. Fiz. Nauk.*, **142**, 99 (1984).
53. E. I. Katz and M. I. Monastyrski, *J. Phys.*, **45**, 709 (1984).
54. H. Kidachi, *Progr. Theor. Phys.*, **72**, 1240 (1984).
55. Y. F. Sun and J. Swift, *Phys. Rev. A: Gen. Phys.*, **33**, 2735 (1986).
56. Y. F. Sun and J. Swift, *Phys. Rev. A: Gen. Phys.*, **33**, 2740 (1986).
57. C. Godrèche and L. de Seze, *J. Phys. Lett. (Fr.)*, **46**, 39 (1986).
58. E. I. Katz and V. V. Lebedev, *Zh. Exp. Teor. Fiz.*, **86**, 558 (1984).
59. H. R. Brand, *Phys. Rev. A: Gen. Phys.*, **33**, 634 (1986).

CHAPTER XII

60. A. L. Tsykalo, *Pis'ma Zh. Tekh. Fiz.*, **6**, 495 (1980).
61. A. L. Tsykalo, *Zh. Fiz. Khim.*, **54**, 3014 (1980).
62. A. L. Tyskalo and A. D. Bagmet, *Kristallografiya*, **27**, 941 (1982).
63. A. L. Tsykalo and A. D. Bagmet, *Zh. Tekh, Fiz.*, **52**, 961 (1982).
64. A. L. Tsykalo, *Doklady AN SSSR*, **249**, 1348 (1979).
65. A. L. Tsykalo and A. D. Bagmet, *Kristallografiya*, **25**, 568 (1980).
66. A. L. Tsykalo, E. E. Nyankina, and A. D. Bagmet, *Zh. Tekh. Fiz.*, **50**, 2365 (1980).

67. B. J. Berne and P. Pechukas, *J. Chem. Phys.*, **56**, 4213 (1972).
68. W. J. A. Goossens, *Phys. Lett. A*, **31**, 413 (1970).
69. R. Eppenga and D. Frenkel, *Mol. Phys.*, **52**, 1303 (1984).
70. J. Billard, Proc. European Conf. on Liquid Crystals, Garmish, Springer Verlag (1980), p. 383.
71. A. M. Godquin-Giroud *et. al.*, *J. Phys. Lett. (Fr.)*, **45**, 387 (1984).
72. J. M. Buisine and M. Donan, *C. R. Acad. Sci. Ser. 2*, **303**, 1769 (1986).
73. A. M. Godquin, J. M. Latour and J. C. Marchon, *Inorg. Chem.*, **24**, 4452 (1985).
74. M. Sorai, *Thermochim. acta.*, **88**, 1 (1985).
75. R. Y. Dong, D. Goldfarb, M. E. Moseley, Z. Luz and H. Zimmermann, *J. Phys. Chem.*, **88**, 3148 (1984).
76. A. L. Tsykalo, *Doklady AN SSSR.*, **256**, 1367 (1981).
77. A. L. Tsykalo, *Molec. Cryst. Liq. Cryst.*, **128**, 99 (1985).
78. J. M. Buisine, R. Cayuela, C. Destrade and H. T. Nguyen, *Mol. Cryst. Liq. Cryst.*, **144**, 137 (1987).

PART V.

Mixtures of Liquid Crystals

But if that flow'r with base infection meet,
The basest weed outbraves his dignity.
William Shakespeare, "Sonnet 94"

The variety of structures and physical characteristics of different types of liquid crystals as well as the striving to make the utmost use of the wide complex of their properties for practical purposes have led to the appearance of a large number of investigations concerning mixtures of liquid crystals. Here, very interesting both in the scientific and applied respects, results have been obtained. In particular, new types of state diagrams, reflecting the presence under definite conditions of so-called "induced" phases, which are not observed at all in the starting individual components of the mixture, have been discovered. Interesting features of the temperature induced changes in the pitch of the helix in solutions of liquid crystals (for example, chiral nematics, see Chap. 9) have also been observed. From the practical point of view, the importance of solutions of liquid crystals is due to the possibility of obtaining liquid-crystalline materials, fulfilling to the maximum extent possible the goals formulated, by a quite simple method—forming mixtures with the required composition.

At the same time, this direction presents investigators with quite complicated problems. Since it is impossible to study experimentally the properties of solutions for the entire variety of compositions of binary and, especially, multicomponent mixtures, computational-theoretical methods of investigation assume special importance. However, these methods are complicated, since it is necessary to take into account the features of the interactions of the like and unlike particles comprising the solutions and having a complicated shape. In addition, the interactions mentioned can in general have different natures.

CHAPTER XIII.

The Structure of the Solutions of Liquid Crystals

We in this world, as in a circle,
Live and create one in another.
Johann Wolfgang von Goethe, *Faust*

13.1. EXPERIMENTAL INVESTIGATIONS OF STRUCTURAL FEATURES

Investigators have devoted their greatest attention to the study of the pitch of the helix of mixtures of CLC or mixtures CLC–NLC, in which the pitch is highly sensitive to external actions. Inasmuch as the pitch of the helix is related in a known manner to the wavelength of light selectively reflected by the liquid-crystalline substance,[1] this feature, on the one hand, opens up possibilities of different practical applications and, on the other, simplifies investigatory work.

Some of the results of the classical investigation by Fergason *et al.*[2] are presented in Figs 13.1–13.2. It is evident that the temperature dependence of the wavelength of the peak in selective reflection for solutions of CLC can be observed at a different temperature level and has a different curvature. For this reason, by varying the nature of the components and the composition of the mixture, it is possible to select a mixture that has the required working range of "color play" temperatures and the appropriate slope of the working characteristic, on which the accuracy with which it is possible to judge the temperature according to the color of the liquid-crystalline material, depends. Fergason investigated in detail the effect of separate components on the selective reflection of light, as well as the characteristics of this reflection as a function of the polarity of the solvent.

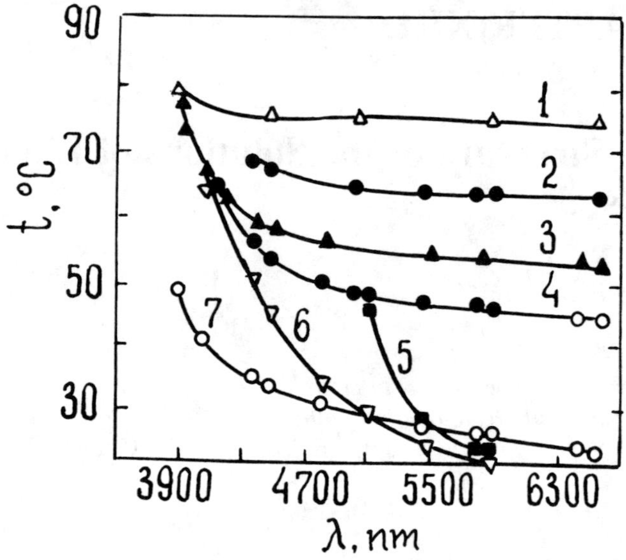

Figure 13.1. Dependences of temperature on wavelength of maximum of selective reflection of light for binary mixtures of cholesteryl nonanoate and esters of cholesterol:[2] 1—pure cholesteryl nonanoate; 2—20% of cholesteryl hydrocinnamate; 3—20% of cholesteryl butyrate; 4—20% of cholesteryl propionate; 5—20% of cholesteryl chloride; 6—20% of cholesteryl acetate; 7—20% of cholesteryl methylcarbonate.

In recent years, the pitch of the helix of solutions of liquid crystals has been studied by many investigators. Here, both the method of infrared spectroscopy[3] and the usual optical method[4–9] were used. Together with the addition of experimental data required for practical and theoretical generalizations, a series of new interesting phenomena have been discovered. For example, the appearance of double inversion of the direction of the pitch of the helix of the binary mixture cholesteryl chloride (CC)—cholesteryl-2-(2-etoxy-etoxy-ethyl)-carbonate (CEEC), both of whose individual components are characterized by dextrarotary spiral structure, was observed in Ref. 4. The unusual change in the pitch of the helix in the mixture CEEC–MBBA when the concentration of the first component changed (decrease and then sharp increase to infinity) was discovered in Ref. 5.

The concentration dependences of the wavelength of selective reflection of a wide range of binary mixtures of nematic and cholesteric liquid crystals were found in Ref. 7. Here, the deviations of the concentration dependence 1/λ from a linear dependence in connection with the features of the phase diagrams of the mixtures were examined

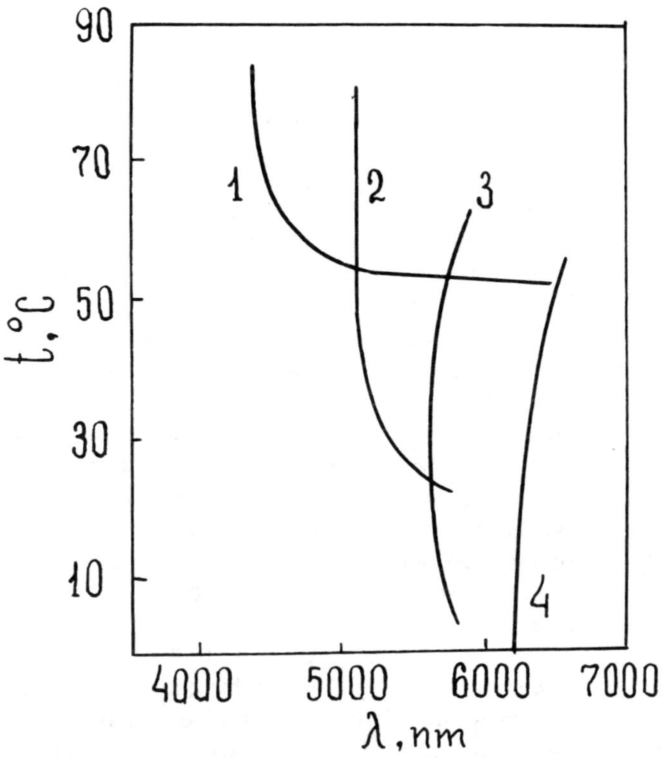

Figure 13.2. Dependence of temperature on wavelength of selective reflection of light for binary mixtures of cholesteryl nonanoate (I) and cholesteryl chloride (II): 1–1–90%; 2–1–80%; 3–1–75%; 4–1–70%.[2]

and discussed in detail. Deviations from a linear dependence are observed in mixtures with CLC having short side chains. It was established that the largest deviations are observed in mixtures for which the temperature of the phase transition from the smectic phase to the cholesteric phase is higher than for the individual components, while the smallest deviations occur in mixtures where the smectic phase is not observed. Inversion of the pitch of the helix was observed in nematic–cholesteric mixtures and discussed in connection with the chemical structure of molecules (number of chemical bonds: chiral center-aromatic ring) in Refs. 8 and 9. The review by Gajewska *et al.*[10] is concerned with examining the methods for measuring the pitch of the cholesteric structure and contains data obtained by many measurements of concentration dependences of the pitch in cholesteric and cholesteric–nematic mixtures of liquid crystals.

The dependences of λ on temperature and concentration in binary mixtures MBBA—cholesteryl chloride, cholesteryl oleylcarbonate, and cholesteryl pelargonate were obtained and approximated by simple linear relations in Ref. 11. The linear dependence of λ on temperature was also established in Ref. 12, which can be explained by Keating's[13] phenomenological theory. It was shown in Ref. 14 that it is possible to obtain a wide temperature range of the existence of the cholesteric mesophase and of the "color variation" in binary mixtures of cholesterol esters. The working temperature range can be narrowed by creating three and four component mixtures.

References 15 and 16 are devoted to the experimental study of the dependence of the pitch of the helix of binary mixtures on their composition and temperature and the structural changes caused in mixtures by mechanical displacement and pressure are investigated in Refs. 17 and 18. It was shown that for large displacements, a homeotropic texture appears.[17] As far as the pressure is concerned, as it increases, the magnitude of the pitch increases, which reaches infinity at a pressure corresponding to the CLC–SLC transition.[18] It was possible to establish a linear relation between the pressure and the transition temperature.

The dependence of the spiral pitch of NLC–CLC mixtures on the thickness of a sample and that of molecular interaction peculiarities characterizing the twisting force were studied in paper 19. In this work special attention was paid to cases where the spiral pitch was much larger than the thickness of a sample and long axes of nematic molecules were oriented nearly perpendicular to the substrate. The conditions were found under which small amounts of cholesteric admixtures made it possible to obtain homeotropic NCL orientation.

Textural changes, induced in mixtures by an electric field, have also been studied in detail.[20–24] Thus it was possible to observe the transition from the homeotropic texture to a planar texture under the action of a high-frequency electric field.[20] When a field is imposed on a planar texture, a confocal texture appears, which can remain throughout the "memory" period varying from several seconds to several weeks (depending on the pitch of the helix and the thickness of the specimen).[21] If the intensity of the field reaches the critical value, the helix "untwists" and a homeotropic nematic texture forms.

The study of electrooptic properties of NLC–CLC mixtures enabled the threshold voltage of the nematic–cholesteric transition of mixture V_t to be defined, while experimental data revealed the availability of the power dependence of $V_t \sim d^\alpha$, where d = the sample thickness and α depends on the concentration of cholesteric component.[25]

Nonmesogenic additives (NMA) can have a considerable effect on the structure of liquid-crystalline mixtures.[26-30] Thus it has been observed that the introduction of NMA lowers the temperatures of phase transitions and, correspondingly, displaces the temperature characteristics of the mesophase.[26] It has been proposed that the change in the concentration factors in mixtures can be considered as being equivalent to temperature changes.[29] The temperature dependences of the pitch of the helix of CLC with NMA are studied in Ref. 30, where a mechanism for the effect of NMA on the structure of a liquid crystal was proposed. Penetrating into the volume occupied by the system of molecules of the specimen, the molecules of the NMA weaken the anisotropic interaction of mesogenic particles, facilitating oscillatory motions, i.e., effects analogous to the effect of an increase in temperature, arise.

A number of papers[31,32] are concerned with the investigation of the microstructure of mixtures of CLC. Thus the microscopic investigation of "aging" of mixtures of CLC (study of the textures and effect of aging on the intensity and uniformity of the reflected light) was performed in Ref. 31. The temperature dependence of the pitch of the helical structure in the region of the CLC–IL phase transition in six multicomponent CLC, containing cholesteryl acetate, cholesteryl benzoate, cholesteryl pelargonate, and cholesteryl oleate, was specially investigated in Ref. 32. It was shown that this first-order transition can be considered as resulting from the vanishing of anharmonic interaction among the internal vibrations of molecules.

The submicrostructure of cholesteric–nematic mixtures was studied in detail in Refs. 33 and 34 by the method of low-angle x-ray diffraction. Mixtures of cholesteryl myristate and *n*-hexyloxyphenyl ester of *n*-butyloxybenzoic acid (10, 20, 30 and 40% NLC) during the cooling process were studied.[33] Intervals of existence of mesophases, as well as some other results of this investigation, are presented in Table 13.1. It was shown that as the concentration of NLC increases, the rate of growth of the lengths of the crystalline section drops, inasmuch as the introduction of the molecules of the nematic substance weakens the ordering of the molecular structure of the cholesteric. After the transition of the cholesteric phase into the smectic phase (which is clearly observed on the Bragg scattering curves), the process of ordering of the smectic lamellar structure stretches out in temperature, which is related to the role of the molecules of the

Table 13.1

Temperatures of phase transitions and structure features of binary system cholesteryl myristate-*n*-hexyloxyphenyl ester of *n*-butoxybenzoic acid[33]

Content of nematic substance	Temperatures of transitions, °C		t, °C	Coefficient of crystallinity	Length of crystalline part, Å	Length of amorphous part, Å
	CLC	SLC				
0.10	82.0	75.0	78.5	0.72	24.1	9.4
			76.0	0.80	26.2	6.6
			72.5	0.9	28.4	3.1
			55.5	1.0	32.3	—
0.20	80.5	62.5	72.5	0.68	22.3	10.5
			68.0	0.73	23.6	8.7
			59.0	0.84	26.1	5.0
			51.5	0.89	27.1	3.4
0.30	80.0	62.0	76.0	0.66	21.2	11.0
			66.0	0.79	24.7	8.9
			59.0	0.90	26.7	3.0
			49.0	0.93	26.9	2.0
0.40	80.0	—	76.0	0.64	20.1	11.3
			62.0	0.73	22.7	8.4
			51.5	0.82	25.3	5.5
			46.5	0.85	25.8	4.6

nematic, which are situated in amorphous segments and inhibit the formation of layers. In a mixture with 40% nematic, the disordering of the lamellar structure reached a critical value and the smectic phase vanished. Thus, in cholesteric–nematic mixtures, it is not only possible to control intensity and characteristics of the process of ordering of the structure of the cholesteric phase and obtaining it with quasi-smectic features, but it is also possible to obtain a smectic phase with a considerably more ordered structure than the smectic of ordinary cholesteric substances.

These data were supplemented by the results of the work[34] in which cholesteryl myristate was used as cholesteric component of nematic–cholesteric system.

One of the effective methods for studying the orientational ordering of chiral compounds, dissolved in NLC and thereby inducing the solution of a cholesteric type, is the study of the spectra of the Cotton transmission and rotation effects.[35] The orientation of menthone and other substances, dissolved in a nematic, was studied using this method. It was shown that C=O group is oriented perpendicular to the director and, in addition, different chiral compounds can lead to a left-handed as well as a right-handed helix in the same nematic. The relationship between the features of the twisted structure and the nature of the cholesteric giving rise to this twisting was also studied by the method of induced circular dichroism.[36] The problem of compensation of the temperature dependence of transient characteristics of twisted nematic displays by means of cholesteric additives is discussed in Ref. 37. The effect of optically active compounds on amphiphilic NLC is interesting. It has been established that chiral additives cause all such systems to transform into the cholesteric mesophase.[38]

A new trend concerning the application of the nuclear magnetic resonance method shows great possibilities of studying the structure.[39-41] Possibilities of this method for mixtures of nematic substances with opposite diamagnetic anisotropy as well as NLC and CLC mixtures were studied in paper 39. Orientational order of components for mixtures containing partially oriented liquid crystals was studied on the base of spectrum analysis of NMR.[40,41]

The structure of solutions exhibiting smectic phase is less studied.The concentration dependence of the interplanar distances in the smectic B phase was obtained based on data from an x-ray diffraction analysis of the binary mixture.[42] The structure of the B phase of these two compounds differed: in one of them, the long molecular axes were perpendicular to the layers, while in the other, they were tilted with respect to the layers. Liquid crystals exhibiting a smectic C phase have been studied by the method of polarization microscopy and isomorphic displacement and it has been established that the addition of chiral molecules leads to the appearance of optical rotation in this phase.[43] It has been shown that some solutions exhibiting a smectic phase can serve as a matrix for obtaining oriented molecules of dissolved substance in studying spectra.[44] For this purpose, the isotropic melt is placed into a magnetic field, as a result of which, the optical axis of the liquid crystal turns out to be oriented along the applied field.

An interesting investigation of "blue phases" (BP) in a CLC–NLC mixture [cholesteryl myristate (42.37% mole)—*p*-pentyl-phenyl-2-chloro-4-(*p*-pentylbenzoyloxy)-benzoate] was performed in Ref. 45. Two blue phases (BP-I and BP-II), whose textural characteristics and transition temperatures are presented in Table 13.2, were observed in this mixture.

Table 13.2

Temperatures of phase transitions and textures (in heating and cooling) of mixture cholesteryl myristate (42.37% mole)—*p*-pentyl-phenyl-2-chloro-4-(*p*-pentylbenzoiloxy)-benzoate[45]

Phases	Cholesteric	Blue I	Blue II	Isotropic
Temperatures in transitions K	369.20	369.45	369.80	
Heating	Planar (blue)	Planar (green-blue)	Planar (green-blue)	
Heating	Focal-conical	Laminar (green-yellow)	Laminar (green-blue)	
Cooling	Focal-conical supercooling ←	Laminar (green-yel-low-red)	Laminar (green-blue)	

13.2. COMPUTATIONAL-THEORETICAL STUDIES OF STRUCTURE

Many theoretical investigations of helical cholesteric structures and temperature and concentration dependences of their pitch are based on Keating's work,[13] in which the cholesteric twisting is considered as the rotational analog of thermal expansion. Thus, in Ref. 46, the following relation is proposed for the pitch of the helix h of mixtures of CLC and NLC

$$h = \left(\sum_i \alpha_i \theta_i\right)^{-1}, \tag{13.1}$$

where α_i is the weight concentration of the components and θ_i is the effective twisting capability (optical activity) of the i-th component. The same authors extended this dependence to mixtures containing an arbitrary number of cholesteric components[47] and the noted that the direction of rotation of the polarization plane is determined by the sign of the sum $\Sigma\alpha_i\theta_i$.

Later, more complicated forms of the concentration dependence of the pitch were proposed. Thus, a quadratic dependence[48]

$$(2h)^{-1} = \left(\sum n_i\right)^{-1}\left\{\sum \beta_i n_i^2 + 2\sum \beta_{ij} n_i n_j\right\}, \tag{13.2}$$

where n_i is the number of moles of the i-th component and β_i is the "twisting force," having the dimensions of area (see Ref. 49 for its definition), has been proposed. For example, the expression for the binary mixture

$$(2h)^{-1} = (n_A + n_B)^{-1} \{\beta_A n_A^2 + \beta_B n_B^2 + 2\beta_{AB} n_A n_B\}.$$

The last term in braces, as shown in Ref. 49, can make a large contribution. Other relations, convenient for practical purposes, for h were examined in Ref. 48, for example,

$$(2hM)^{-1} = N\{\beta_A m_A^2 + \beta_B(1 - m_A)^2 + 2\beta_{AB} m_A(1 - m_A)\}, \quad (13.3)$$

where $m_A = n_A/(n_A + n_B)$ is the molar fraction of the cholesteric component A, $M = [(M_A/d_A - M_B/d_B)m_A + M_B/d_B]^{-1}$ is the molar density of the mixture, M_A and M_B are the molecular weights of the components, d_A and d_B are their density, and N is Avogadro's number. If it is assumed that

$$2\beta_{AB} \equiv \beta_A + \beta_B + \delta\beta_{AB}, \quad (13.4)$$

then expression (13.3) can be rewritten in the form

$$(2hM)^{-1} = N\{(\beta_A - \beta_B)m_A + \beta_B + \delta\beta_{AB} m_A(1 - m_A)\}. \quad (13.5)$$

The computational results were compared with experimental data for binary mixtures of CLC as well as for triple mixtures, containing two cholesteric and one nematic component. The computed and measured[50,51] values for three mixtures are compared in Fig. 13.3 (mixtures of cholesteryl chloride (CC)—cholesteryl-2-(2-etoxy-etoxy)-ethylcarbonate (CEEC), 4-cyanobenzal-amino-cinnamate (CBAC)–CC and CBAC–CEEC).[48] The agreement is good. Excellent agreement between the computed and experimental values is also observed for triple mixtures, each of which contains two CLC (CC, cholesteryl nonanoate CN), and NLC (MBBA). The molar "twisting" forces, found from the experimental data, are presented in Table 13.3.

An extended model of Priest and Lubensky[52] likewise makes it possible to interpret changes in the pitch of the cholesteric mixtures.[53] This method is based on minimizing the Hamiltonian of the mixture of liquid crystals and determining the corresponding pitch of the helical structure, which is expressed in terms of the concentration of components, the parameter of the nematic order, and the energy

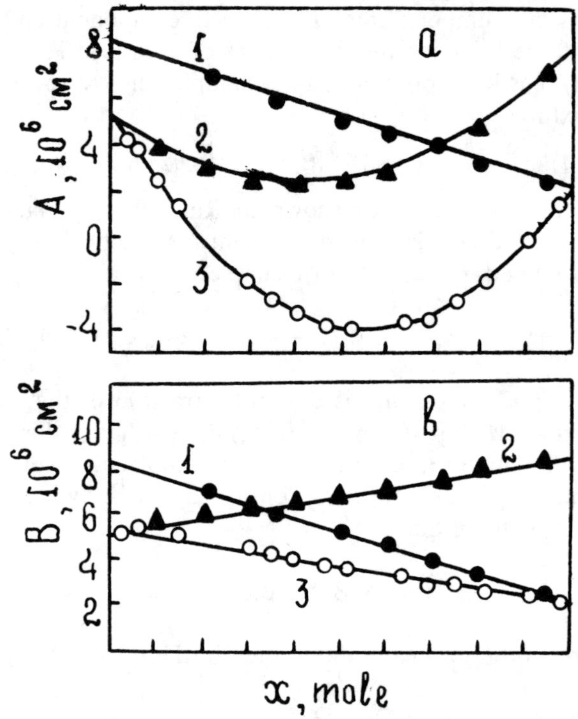

Figure 13.3. a—Dependences of quantity $A = (2\ \mathrm{hM})^{-1}$ on molar content of CEEC in CC (1), CC in CBAC (2) and CEEC in CBAC (3); b—verification of validity of equation (13.5): dependence $B = (2\ \mathrm{hM})^{-1} - N\,\delta\beta\, m_A\,(I - m_A)$ on x_A.[48]

Table 13.3

Molar twisting forces of CLC $N\beta_A$ and $N\beta_{AB}$ (10^6 cm^2) in mixtures containing CLC and NLC[48]

Cholesteric substances*	Studied pairs	$N\beta_A$	$N\beta_{AB}$
	Cholesteric-cholesteric		
CC (405)	CBAC-CC	+8.4	−1.9
CEEC (547)	CC-CEEC	+2.2	+5.4
CBAC (346)	CBAC-CEEC	+5.3	−11.4
	Cholesteric-nematic		
CC (405)	CC-MBBA	+9.1	−7.0
CN (527)	CN-MBBA	−9.2	−11.5

* Molecular weights are given in brackets.

constants introduced in Ref. 52. Mixtures of cholesterics (dextro and levorotary CLC), as well as cholesteric–nematic mixtures can be investigated in this manner. In the first case, the pitch is independent of the temperature, and when the concentrations of both cholesterics are equal, it can become infinite. In the second case, the model predicts the appearance of cholesteric twisting in the solution and, in addition, the pitch depends on the concentration nonlinearly.[54]

The use of the Frank–Oseen approximation makes it possible to minimize the free energy of the solution taking into account the cross terms, related with the CLC–NLC interaction and to determine the pitch of the helix.[55] As a result, the interesting conclusion is obtained that although the pitch of the helix in general depends on the concentrations of CLC and NLC, the helical structure arises even with very low concentrations of the cholesteric, which agrees with Ref. 54 and experimental data.

The structure of liquid-crystalline mixtures can be investigated not only based on the phenomenological theory[52–55] but also based on the extended molecular-statistical theory of cholesterics, proposed by Goossens.[56] This method is based on writing the interaction energy of the particles located in neighboring quasilayers (including also particles of different type) and minimizing this energy. In this way, a relationship is found between the angle of rotation θ and the molar fraction of the cholesteric.[57,58] Comparison with experimental data shows that the dependence obtained $\theta(x)$ correctly predicts the spiral inversion effect, when with equal concentrations of dextro- and levorotary CLC, the solution behaves like a nematic, and it also correctly describes the observed minima on the curves $\theta(x)$ for definite values of the concentrations x. It is interesting to note that the twistedness of the structure can be caused not only by dipole–quadrupole interactions (according to Goossens' theory[56]), but also by the long-range dipole–dipole forces. Thus, it was shown in Ref. 59 that the dipole–dipole interactions can lead to the formation of screw structures in nematics, depending on the dimensions of the specimen, magnitude of the spontaneous polarization, and the elastic moduli.

In spite of its relative simplicity and clarity, Goossens' theory is not entirely satisfactory and has a number of limitations. It was shown in Ref. 60 that in the case of mixtures exhibiting strong deviations of concentration dependences of the pitch from linear dependences, it is necessary to take into account the detailed feature of the shape of the interacting particles (in particular, the presence of functional groups, which can cause steric difficulties under definite mutual positions of the molecules). It has also been suggested that the specific steric repulsion of molecules (neglected in Goossens' theory) play a role in the appearance of the excess twisting in solutions.[61] These effects were included for cholesterics and cholesteric–nematic mixtures in Ref. 62, leading to the following working expression:

$$\theta = \sum_i \sum_j \theta_{ij}(T) w_i w_j,$$

where $\theta = 2\pi d h^{-1}$, d is the distance between quasinematic layers, w are the weight fractions of components, θ_{ij} are the mutual twistabilities of components with pairwise interaction, and, in addition, for two component nematic–cholesteric solutions $\theta_{22} = 0$. The computational results[62] coincided with the experimental data[60] within accuracy of the latter. At the same time, it was determined that the interaction potential[63] is not realistic enough.[62,64]

In investigating the specific steric repulsion forces between molecules of cholesteryl derivatives, it was found that the steric factors of anharmonicity of rotational vibrations have a large effect.[65] The contribution of this effect to the temperature dependence of θ has the following form in the first approximation:

$$\Delta\theta = a[\langle P_2\rangle(T/T_I)],$$

where a is a constant, characterizing the intensity of the anharmonicity, $\langle P_2\rangle(T/T_I)$ is the orientational order parameter, and T_I is the temperature of the transition to the isotropic liquid. Using this approach and the theory in Ref. 62, it was possible to obtain the dependence of the pitch on temperature and composition for different concentrations,[65,66] which agrees well with the experimental data in spite of the quite complicated nature of these dependences (Fig. 13.4).

A special investigation of the cholesteric twisting of NLC with small additions of CLC (weakly doped nematic[67]) were performed in Ref. 68 and, in addition, in order to perform the calculations, it was necessary to have information on the dependence of the orientational order on temperature and pressure. This information can be obtained from experimental data or based on an appropriate theoretical model. It was shown that the temperature dependence of the pitch of the helix near the line of the transition to the isotropic liquid is determined by the heat capacity of the "background" nematic and the heat of the transition to the isotropic liquid state. The computed and experimental data agree well for a number of doped nematics.[68]

In conclusion we shall note that the results of the above-mentioned

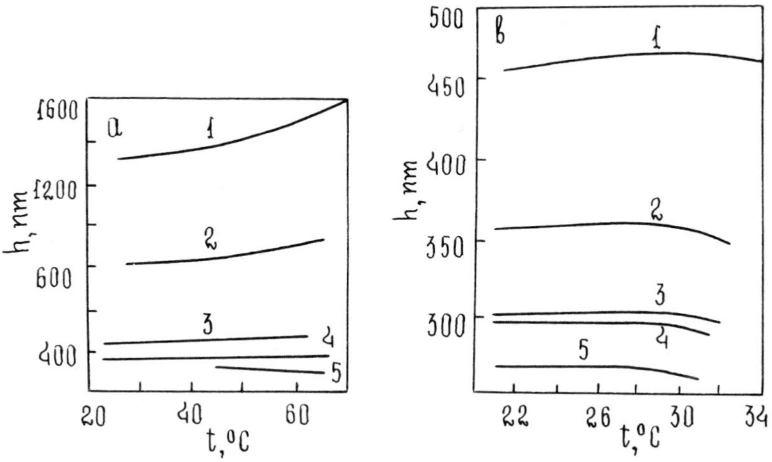

Figure 13.4. Temperature dependence of helix pith of mixtures of nematic and cholesteric liquid crystals:[66] a—cholesteryl formiate—4-butyl-4′-methoxyazoxybenzene (BMAOB) system: 1—x_{NLC} = 0.95; 2—0.90; 3—0.80; 4—0.73; 5—0.20; b—cholesteryl oleate—cholesteryl formiate—MBBA: 1—x_{NLC} = 0.70; 2—0.60; 3—0.50; 4—0; 5—0.10.

papers[40,41] were explained in terms of the mean field theory for binary solutions of liquid crystals with good qualitative agreement. The correlation of order parameter which was found theoretically for the model of cylindrically symmetrical molecules was discussed, as well as order parameters of flexible molecules fragments of LC measured by NMR method.[40] Temperatures of isotropic–nematic transition were calculated for the case of solutions of small rigid molecules in NLC 5CB and it was shown that the relationship between order parameters of small molecules and 5CB order parameter is defined by the same universal curve.[41] Detailed discussion of regularities for spiral structure induction in mixtures containing optically active unmesomorphic admixtures (more than 30 admixtures) was given in the above-mentioned Ref. 20, Part III.

CHAPTER XIV.

Thermophysical Properties of Solutions of Liquid Crystals

Something there is that doesn't love a wall,
That sends the frozen-ground-swell under it
And spills the upper boulders in the sun,
Ana makes gaps even two can pass abreast.
Robert Frost, "Mending Wall"

14.1. PHASE DIAGRAMS AND REGIONS OF EXISTENCE OF LIQUID CRYSTAL MESOPHASES

We shall begin the review of works concerning solutions of liquid crystals with an examination of phase diagrams, which contain in a compact form information concerning the regions of existence of different liquid crystalline mesophases.

Phase diagrams have been used to investigate types of mesophases since 1959. The principle of the method is based on the indication of miscibility of two mesophases and are illustrated in Fig. 14.1.[69] Substance I in a liquid crystalline state mixes with a high-temperature mesophase of substance II for all concentrations (phases of a single type A in Fig. 14.1a). The region of the low-temperature phase B of substance II is situated lower. Figure 14.1b shows the region of complete mixing of substances II and III (phase B). However, in solutions of substances I and III (Fig. 14.1c), immiscibility of phases A and B is observed (a zone of immiscibility, which is shaded in Fig. 14.1c, is observed between regions of existence of phases A and B).

State diagrams are constructed primarily by the methods of optical microscopy, differential scanning calorimetry, x-ray diffraction analysis, differential thermal analysis, as well as volumetric measurements. Interesting experimental material has already been accumu-

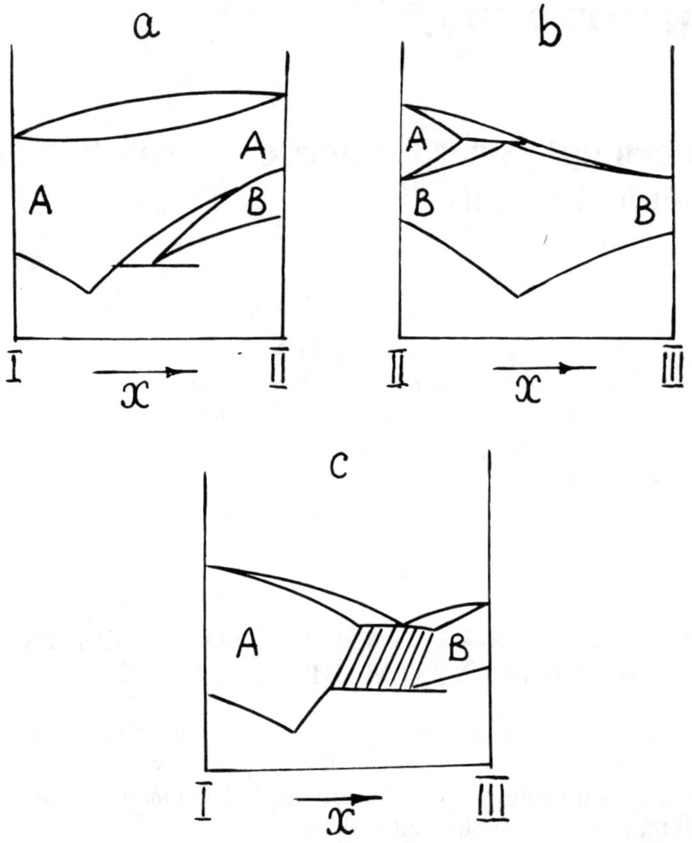

Figure 14.1. Scheme illustrating the application of phase diagrams for study of mesophases of liquid crystals[69] (elucidations in text).

lated in this area. Thus the study of state diagrams of binary mixtures of liquid crystals led to the discovery of eight types of mesophases: nematic (N) and smectic A, B, C, D, E, F, and G types. A special notation is usually not used for the cholesteric mesophase, which mixes with the nematic phase and is observed in substances exhibiting many other liquid crystalline mesophases also.

The types of mesophases mentioned above are encountered both in individual substances and in solutions. The known variants of polymorphic forms are presented in Table 14.1 according to data in Ref. 69. The sequence in the combinations presented from right to left corresponds to the sequence in which the phases appear with de-

Table 14.1
Types of polymorphism of liquid crystals[69]

Character of polymorphism	Observed combinations of liquid crystal mesophases
Dimorphous	AN BN CN GN GA BA EA CD BC EB
Trimorphous	BAN CAN GAN BCN EBA GBA BCA GCA CDA
Tetramorphous	BCAN EBAN GBAN GFCA
Pentamorphous	GBCAN

creasing temperature. In this case, all variants of combinations in Table 14.1 correspond to the sequence

$$\begin{smallmatrix}E\\ \\ G\end{smallmatrix}B\ F\ C\ D\ A\ N,$$

and in addition, mixing of the liquid crystals types in this sequence has not yet been observed. As far as the order E–G or G–E is concerned, it is not determined, since these two types of phases have not yet been observed in a single substance simultaneously.[69]

The phase diagrams of a large number of binary mixtures of liquid crystals were obtained by Arnold and Sackmann.[70] These authors established that NLC are mixed in the entire range of compositions of mixtures, which was confirmed by further investigations. Thus mixtures of homologues in the series di-alkoxyazoxybenzenes (mOAB) were studied in Ref. 71. The diagrams obtained turned out to be analogous to the phase diagrams with the binary eutectic (Fig. 14.2 and 14.3).[71,72] Park *et al.*[73] studied the effect of molecular complexing on the properties of binary mixtures of NLC. The phase diagram of MBBA with *n,n'*-cyanopentylbiphenyl (CPB) (Fig. 14.4) was studied with the help of the methods of optical microscopy and differential scanning calorimetry.[73] It was shown that the dependence of the temperatures of transitions from solid-crystalline to nematic phase ($T_{S\text{–}N}$) has a complicated character and has two eutectic points for CBP concentrations of 15% and 70%. The concentration dependence of the temperatures of transitions from nematic to isotropic phase is nonlinear and is characterized by the presence of a peak, which, as the peak in $T_{S\text{–}N'}$ is reached in the equimolar mixture. This result is explained by the formation of weakly bound complexes and charge transfer between MBBA (proton-donor) and CPB (proton-acceptor).

MBBA is a good solvent in a mixture with nonmesogenic substance, which was established in the first investigation of such systems.[74] It follows from the diagram obtained (Fig. 14.5)[74] that nematic ordering is remained with a considerable concentration of benzene (up to 16%)

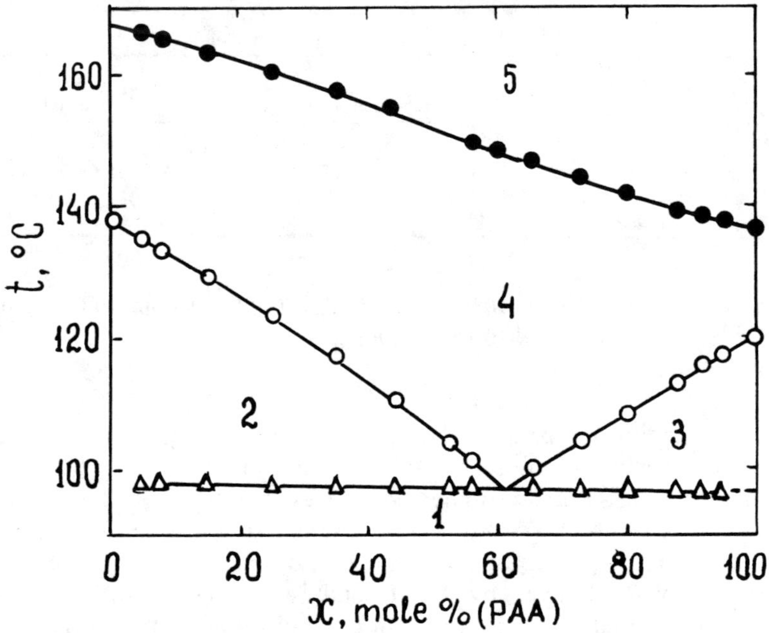

Figure 14.2. Phase diagram of *n*-azoxyanisole (PAA)—*n*-azoxyphenetole (PAP) system:[71] 1—solid phase; 2—solid PAP + NLC; 3—solid PAA + NLC; 4—NLC; 5—isotropic liquid.

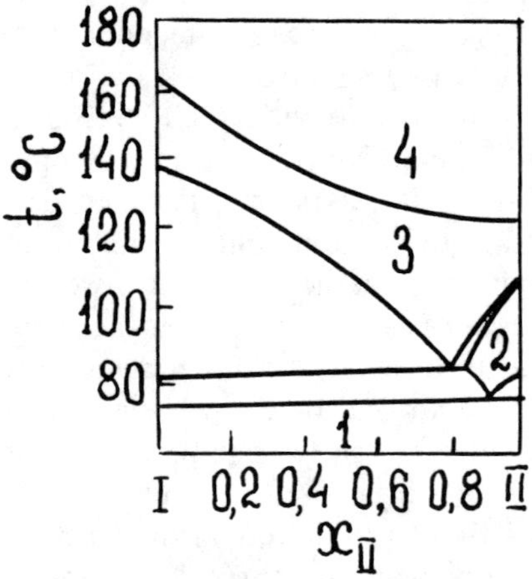

Figure 14.3. Phase diagram of *n*-azoxyphenetole (I)—*n*-octoxyazoxybenzene (II) system:[72] 1—solid phase; 2—smectic phase; 3—nematic phase; 4—isotropic liquid.

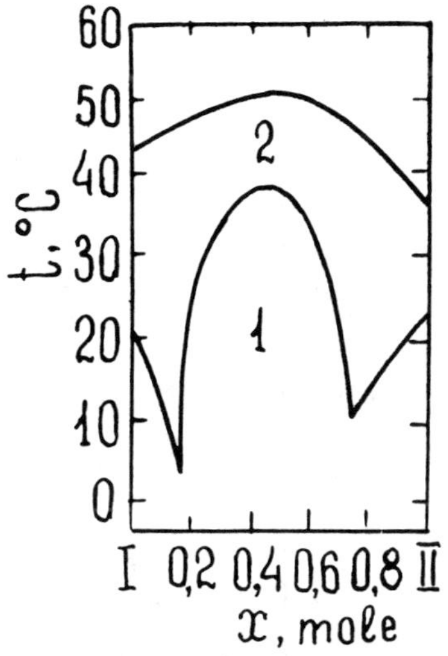

Figure 14.4. Diagram of states of system *N*-(*p*-methoxybenzylidene)-*p*-*n*-butylaniline (MBBA) (I)—*n*-*n*′-cyanopentylbiphenyl (CPB) (II):[73] 1—solid, 2—NLC, 3—isotropic liquid.

with a corresponding narrowing of the temperature region of existence of the liquid crystalline phase. The region of coexistence of NLC and IL turned out to be very narrow (in Fig. 14.5, this region is shaded). This apparently indicates that the NLC is closer to an isotropic liquid than to a solid. The relative magnitudes of the heats of transitions agree with this[71]: the ratio of ΔH_{SC-N} to ΔH_{N-I} equals 41.6 for *n*-azoxyanisole and 12.4 for *n*-azoxyphenetole. For this reason, the corresponding region is not indicated in Figs. 14.2 and 14.3. According to estimates,[75] the extent of the heterogeneous region is 0.4–0.5% mol., while the change in temperature, corresponding to the transition of the anisotropic solution to the isotropic liquid, does not exceed 0.4°C.

In the case of a mixture of nematics, the temperature region of existencc of the liquid crystalline phase can be quite wide. It is remarkable that such mixtures can be supercooled to a larger degree than the pure components. This result was obtained in Ref. 76 in connection with the study of the state diagram of mixtures consisting of the first and second homologues of the series of *n*-alkoxybenzylidene-*n*′-butylanilines.

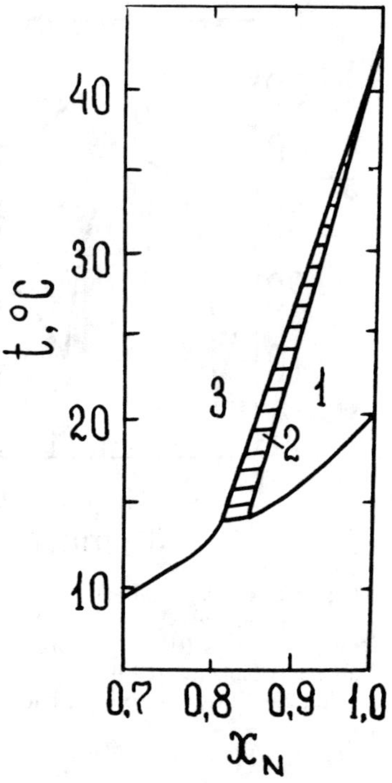

Figure 14.5. Diagram of states of MBBA—benzene system[74] (x_{NLC}—mole fraction of MBBA): 1—NLC; 2—NLC + isotropic liquid; 3—isotropic liquid.

Mixtures of cholesteryl erucylcarbonate (CEC) and cholesteryl nonanoate (CN) also exhibit a greater capability for supercooling than the pure components.[75] The study of such noncrystallizing systems clarified the features of the phase diagram. It was shown that CEC and CN form a continuous series of liquid crystalline solutions. However, their ability to crystallize can differ. The diagram of phase transformations has the form

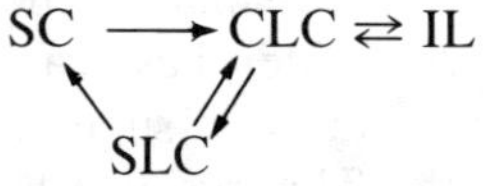

It is evident from the data in Fig. 14.6 that there are two mesophases in the entire region of the noncrystallizing compositions: the smectic mesophase at low temperatures and a cholesteric mesophase at high

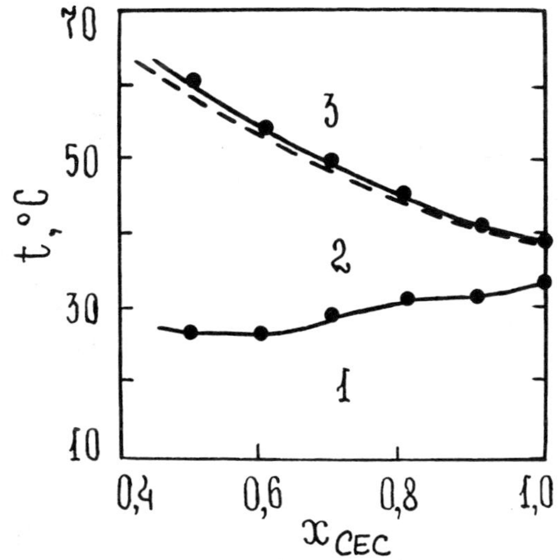

Figure 14.6. Phase diagram of solutions of cholesteryl erucylcarbonate (CEC) and cholesteryl nonanoate (CN):[75] 1—smectic (or solid) phase; 2—cholesteric phase; 3—isotropic phase.

temperatures. The region of the cholesteric phase expands as the concentration of CN increases and reaches 31°C for an equimolar composition; in this case, the temperatures of the smectic–cholesteric phase transitions decrease, while the temperatures of transitions to the isotropic liquid increase.

Binary liquid crystalline systems (mixtures of CLC) were also studied in Ref. 77. The phase transition temperatures were determined and the phase diagrams of the systems were constructed by the method of differential thermal analysis: a) cholesteryl nonanoate–cholesteryl propionate (CN–CP); b) cholesteryl nonanoate–cholesteryl oleate (CN–CO). The first system exhibits complete miscibility in the isotropic liquid and cholesteric states (Fig. 14.7a and b), while the solid crystals CN and CP do not mix at all. At point of equilibrium of the cholesteric phase with solid crystals CN and CP (three-phase equilibrium) exists at 46°C. As thc specimen is cooled, a smectic phase appears in CN and its mixtures, containing up to 35% CP. As the CP content continues to increase in the mixture, CP crystals begin to grow out of the cholesteric phase, and the cholesteric transforms completely into the solid phase only with strong supercooling. This transition occurs very slowly: mixtures close to eutectic mixtures

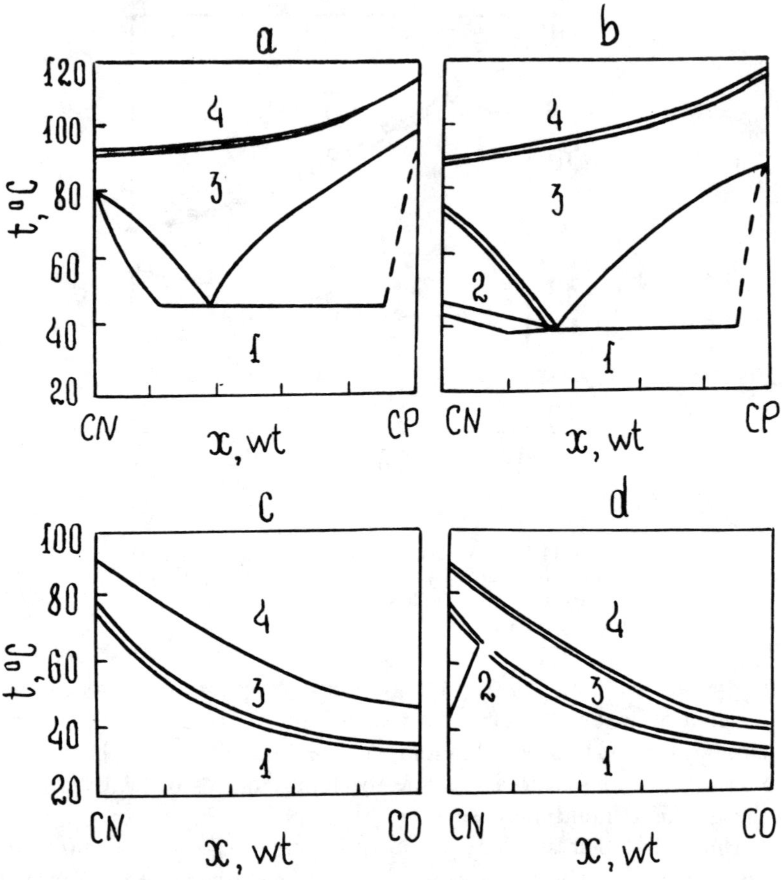

Figure 14.7. Phase diagrams of binary systems:[77] a—cholesteryl nonanoate–cholesteryl propionate, heating; b—the same, cooling; c—cholesteryl nonanoate—cholesteryl oleate, heating; d—the same, cooling; 1—solid crystal; 2—SLC; 3—CLC; 4—isotropic liquid.

transform completely to the solid state within days. In Fig. 14.7b, this transition is denoted by the dashed line.

The components of the system CN–CO mix in all ratios (Figs. 14.7c and d). The monotropic smectic phase exists in a small range of concentrations (up to 40% by weight of CO) and a sharply decreasing temperature interval (ΔT = 1°C already with 10% CO). A small heterogeneous region (not more than 0.3°C) may not always be recorded experimentally with the transition of the CLC to the solid state. The complete miscibility is indicated by the small heterogeneous region with all transitions except the transition of CLC into the solid state (and vice versa) in the system CN–CP.[77] Total miscibility of components in the cholesteric phase and the formation of noncrystallizing solutions permit using the mixtures investigated as temperature indicators.[76,77]

The unlimited mutual solubility both in the liquid crystalline and in the solid phases was observed for mixtures of cholesteryl myristate (CM)–cholestanyl myristate (CNM).[78] The phase diagram was studied by methods of differential scanning calorimetry and polarization microscopy and the formation of two mesophases was discovered at small and medium concentrations of CM. As far as the system cholesteryl oleate (CO)–CNM is concerned, this system belongs to a simple eutectic type.

It should be noted that as the range of experimental data on phase diagrams of mixtures increases, in many cases, data appear which exhibit poor mutual agreement or are difficult to interpret taking into account the conditions of thermodynamic equilibrium.[76,79–81] This could be explained by inadequacies of the experimental techniques used or in the methods used to analyze the data. Thus the observation or photographing of specimens, as a rule, does not give unique information on the type of phase diagram due to the absence of a solidus line. Data on the temperatures of crystal–nematic phase transitions could be far from equilibrium data even with slow cooling and intense mixing, inasmuch as the high viscosity and low thermal conductivity facilitate considerable supercooling. When recording heating curves for mixtures, the temperatures of phase transformations can depend considerably on the past history of the specimens.

The difficulties noted above in studying the low-melting liquid crystalline systems are a result of the features of mixtures of liquid crystals. This makes it necessary to use the lowest possible rates of change of temperature in order to obtain reproducible results. The entire range of phase transformations often constitutes only several degrees, so that when an experiment is not set up carefully enough, the points of the onset and termination of a transition may not be fixed.

To overcome the difficulties mentioned above, a special technique was proposed in Ref. 81 in order to investigate phase equilibria in low-melting liquid crystalline systems, based on recording the differential heating curves. The system MBBA–EBBA was investigated using this technique. It follows from the state diagram (Fig. 14.8) that a weak associate with composition 2:1, giving solid solutions with components, can be formed in this mixture: α, forming with a max-

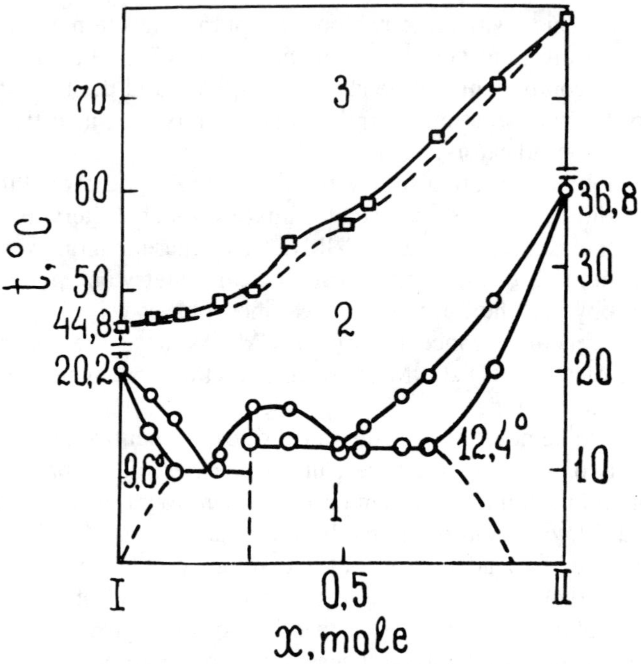

Figure 14.8. Phase diagram of system MBBA (I)—EBBA (II):[81] 1—solid crystal; 2—NLC; 3—isotropic liquid.

imum concentration of 12% mole EBBA at 9.6°C, and β, forming in the region of 70–100% EBBA with minimum concentration at 12.4°C. The transition from NLC to IL is characterized by a narrow two-phase region: the temperatures of the onset of the transition (dashed line) and its termination (continuous line) differ by 0.5–1.0°C. The phase diagram obtained is close to the one found in Ref. 76: the lines of the liquidus practically coincide. However, the absence of a solidus line does not permit judging uniquely the type of interaction of the components. As a result, it is impossible to make comparisons with the results in Refs. 79 and 80; moreover, components containing impurities could have been used in these works. The latter circumstance, apparently, led to measurements that were too low along the coordinates of the eutectic. For this reason, the noted existence of the associate[81] must be confirmed by results obtained with the help of other methods.

As a rule, binary liquid crystalline systems form eutectics with a molar ratio of components 1:1 or 1:2.[82–84] An example is the phase

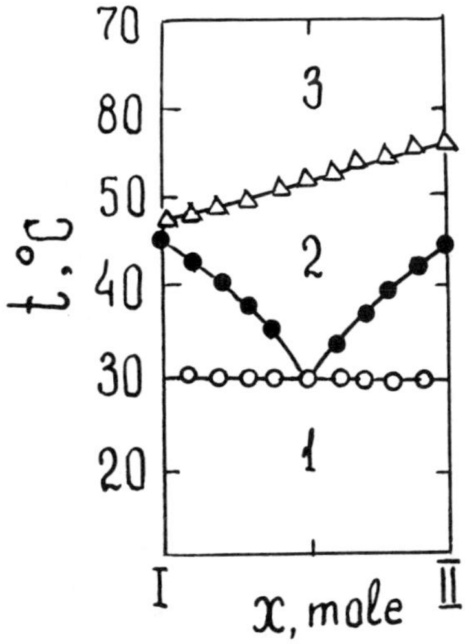

Figure 14.9. Phase diagram of *p*-cyanophenyl ester of *p*-*n*-hexylbenzoic acid (I)—*p*-cyanophenyl ester of *p*-*n*-heptylbenzoic acid (II):[84] 1—solid crystal; 2—NLC; 3—isotropic liquid; open circles—melting of eutectics; black circles—transition to NLC; triangles—transition to isotropic liquid.

diagram of *p*-cyanophenyl ester of *p*-*n*-hexylbenzoic acid—*p*-cyanophenyl ester of *p*-*n*-heptylbenzoic acid,[84] presented in Fig. 14.9.

The phase diagrams of double mixtures with points of coexistence of nematic and smectic A and C phases (with triple points) were discovered in Ref. 85 and investigated in Refs. 86 and 87. A mixture of octyloxycyanobiphenyl and heptyloxyaminofluorenone was examined.[86] The diagram obtained by the method of optical microscopy is shown in Fig. 14.10. The curves exhibit a monotropic transition from some phases to other phases, but it was impossible to observe specimens at temperatures below 50°C due to spontaneous crystallization. Figure 14.11 shows a qualitative comparison of the theoretical temperature dependence of the coefficient of clasticity C_1(a) and the experimental dependence $x - x_{NAC}$ on temperature (b).[86]

A different type of phase diagram, containing a triple point (Fig. 14.12), was observed.[88] There are indications that such systems occur more frequently than previously thought.

A very detailed phase diagram was obtained for binary systems

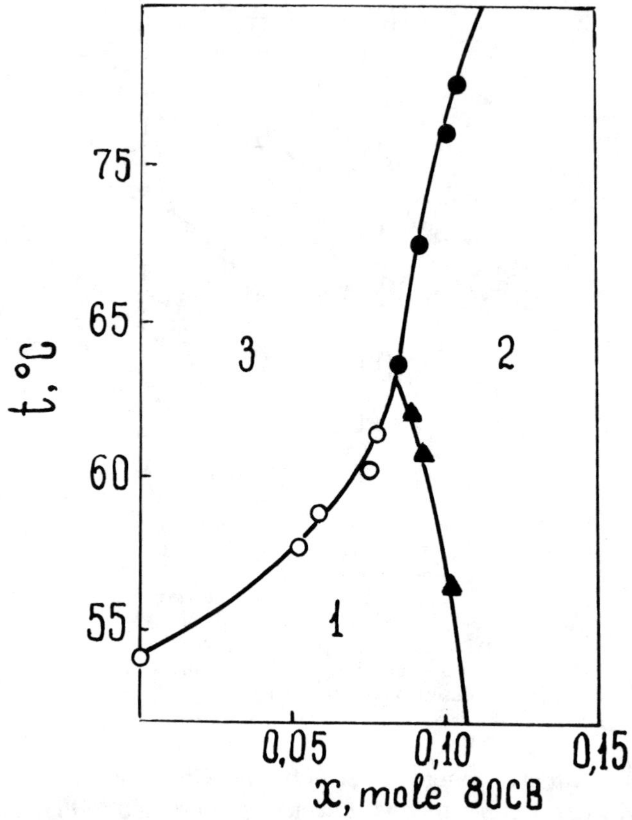

Figure 14.10. Monotropic part of diagram of binary system heptyloxyaminofluorenone—80 CB including the triple point;[86] 1—SLC C; 2—SLC A; 3—NLC.

4,4′-di-(*n*-heptyl)-azoxylbenzene (7AB) and 4-*n*-pentyl-4′-cyanobiphenyl (5CB) (Fig. 14.13).[89] The form of this diagram is quite complicated: it contains regions with two nematic and three smectic phases—A, B, and E. A phenomenon that can only occur in solutions of liquid crystals is distinctly manifested here: the appearance of types of liquid crystalline mesophases which are not observed at all in the starting pure components. There have been a large number of experimental investigations of such "induced" mesophases. For example, the formation of NLC in a mixture of dextro- and levorotary CLC at well-defined temperatures was observed in Ref. 90. Later, this phenomenon was studied in detail (see Secs. 13.1 and 13.2). A great deal of attention was also devoted to the investigation of smectic mesophases,

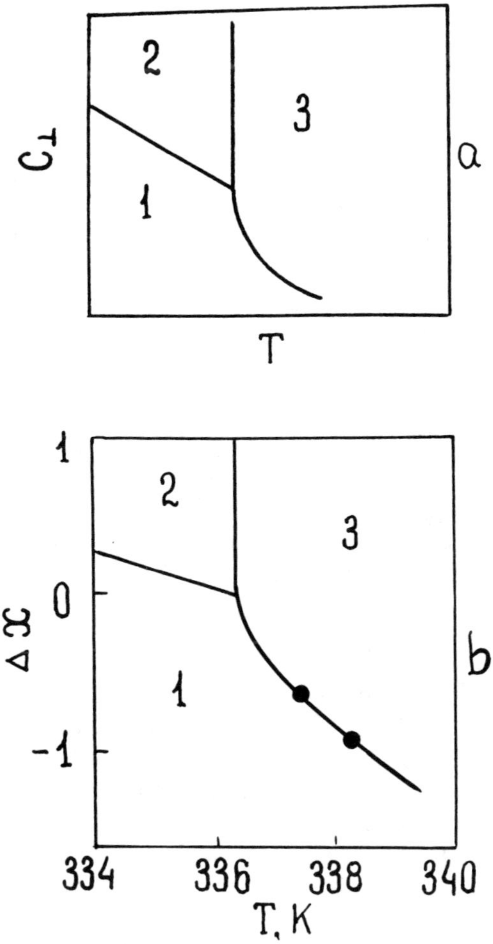

Figure 14.11. Comparison of theoretical diagram of elasticity coefficient $C_\perp$ (a) with experimental results ($\Delta x = x - x_{NLC}$) (b):[86] 1—SLC; 2—SLC A; 3—NLC.

induced in mixtures of nematic liquid crystals[91–96] which, at the suggestion of Ref. 71, must form only phase diagrams of the usual eutectic type. Detailed results of investigations of two-component mixtures of enantiotropic nematic mesophases, giving SLC, are presented in Refs. 92 and 93.

It was discovered in Ref. 92 that mixtures of *p*-heptyloxybenzylidene-*p*′-aminobenzonitrile and *p*-methoxy-*p*′-hexasiloxy-azoben-

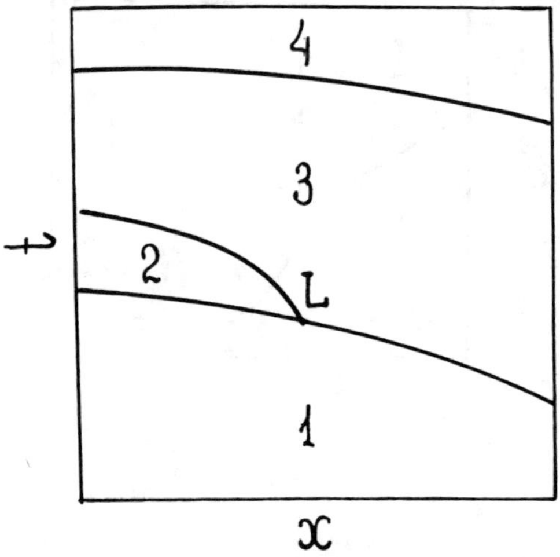

Figure 14.12. Quality view of phase diagram of binary system of liquid crystals demonstrating smectic C (1), smectic A (2) and nematic (3) mesophases:[88] 4—isotropic liquid, *L*—multicritical point.

zene give a monotropic smectic mesophase at well-defined temperatures. Later,[93] 12 binary systems of alkoxybenzilideneamino-benzonitriles with compounds characterized by nearly the same lengths of molecules (the maximum difference in the molecular lengths was about 5 Å) were examined in Ref. 93. A smectic mesophase, which remained at room temperatures for several hours, formed in all mixtures when cooled (Fig. 14.14). Thus the proposition[92] that the formation of a smectic mesophase accompanying mixing of nematics must be expected in compounds with approximately identical molecular lengths is, apparently, confirmed.

The structure and electrical properties of induced phases are of considerable interest.[93,94] It was possible to observe the minimum of the dependence of the transparency of samples on the electric field intensity *E*. This means that transparency of a thin smectic layer can be increased both by increasing and decreasing *E*.[93] The increase in transparency is accompanied by a transition of the fine-grain polygonal texture to a homeotropic texture.[94]

The phase diagram of binary mixtures of 4-*n*-alkyl- and 4-*n*-alkoxy-4′-cyanodiphenyls with homologues of 4,4′-di-*n*-alkyloxyazoxybenzenes containing an induced smectic A phase, was investigated in

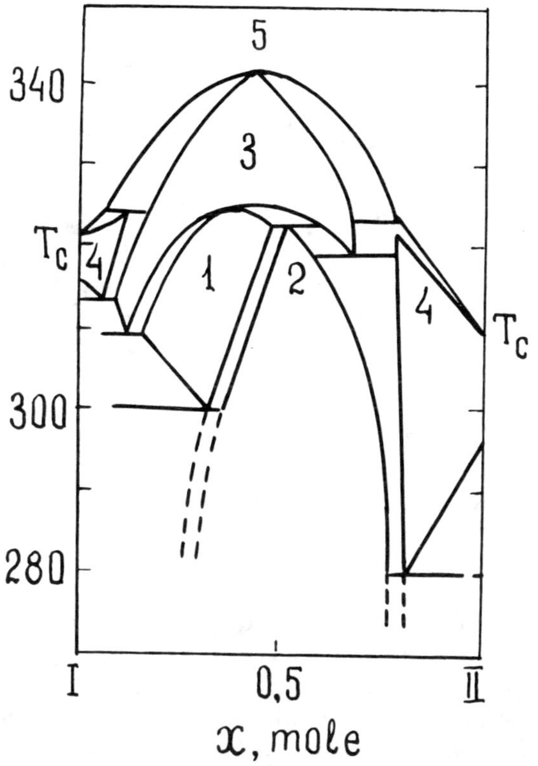

Figure 14.13. Isobaric phase diagram of binary system 4,4′-di(*n*-heptyl)-azoxybenzene (7AB,I)—4-*n*-pentyl-4′-cyanobiphenyl (5CB,II):[89] 1—SLC B; 2—SLC E; 3—SLC A; 4—NLC; 5—isotropic liquid.

Ref. 95. In this induced phase, the temperature of the transition to the isotropic liquid is higher than for the pure components.

Analysis of the results of investigations shows that induction is observed in mixtures for which one nematic liquid crystal contains molecules with polar terminal groups and the other contains nonpolar groups. Such binary mixtures were investigated in detail[96] with the help of the contact method.[97] Eleven liquid crystalline compounds, containing nonpolar terminal groups (N) and eight compounds with polar terminal groups (P) were studied. As a result, the induced smectic phases were observed in mixtures of nematics containing nonpolar terminal groups with nematics containing polar terminal groups (mixtures of the type N/P), but were not observed in mixtures

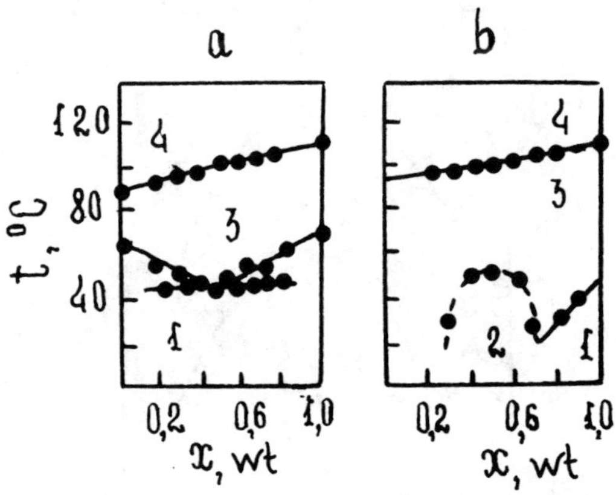

Figure 14.14. Phase diagram of system alkoxybenzylidene-amino-benzonitrile (BN, $n = 7$)—compound I:[93] a—heating; b—cooling; 1—solid crystal; 2—SLC; 3—NLC; 4—isotropic liquid; compound I: CH_3O—⟨○⟩—N═N—⟨○⟩—$OCO(CH_2)_4CH_3$.

N/N and P/P. The typical phase diagram is shown in Fig. 14.15.[96] Induced smectic B phases (at higher temperatures) and *E* phases (at lower temperatures) were observed in a number of systems. The induced smectic mesophases with tilted director were not observed.

The SLC-C phase, contained in some nonpolar substances, stabilized relative to the nematic phase with the addition of compounds containing polar terminal groups. However, as the concentration of the liquid crystalline compound with polar terminal groups increased, the temperature of the transition to the induced CLC-A phase decreased (Fig. 14.16).[96]

The concentration dependence of the interlayer distance of the smectic phase was determined for the system HEPTOAB-CBOOA (Fig. 14.17),[96] and in addition the minimum values of this distance were observed for mixtures containing about 60% CBOOA. For smaller concentrations of the component with polar terminal groups, the interlayer distances varied linearly along a straight line connecting the lengths of molecules of both components L_1 and L_2 (additivity rule):

$$d = x_1L_1 + x_2L_2. \tag{14.1}$$

At higher concentrations of the component with polar terminal groups,

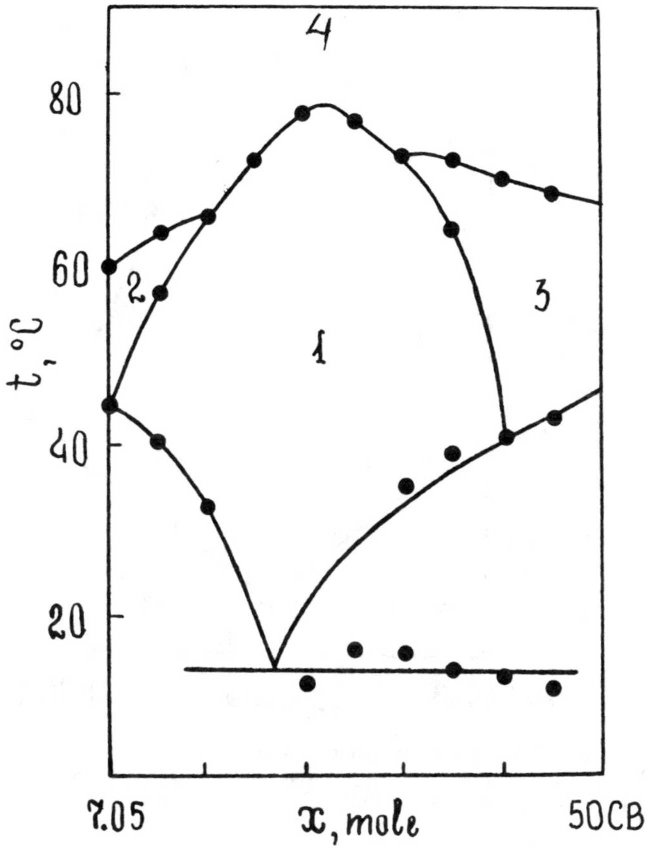

Figure 14.15. Phase diagram of system 4-*n*-pentylphenyl-4′-*n*-heptyloxybenzoate (7.05)—4-*n*-pentyloxy-4′-cyanobiphenyl (5OCB):[96] 1—SLC A; 2 and 3—NLC; 4—isotropic liquid.

the thicknesses of the layers increase with an increase of these concentrations.

The interlayer distances were also studied for systems investigated in Ref. 89 (Fig. 14.18) and the results correspond to the data on Ref. 96. In particular, the increase in *d* on the right side of the dependence in Fig. 14.18 can be explained by the displacement of the equilibrium mixture of monomers and dimers toward dimers with increasing concentration of the polar component, so that there is a tendency toward the formation of a "double layer" structure.[89] The structure, characterized by the presence of a double layer, of liquid crystals con-

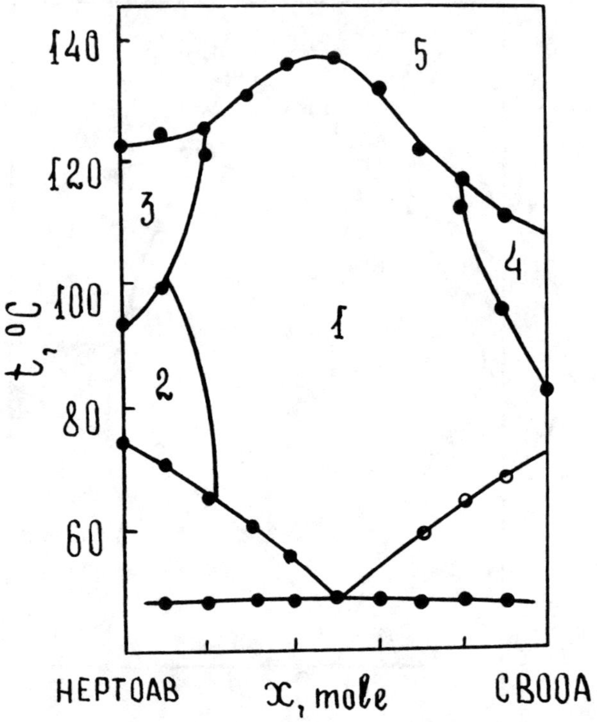

Figure 14.16. Phase diagram of system 4,4′-di-*n*-heptyloxyazoxybenzene (HEPTOAB)—4-cyanobenzylidene-4′-*n*-octyloxyaniline (CBOOA):[96] 1—SLC A; 2—SLC C; 3 and 4—NLC; 5—isotropic liquid.

taining polar terminal groups, giving rise to antiparallel pairwise arrangement of molecules. Such an interaction facilitates the formation of a nematic phase, and not a layered smectic structure. On the other hand, the addition of a nematic, containing nonpolar terminal groups, permits separating the dipoles. As a result, smectic layers, which are stable relative to the component containing the polar terminal groups, are formed.[89]

The method of infrared spectroscopy can be used to study complexes of this type. The complexes forming in a system containing 4,4′-diacyldiphenyl and *N*,*N*′-dialkyl-4,4′-diaminodiphenyl, were studied in this manner.[98] Three types of smectic phases were found to exist in this system, and it was established that hydrogen bonds play an important role in the formation of the observed complexes.

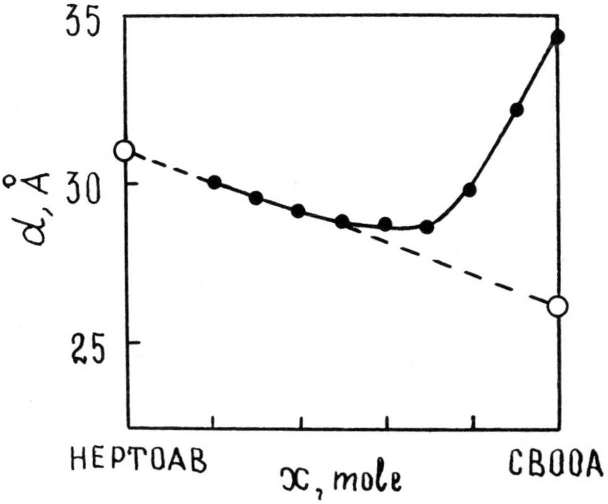

Figure 14.17. Interlayer distance of smectic phase of system HEPTOAB–CBOOA at 77°C.[96]

The effect of induction of smectic mesophases limits the number of possible combinations of NLC in preparing nematic mixtures existing in a wide temperature range.[99] In addition, this effect as well as pretransitional phenomena give rise to the features of electrical and electrooptical properties of mixtures of nematics.[100] It is remarkable that mixtures with induced smectic phases are interesting objects for investigating second-order phase transitions.[86,101]

Phase diagrams of mixtures of chiral nematics have a unique character (see Part 3). Figure 14.19 shows the qualitative form of the phase diagram of this type and the typical dependence of the temperature range of the "color play" effect. A characteristic peculiarity of such solutions is the existence of liquid crystalline state (including the capability of selective reflection in the visible range) down to low temperatures.[102,103]

Attempts to use the method of mixing for investigating mesophases in such systems were already undertaken in the first period of study of liquid crystals consisting of disc-like molecules.[104,105] One of the first and most complete works in this area concerned benzene-hexa-*n*-alkanoates.[104] The data obtained, together with some simplifying assumptions, permit predicting the nature of the isobaric phase diagrams of any pair of terms in this series. Thus, the lyotropic meso-

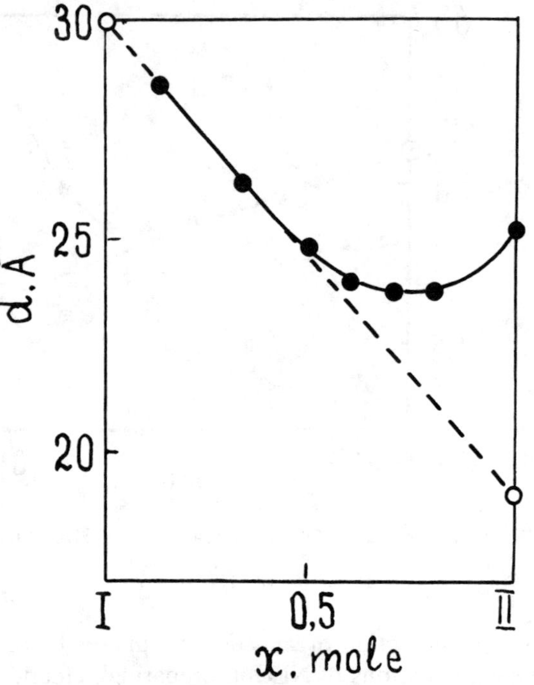

Figure 14.18. Interlayer distance in solution 7AB–5CB:[89] solid line corresponds to experiment; dashed line—to results of calculation on eq. (14.1).

morphism of the liquid crystalline mesophase was established. The obtained isobaric phase diagrams of solutions are presented in Figs. 14.20–14.23. It was possible to show that the temperatures of the real and virtual mesophase–isotropic liquid transitions depend linearly on the molecular weight, while the temperatures of the solid crystal–isotropic liquid transition exhibit a minimum depending on the molecular weight (Fig. 14.24). As far as solutions of the liquid crystals consisting of disc-like molecules (BHA, $p = 5$) with non-mesogenic substance (benzene) are concerned, the region of existence of the mesophase rapidly decreases with decreasing concentration of the mesogenic substance and constitutes about 2°C in the middle range of concentrations (Fig. 14.25).

The isobaric phase diagram of the mixture hexa-alkoxy-derivatives of triphenylene ($n = 5$ and 7) was obtained by methods of calorimetric measurements and nuclear magnetic resonance (Fig. 14.26).[105] This

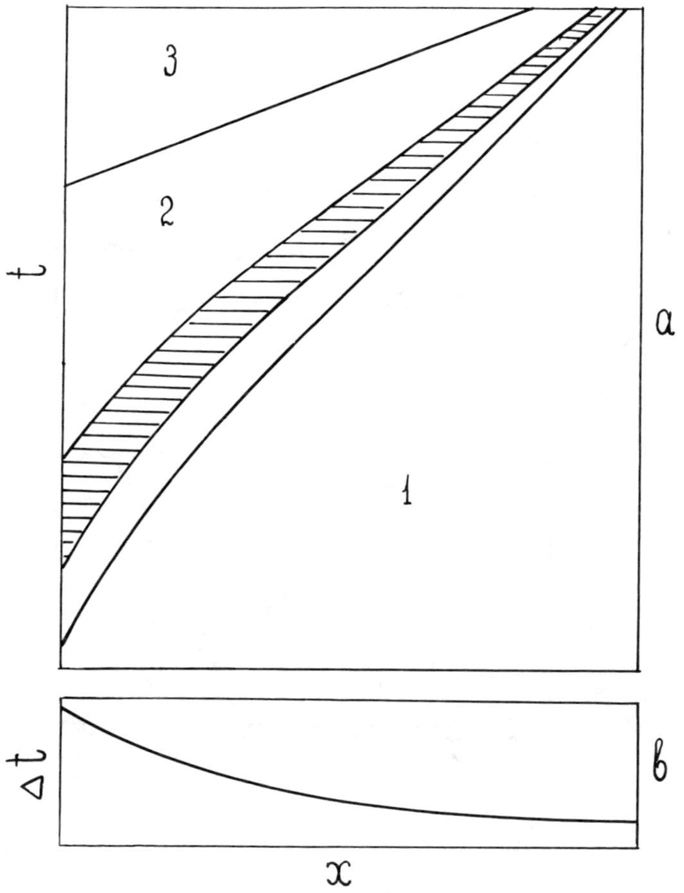

Figure 14.19. Quality view of phase diagram of mixture of chiral nematics (a) and typical concentration dependence of temperature interval Δt of "colour play" of this mixture (b):[102,103] 1—SLC; 2—CLC; 3—IL. The region of "colour play" is shaded. The scale in the figure is not observed. The lower boundary of the "colour play" region can reach −30°C and lower; characteristic values of Δt are between 0.5 and 5°C.

diagram is qualitatively similar to the phase diagram of BHA solutions (p = 6 and 7, Fig. 14.21).

A special study of interactions of solvent–solute in mesophases, formed by diskotic particles, was formed in Ref. 106. In particular, mixtures containing benzene-hexa-n-heptanoate and benzene-hexa-n-octanoate were investigated. The phase diagram of this mixture is

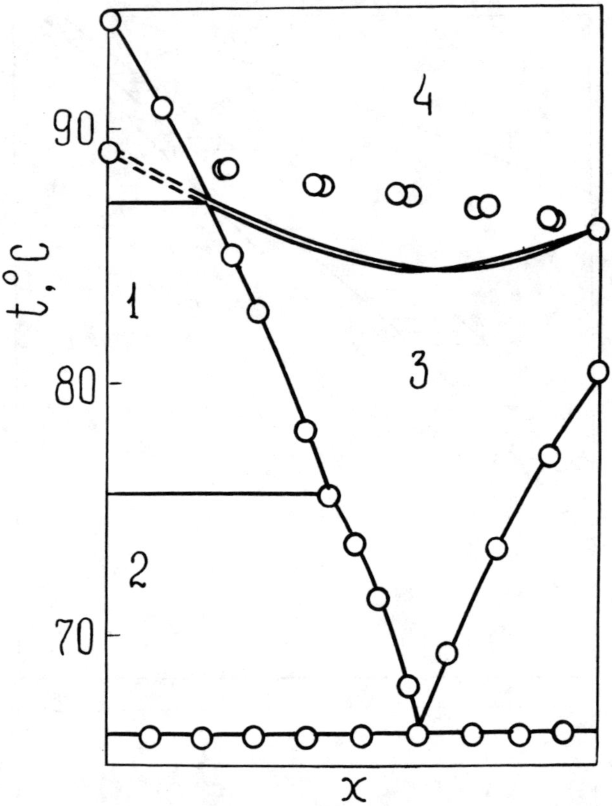

Figure 14.20. Phase isobaric diagram of mixtures of substances BHA $p = 5$ (to the left) and $p = 6$ (to the right):[104] 1 and 2—solid crystal; 3—liquid crystal, 4—isotropic liquid.

presented in Fig. 14.27. The enthalpies of transitions solid–mesophase and mesophase–isotropic liquid were studied in detail (Fig. 14.28). Benzene, ethanol, and other molecules were used as the small nonmesogenic molecules, soluble in this system. Detailed data were obtained on the concentration dependences of the transition temperatures (Figs. 14.29–14.30). It was established that diskotic mesophases have a strong ordering effect on the dissolved substances.

An example of the effectiveness of using the mixing method in searching for and investigating of new types of mesophases is the study of mixtures of hexa-substituted triphenylene.[107] One of the results of this investigation is the phase diagram presented in Fig.

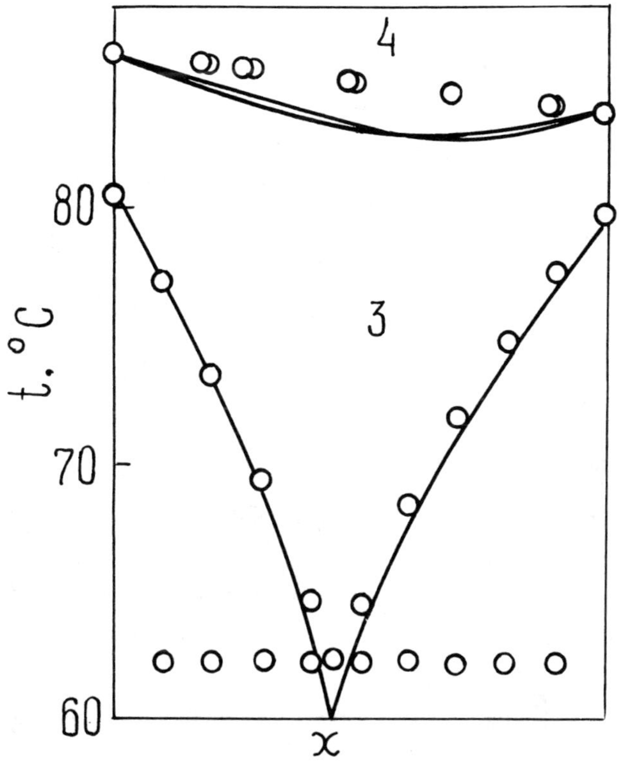

Figure 14.21. Phase isobaric diagram of mixtures of substances BHA $p = 6$ (to the left) and $p = 7$ (to the right):[104] designations see in Figure 14.20.

14.31. This diagram contains regions of diskotic phases, denoted by the authors by D, D_F, and D_B. The phase diagram of the mixture di-isobutylsilanediole and hexa-*n*-heptanoate (Fig. 14.32)[108] on which the diskotic mesophase occupies a considerable region encompassing the entire range of concentrations is also interesting.

Investigations of phase equilibria in liquid crystals which were recently carried out have a great methodical importance.[109–112] Thus, in paper 109 typical constructions of experimental equipment were described, which made it possible to do the study in the range of pressure up to 300 MPa. The results of theoretical calculations of phase equilibria in mixtures were then discussed. Data of the wide thermodynamical investigation of phase equilibria in solutions of liquid crystals and the method of equilibria conditions calculations were

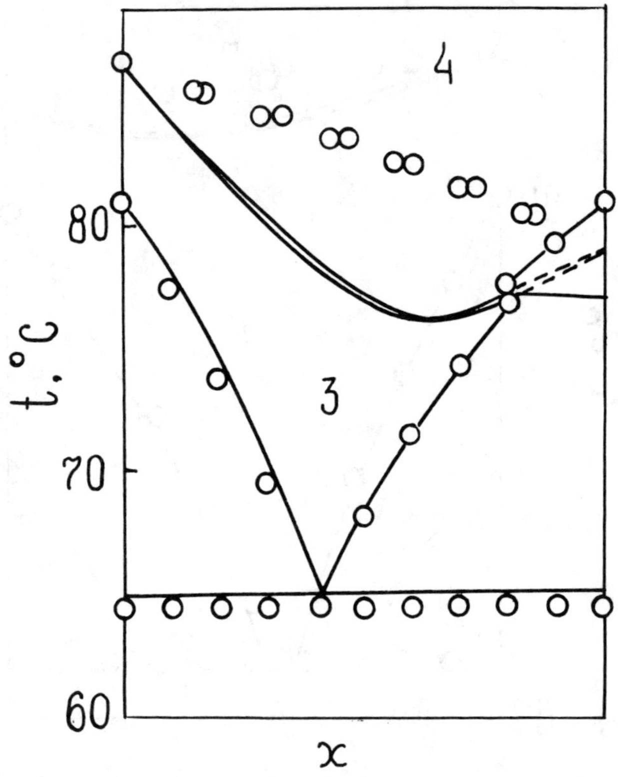

Figure 14.22. Phase isobaric diagram of mixtures of substances BHA $p = 6$ (to the left) and $p = 8$ (to the right):[104] designations see in Figure 14.20.

given in paper 110. The application of thermodynamic properties (heat capacity, excess Gibbs energy, etc.) as criteria made it possible to propose a classification of phase diagrams for mixtures of liquid crystals including ideal, non-ideal and reentrant phases.

The results characterize the calculation of phase diagrams demonstrating smectic phase A,C,F,G, and the reentrant phase, as well as Gibbs excess energy and chemical potentials versus concentrations and temperature. The availability of numerous reentrant nematic phases in systems of molecules having aromatic cores and highly polarized end groups (CN or NO_2) was noted in paper 111. At the same time it was noted that there were some types of smectic A phases with different kinds of ordering of the layer packing in such

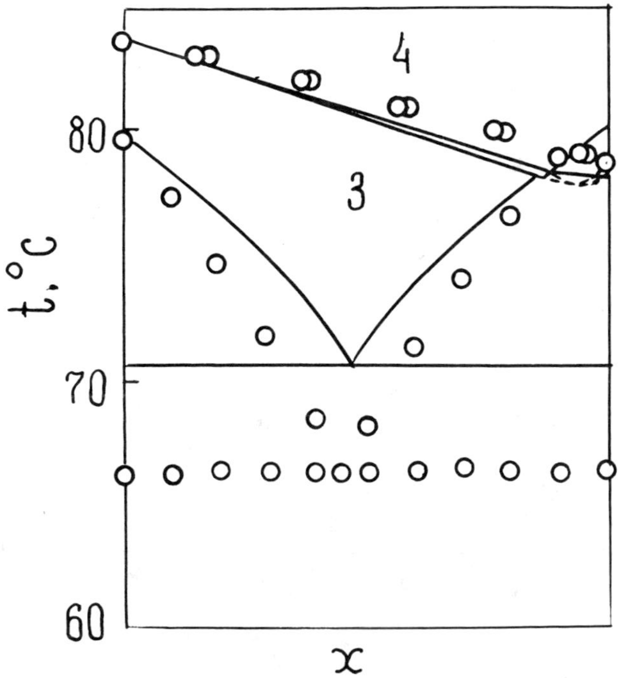

Figure 14.23. Phase isobaric diagram of mixtures of substances BHA $p = 7$ (to the left) and $p = 8$ (to the right):[104] designations see in Figure 14.20.

systems. The appearance of "nematic gaps" in mixtures of smectic is not only of theoretical, but also of great practical importance, as there appears to be a possibility of obtaining nematics with low viscosity from smectic phases. In addition it becomes possible to prevent the formation of the smectic phase in mixtures at low temperatures.[112]

Recent theoretical studies of phase diagrams for solutions of liquid crystals are of great interest.[113] Paper 113 gives the interpretation of phase diagrams for mixtures of cholesterol esters with different nematogens from the viewpoint of specific interactions between molecules of various components. The increasing of temperature and the appearance of the smectic-A phase was shown to be related to the ratio of molecular lengths of components mixed.[114] Generalization of the lattice model of the nematic phase for the case of mixtures of rod-like molecules was given in paper 115. The authors attributed

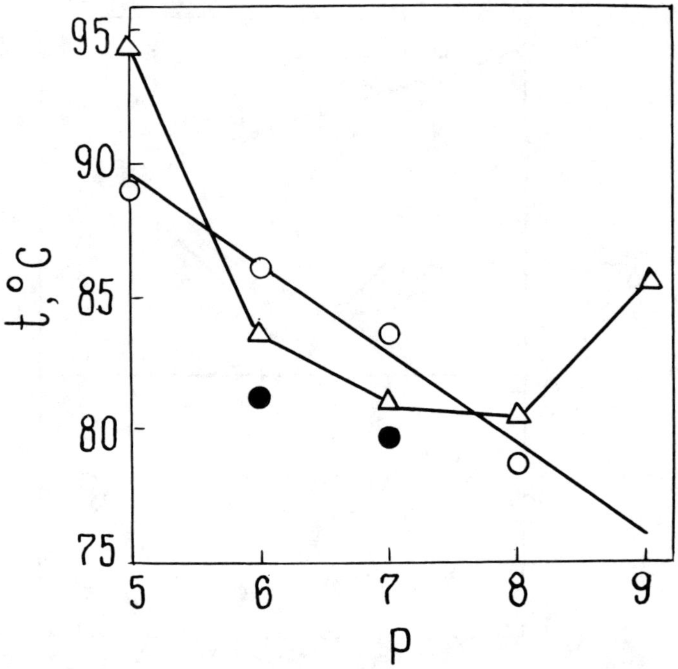

Figure 14.24. Temperature dependences of transitions of substances BHA versus *p*:[104] triangles—solid crystal—isotropic liquid transition; black circles—solid crystal—liquid crystal transition; open circles—liquid crystal—isotropic liquid transition.

the difference between theoretical and experimental data to the role of rotational diffusion in nematics, resulting in the quick formation of domains of large sizes. The method of the mean field approximation[116] was generalized for the case of a binary mixture of nematic liquid crystals.[117] This method enabled them to draw the phase diagram in coordinates: tempeature-concentration, as well as to study the behavior of uniaxial and biaxial order parameters. A more complicated case was also considered in Ref. 118, when a semi-flexible polymer was one of the components of the system. Experimental data for such a system were compared with the results obtained by means of the simplest variants of the Flory theory of the mean field. It appeared that good agreement between experimental and theoretical data could be reached if the temperature dependence of geometrical parameters of molecules was taken into account.

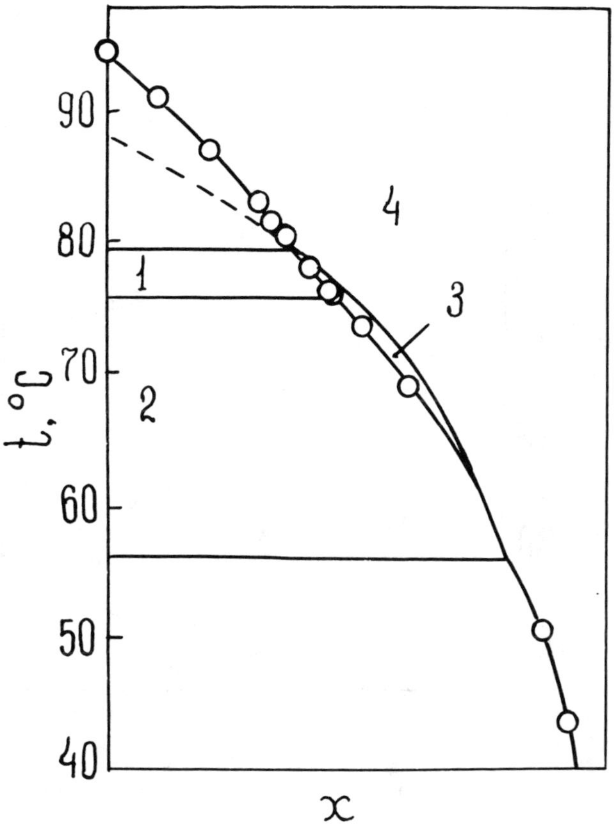

Figure 14.25. Phase diagram of mixtures of substance BHA $p = 5$ (to the left) and benzene (to the right);[104] see designations in Figure 14.20.

14.2 THERMOPHYSICAL PROPERTIES OF SOLUTIONS OF LIQUID CRYSTALS AND COMPUTATIONAL-THEORETICAL METHODS FOR DETERMINING THESE PROPERTIES

A well-defined range of data on the properties of mixtures of liquid crystals has not been accumulated. These data concern the temperatures of phase transitions, density, mixing enthalpies and volumes, etc. On the whole, however, there are still few such data.

The results of experimental investigations of the temperature de-

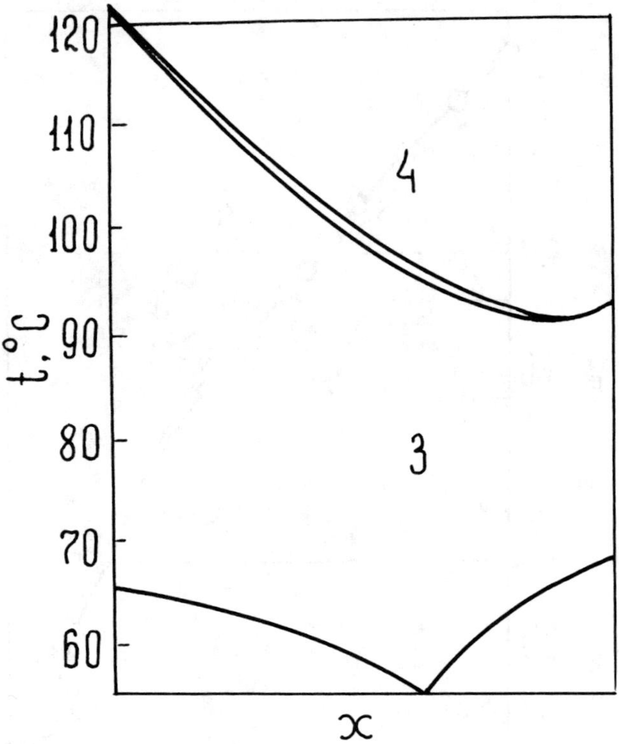

Figure 14.26. Phase isobaric diagram for the mixture of hexa-alkoxy—derivations of triphenyl $n = 5$ (to the left) and $n = 7$ (to the right);[105] see designations in Figure 14.20.

pendences of the density of mixtures of liquid crystals are presented in Refs. 119–121. A characteristic example is the solution PAA–cholesteryl propionate. Analogously to the case of pure substances, the nature of the isotherms of the density of mixtures varies considerably in the region of the transition liquid crystals–isotropic liquid.[119] Far from the point of the transition, the density of the mixture can be determined using the assumption of additivity from the densities of the individual substances.[120]

A detailed study of the thermodynamic properties of binary mixtures consisting of 4,4′-dipentylazoxybenzene (DPAB), 4-ethyl-4′-pentylazoxybezene (EPAB), and 4-*n*-pentyl-4′-cyanobiphenyl (PCB), is performed in Ref. 121. Based on the experimental data obtained, the isothermal mixing volumes V^S (coinciding with the excess volumes V^E in cases when both pure components and the solution are in the

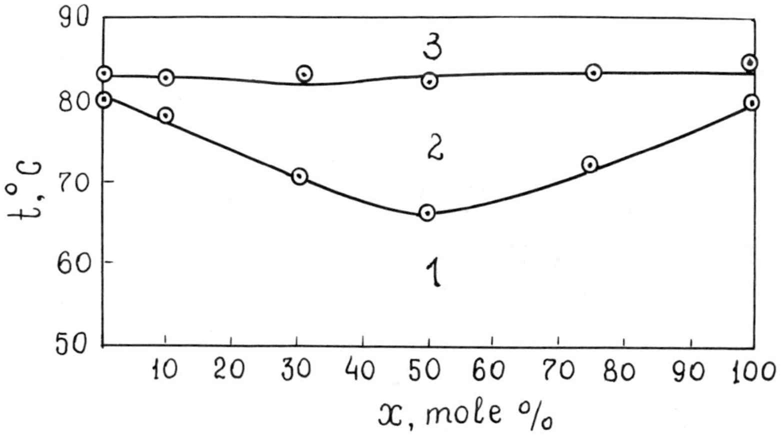

Figure 14.27. Phase diagram of binary system consisting of benzene-hexa-*n*-heptanoate and octanoate:[106] 1—solid crystal; 2—liquid crystal; 3—isotropic liquid.

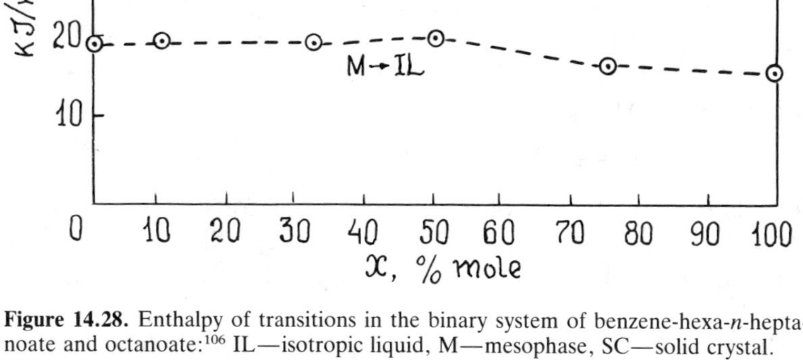

Figure 14.28. Enthalpy of transitions in the binary system of benzene-hexa-*n*-heptanoate and octanoate:[106] IL—isotropic liquid, M—mesophase, SC—solid crystal.

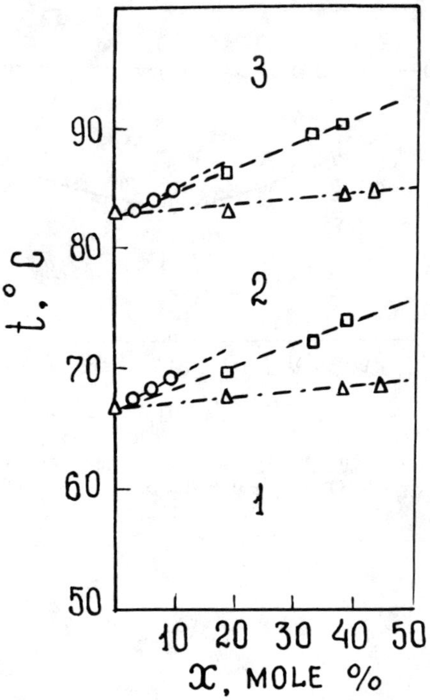

Figure 14.29. Behavior of solutions of benzene-hexa-*n*-heptanoate and octanoate (1:1) in equimolar systems with solvents:[106] circles—dimetylbenzene, squares—benzene, triangles—ethanol; 1—solid crystal, 2—liquid crystal, 3—isotropic liquid.

same phase) were calculated. These results are presented in Table 14.2 and 14.3 in Fig. 14.33a and 14.33b (the phase diagrams of the mixtures are also presented in the figures). Characteristically, in all cases, $V^S < 0$. Table 14.4 presents the values of $\Delta V^S/\Delta t$, calculated according to the expression

$$\Delta V^S/\Delta t = \Delta V/\Delta t - x_1(\Delta V_1/\Delta t) - (1 - x)(\Delta V_2/\Delta t).$$

The thermal coefficients of isobaric expansion for a mixture containing EPAB and PCB were also determined. The difference between the densities of DPAB and EPAB is apparently due to the role of the group $(CH_2)_3$, which is caused by the difference in the structures of the molecules of these substances. A calculation of the volume of

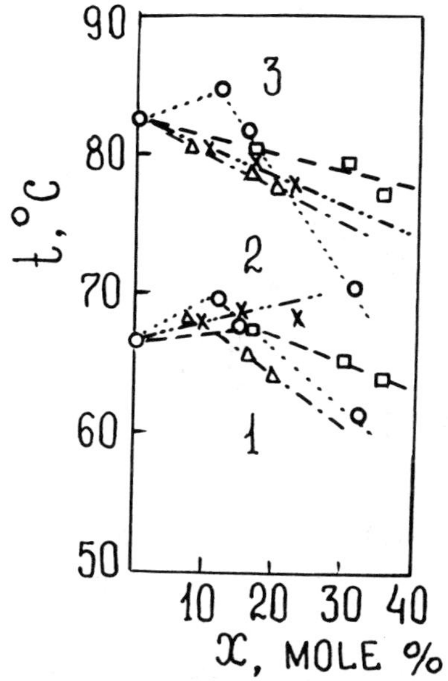

Figure 14.30. Behavior of solutions of benzene-hexa-*n*-heptanoate and octanoate (1:1) in equimolar systems with solvents:[106] circles—anisole, squares—acetate, triangles—nitrobenzene, crosses—acetophenol; 1, 2, 3—see Figure 14.29.

this group, performed in Ref. 121 for the pair DPAB–EPAB and the 4-octyl-4′-ethylazoxybenzene led to a quantity close to 50 cm^3 $mole^{-1}$.

The molar volumes were determined both by a dilatometric method[122] and with the help of gas-liquid chromatography (GLC).[123] The enthalpies and entropies of binary solutions of 42 nonmesogenic substances in two nematogenic solvents were calculated in Ref. 123. As far as the GLC method is concerned, it also was used in Refs. 124 and 125, where in addition to the enthalpies and entropies, the heats of dissolution and the activity coefficients were also determined. The dependences of the activity coefficients f_0^∞ on the inverse temperature $1/T$ for solutions containing one of five homologues of *n*-alkanes in PAA are presented in Fig. 14.34.[72,123] It is evident that as the molecules become longer, the quantities $\ln f_0^\infty$ increase by a factor of 1.2–1.3 and increase abruptly at the point of the transition of PAA

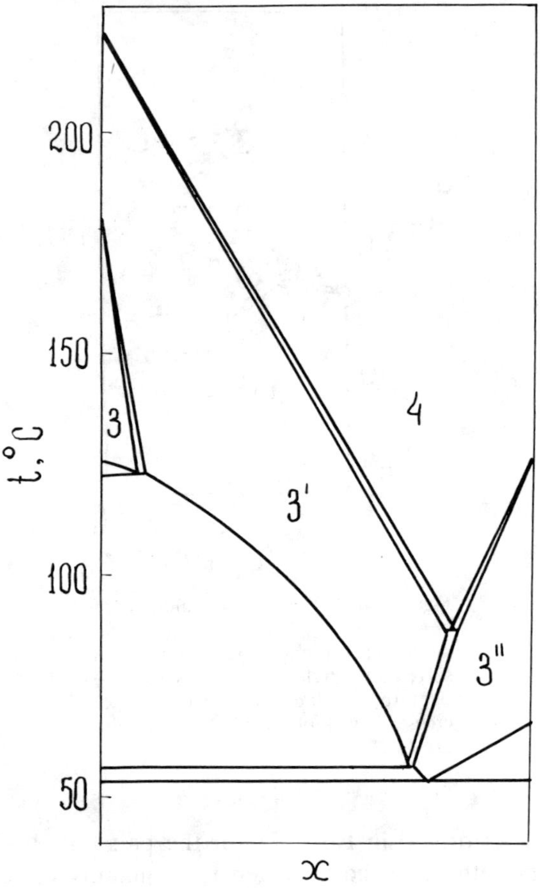

Figure 14.31. Phase isobaric diagram of mixtures of substance 4.1, R_l (Table 11.1) (to the left) and substance 4.1, R_3 (to the right):[107] 3, 3′ and 3″—liquid crystal phases of disc-like molecules, 4—isotropic liquid.

from isotropic to nematic phase. Investigators note that the increase in f_0^∞ could be due to a decrease in the energy of interaction of the dissolved substance with the solvent and to a decrease in the number of possible conformations of the molecule dissolved in the liquid crystal, as well as to a restriction on the freedom of its rotational motions. The dependence of ln f_0 on the concentration of methoxy-benzylidene-propyl-aniline (MBPA) in MBPA–nonmesogen solutions is presented in Fig. 14.35.[124] The change in the nature of the

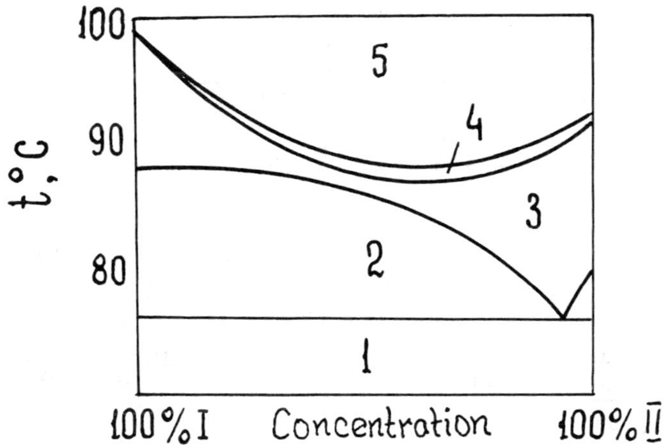

Figure 14.32. Phase diagram of binary mixtures[108] of di-isobutylsilanediole (DIISB) (I) and benzene-hexa-*n*-heptanoate (II): 1—solid crystal, 2—solid crystal + discotic mesophase, 3—discotic mesophase, 4—discotic mesophase + isotropic liquid, 5—isotropic liquid.

concentration dependence accompanying a transition from isotropic phase to anisotropic phase is clearly evident.

The thermodynamic mixing functions were determined not only by the GLC method but also by the method of direct calorimetric measurements.[74,126] Thus the results of measurements of the mixing enthalpies H^M of MBBA with benzene and cyclohexane, presented in Ref. 74, exhibit the considerable endothermic effect of mixing (the largest value $H^M = 1630$ J/mole is observed in the system MBBA–cyclohexane at 45°C). The considerable endothermic effect accompanying the formation of solutions both in the isotropic and in the anisotropic phase was also observed in the case of mixtures of 4-methoxybenzylidene-4′-butylaniline or 4-ethoxybenzylidene-4′-butylaniline with heptane (Fig. 14.36).[126]

The calorimetric method was also used to study the heat capacity of solutions of 4-*n*-octyloxy-4′-cyanobiphenyl (8OCB) and 6OCB. The NLC–SLCA transition is a second-order phase transition which the entropy decreases as the concentration of 6OCB increases. It is remarkable that a mixture containing 27% weight of 6OCB, exhibits a secondary nematic phase and, in addition, the heat capacity does not exhibit critical behavior neither with the transition NLC–SLC-A nor with the transition SLC-A–secondary nematic phase.[127]

Table 14.2

Interpolated values of molar volumes (V) and volumes of isothermal mixing (V^s) of mixtures DPAB-PCB ($cm^3 \cdot mole^{-1}$)[121] *

x, mole DPAB	0.0000	0.3564		0.5040		0.6029		1.0000
t, °C	V	V	V^s	V	V^s	V	V^s	V
	Nematic			Smectic *A*				Nematic
25.00	243.99	274.55	−0.88	287.37	−1.08	304.12	−0.99	332.20
30.00	245.11	275.48	−1.09	288.30	−1.30	305.15	−1.15	333.39
35.00	246.46	276.44	−1.44	289.26	−1.64	306.14	−1.42	334.63
	All components and mixtures are in isotropic phase ($V^s = V^E$)							
70.00	254.06	286.30	−0.47	299.70	−0.61	317.00	−0.64	345.82
75.00	255.03	287.40	−0.50	300.94	−0.57	318.36	−0.57	347.25

* Data[121] are given selectively.

Table 14.3

Interpolated values of molar volumes (V) and volumes of isothermal mixing (V^s) of mixtures EPAB-PCB ($cm^3 \cdot mole^{-1}$)[121]*

x, mole EPAB	0.0000	0.4437		0.4923		0.5856		1.0000
t, °C	V	V	V^s	V	V^s	V	V^s	V
	Nematic			Smectic *A*				Nematic
25.00	243.64	259.15	−2.03	261.03	−2.07	264.85	−1.95	283.17
30.00	244.76	260.09	−2.23	261.92	−2.28	265.80	−2.13	284.33
35.00	246.06	261.08	−2.51	262.92	−2.59	266.80	−2.40	285.57
	Isotropic			Smectic A				Isotropic
40.00	247.9	262.16	−3.25	263.96	−3.37	268.09	−2.92	287.35
	Isotropic			Nematic				Isotropic
45.00	246.92	264.50	−1.97	266.42	−1.98	270.32	−1.77	288.48
50.00	249.91	265.84	−1.66	267.77	−1.65	271.67	−1.45	289.55
52.00	250.30	266.43	−1.48	268.37	−1.47	272.32	−1.23	289.99
	All components and mixtures are in isotropic phase ($V^s = V^E$)							
56.00	251.09	268.05	−0.69	269.98	−0.69	273.76	−0.62	290.87
60.00	251.86	268.94	−0.58	270.86	−0.63	274.64	−0.56	291.72
70.00	253.79	271.05	−0.53	272.99	−0.54	276.77	−0.50	293.89
75.00	254.75	272.09	−0.50	274.04	−0.51	277.83	−0.47	294.97

* Data are given selectively.

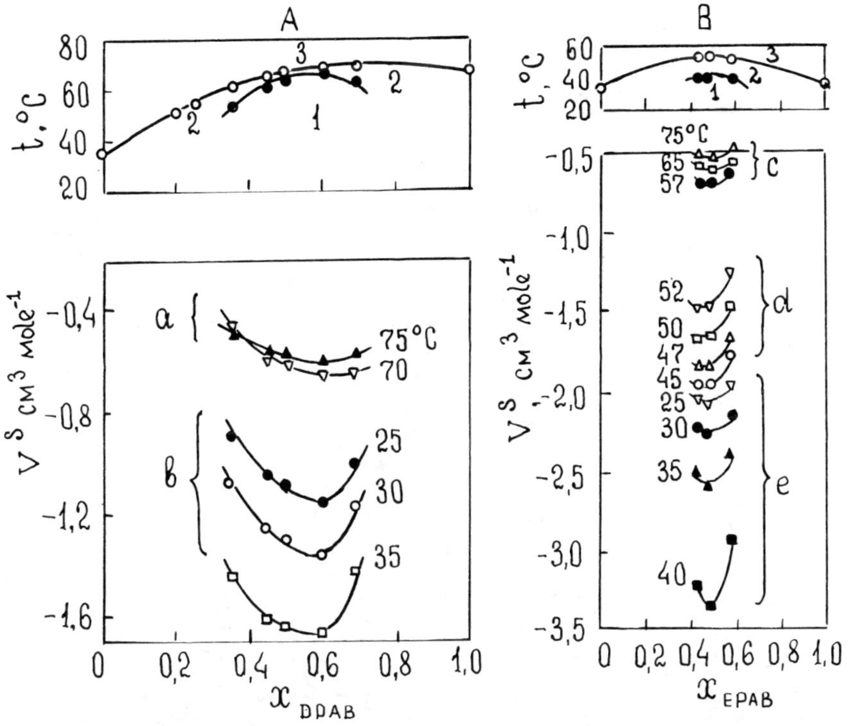

Figure 14.33. Phase diagrams and concentration relationships of V^s for binary systems:[121] A—DPAB (x)—PCB ($1 - x$): a—mixture and components are isotropic liquids ($V^s = V^E$); b—mixtures are SLC A, components are NLC; B—EPAB (x)—PCB ($1 - x$): c—mixtures and components are isotropic liquids ($V^s = V^E$); d—mixtures are NLC, components are isotropic liquids; e—mixtures are SLC A, components are NLC; I—SLC A, 2—NLC, 3—IL.

The concentration dependences of the entropy changes in the system 8OCB–heptyloxyaminofluorenone are presented in Fig. 14.37 based on data in Ref. 86. The changes in the enthalpies of the transitions SLC-E–SLC-A (SLC-B), SLC-B–SLCA-A, and SLC-A–IL are presented in Fig. 14.38 based on results in Ref. 89. The peaks in the concentration dependences of these transitions correspond to concentrations of 5CB equal to 0.5, 0.3, and 0.3–0.5.

A detailed study of the properties of liquid-crystalline mixtures, based on experiments on measuring the density, velocity of ultrasound, and coefficient of absorption, was performed in Ref. 128.

Table 14.4
Thermal coefficients of isobaric expansion of mixture EPAB-PCB (x_{EPAB} = 0.4923 mole) and its components[121]*

	Components				Mixture		
	EPAB		PCB				
t, °C	Phase	$\Delta V_1/\Delta t$, cm³·mole⁻¹·°C⁻¹	Phase	$\Delta V_2/\Delta t$, cm³·mole⁻¹·°C⁻¹	Phase	$\Delta V/\Delta t$, cm³·mole⁻¹·°C⁻¹	$\frac{\Delta V^s}{\Delta t}$, cm³·mole⁻¹·°C⁻¹
25	NLC	0.229	NLC	0.220	SLCA	0.184	−0.040
30	NLC	0.238	NLC	0.237	SLCA	0.189	−0.048
35	NLC	0.277	NLC	0.326	SLCA	0.193	−0.109
45	IL	0.222	IL	0.199	NLC	0.272	+0.062
47	IL	0.222	IL	0.199	NLC	0.268	0.058
56	IL	0.215	IL	0.193	IL	0.227	0.023
60	IL	0.218	IL	0.193	IL	0.216	0.011
65	IL	0.216	IL	0.192	IL	0.211	0.007
70	IL	0.217	IL	0.192	IL	0.210	0.006
75	IL	0.217	IL	0.194	IL	0.212	0.007

* $\Delta V^s/\Delta t = \Delta V_{mix}/\Delta t - x_1\Delta V_1/\Delta t - (1 - x_1)\Delta V_2/\Delta t$, $x_1 = x_{EPAB}$.

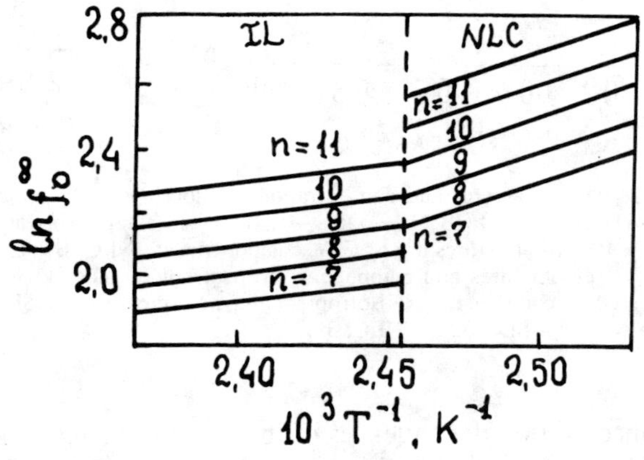

Figure 14.34. The limiting values of activity coefficients f_o^∞ for n-alkanes C_nH_{2n+2} in PAA versus $1/T$.[123]

Special attention was directed toward the region of the transition from liquid-crystalline to isotropic-liquid state. The coefficient of adiabatic compressibility was calculated using the experimental data obtained. The analysis led to the conclusion that Fiksman's well-known theory of critical solutions is applicable to liquid-crystalline mixtures. Based on this theory, it was possible to calculate the temperature

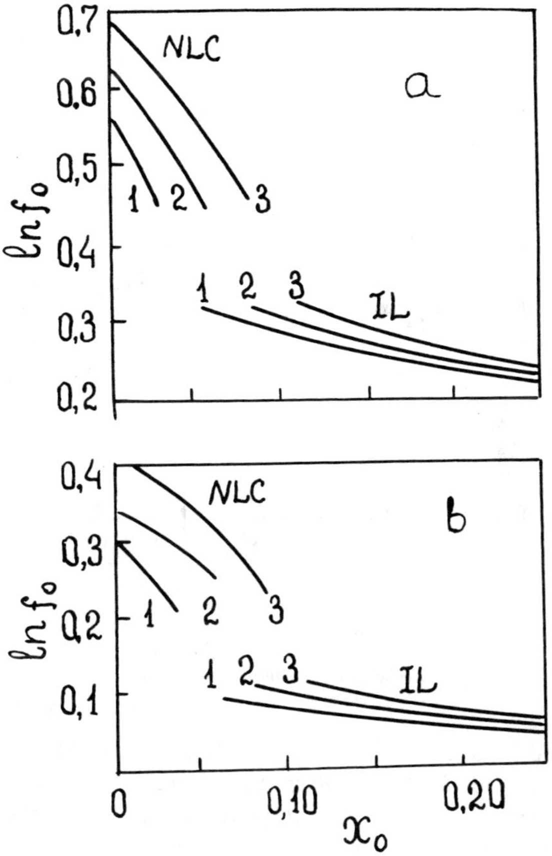

Figure 14.35. Activity coefficients f_o in binary systems NLC–nonmesomorphic substance versus the molar fraction of the latter (x_o):[124] a—system MBPA—CCl_4, b—system MBPA—C_6H_6; 1, 2 and 3 correspond to the temperatures 52.1; 46.3 and 40.1°C respectively.

dependence of the coefficient of absorption of ultrasound, which turned out to agree well with the experimental data. Experiments concerning the investigation of Brillouin zones in mixture of cholesteryl chloride and cholesteryl myristate using an acoustical method are described in Ref. 129. The temperature dependence of the pitch of the helix was studied using the results of the measurements of damping of ultrasound.

The influence of the shape of alkane molecules dissolved in NLC on the thermodynamic properties and orientational order of solutions

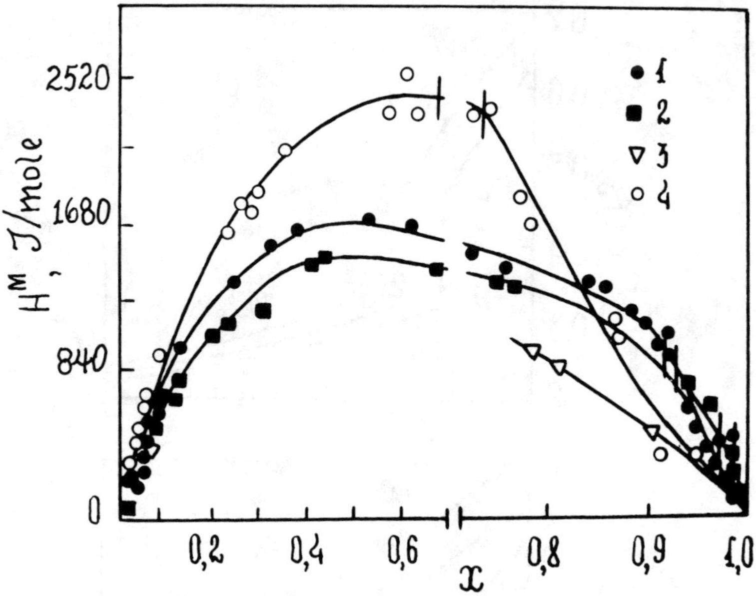

Figure 14.36. Enthalpies of mixing versus mole fraction x of NLC for systems: 4-methoxybenzylidene-4′-butylaniline—heptan at 30°C (1), 40°C (2) and 50°C (3) and 4-ethoxybenzylidene-4′-butylaniline—heptan at 40°C (4).[126]

is analyzed in Ref. 130. Special attention is devoted to the temperatures of the NLC–IL transition and the coefficients of nematic ordering. It was shown that the branched alkanes decrease to a much lesser extent the temperature of the phase transition than the corresponding linear alkanes. The measured changes in the transition temperatures were compared with the calculation for different models of the intermolecular interaction potential. The review in Ref. 131 is devoted to the structure and some properties of liquid-crystalline-phases in systems of amphiphilic compounds.

Ultrasonic behavior of liquid crystalline mixtures of octadecanol and cholesterol (absorption and ultrasound velocity) was studied by means of the impulse method and acoustic interferometer at different concentrations near the temperature of the phase transition isotropic liquid–liquid crystal.[132] One can note a sharp reduction of sound velocity, an increase of sound absorption in the range of isotropic–cholesteric transition, as well as pretransitional phenomena on the side of both isotropic and cholesteric phases.Enthalpy data for the phase transition SLCA–SLCB in the system of 65OBC–PP5CC and temperature dependences of heat capacity of this system and its com-

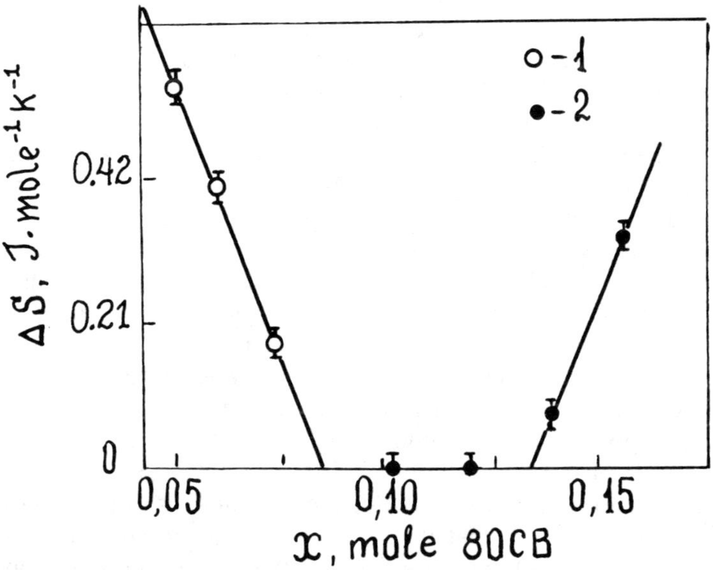

Figure 14.37. Changes of entropy versus mole content of 8OCB:[86] 1—for transition NLC—SLC, 2—for transition NLC—SLC A.

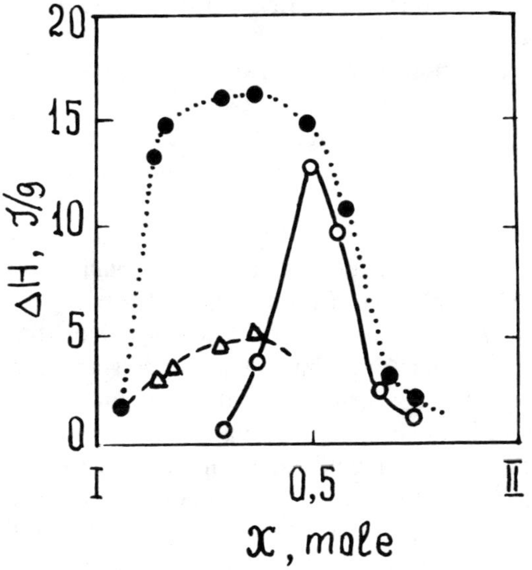

Figure 14.38. Enthalpies of transitions SLC E↔SLC A (SLC B) (solid line), SLC B—SLC A (dashed line) and SLC A—isotropic liquid (points) of system 7AB (I)—5CB (II).[89]

ponent were obtained in paper 133 by polarizational microscopy methods, differential scanning calorimetry and microcalorimetry of high resolution. The phase diagram was drawn and the critical behavior of this system compared to the behavior of the mixture ^{3}He-^{4}He was discussed. In paper 134 densities and heat expansion coefficients for trimers, tetramers and pentamers of a series of *p*-oxybenzoate were determined, and phase transitions of pure components and mixtures were studied.

The range of published experimental data on the thermodynamic properties of solutions of liquid crystals is primarily restricted to the results examined above. As far as the transport characteristics of solutions are concerned, they have been studied to a much lesser extent. Examples are the measurements of the temperature dependences of the viscosity of solutions with different velocity gradients in the regions of phase transitions CLC–IL and CLC–SLC, performed with the help of a rotating viscosimeter.[135] Analysis of these results led to the conclusion that the observed changes in viscosity in the region of phase transitions are due primarily to changes in the pitch of the helix of CLC near these transitions.

Proceeding to a discussion of the computational-theoretical methods for determining the properties of solutions of liquid crystals, we note that it is possible to separate out three basic directions of these investigations.

The first direction is based on the mean molecular field approximation,[135–137] widely used in studying pure liquid crystalline substances (see Part I). It is assumed that the liquid crystalline ordering of particles of the solution is due to the anisotropy of their dispersion interaction. An appropriate choice of the potential function, taking into account the interaction of molecules with the mean molecular force field, permits calculating the thermodynamic functions, their changes with the phase transition, degree of ordering, etc. However, the search for a potential that correctly describes the interaction of molecules with their environment is difficult and requires detailed representations of the structure of the solution.

Mixtures of liquid crystals were investigated from the point of view of the mean molecular field theory in Refs. 138–140. The orientational pseudopotential for a multicomponent mixture of rod-shaped particles[138] was used in calculating the free energy of the model system.[140] In analyzing the stability of phases and predictions of the nature of phase diagrams, it is convenient to use, instead of Gibbs' configurational energy, Helmholtz's free energy, which takes into account the small quantity ΔV in the liquid-crystal–isotropic-liquid transition. The change in this energy within the framework of the mean field model is made up of three terms: a) scalar contribution of the interaction energy; b) mixing entropy (usually assumed to equal the mixing entropy of an ideal solution); and, c) contribution determined by the anisotropic nature of the interaction and leading to phase separation and to formation of a heterogenic system (the solution of an ordinary liquid in NLC and solution of NLC in an ordinary liquid).

The effect of adding a soluble substance consisting of spherical particles into a solution of rods and spheres is demonstrated by the concentration dependences of the Helmholtz function on the molar fraction of the solvent (Fig. 14.39).[140] It is evident that the addition

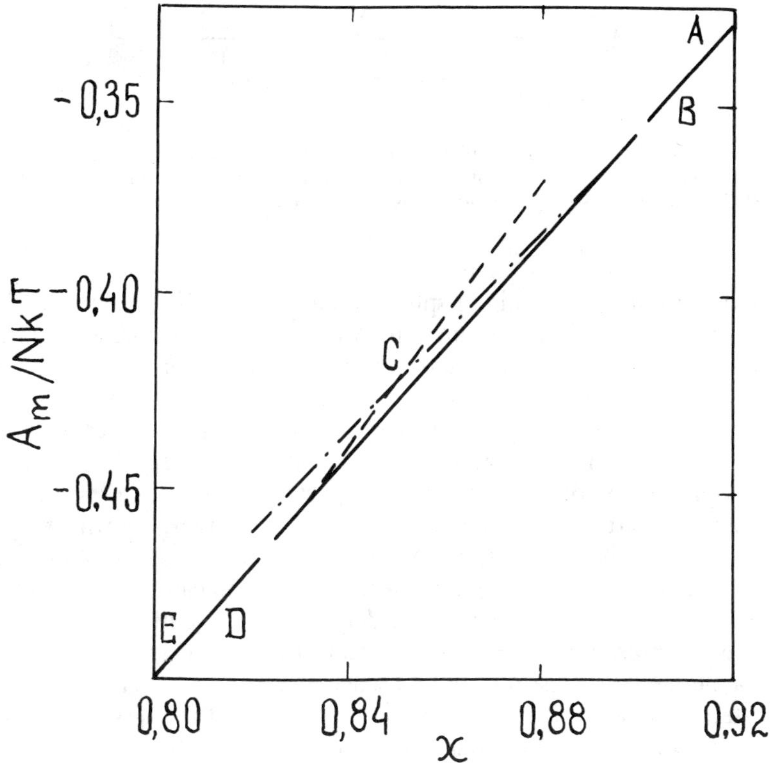

Figure 14.39. Molar Helmholtz function A_m versus mole fraction of solvent x in binary mixture of rods and spheres at reduced temperature 0.852.[140]

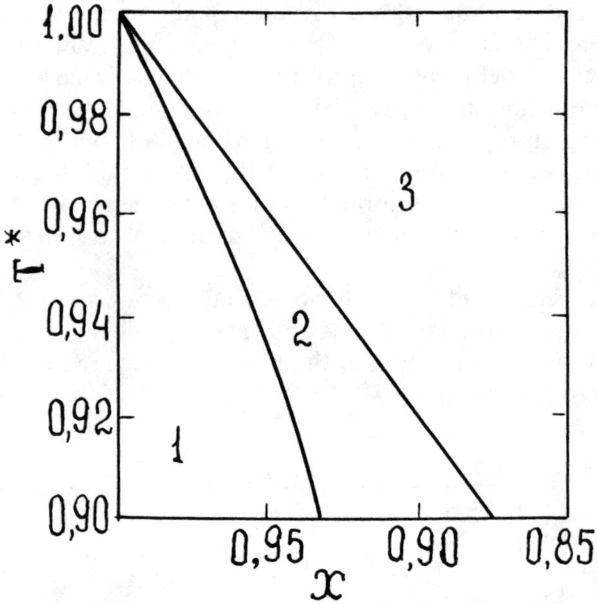

Figure 14.40. Phase diagram of binary mixture of rods and spheres of equal volumes in coordinates: mole fraction of solvent (x) and reduced temperature (T^*):[140] 1—NLC, 2—NLC + isotropic liquid, 3—isotropic liquid.

of a substance consisting of spherical particles to NLC (curve ABC) decreases the Helmholtz function. A decrease of this function is also observed for an isotropic liquid (curve CDE), however, these changes occur at different rates. The point of intersection of the curves corresponds to the liquid-crystal–isotropic-liquid transition, caused by the corresponding change in concentration. However, since Helmholtz's function of the system with separated phases (heterogeneous system) is situated lower (the line BD), the curves, shown by the dot–dashed lines, correspond to states, that are metastable relative to the nematic and isotropic phases. This method can be successfully used to obtain the complete phase diagram of the solution. The phase diagram presented in Fig. 14.40 (Ref. 140) was calculated in this manner and demonstrates the presence of a two-phase region, whose extent increases with increasing molar fraction of the dissolved substance consisting of spherically symmetrical molecules.

The calculations indicate[140] that for values of x close to 0.9, the bounding curves become non-linear and, in addition, the slope of the line bounding the nematic phase is so large at low temperatures that the two-phase region again becomes stable. The section of the phase diagram demonstrating the appearance of the secondary region of phase separation at low temperatures is presented in Fig. 14.41.[140] This effect is related with the change in the ratios of contributions to the free energy due to the increase in the orientational order and relative increase of the corresponding component of the energy and relative decrease in the component related to the mixing energy.

It is remarkable that the results of a direct comparison of the computed and experimental phase diagrams of the solutions of 4,4′-di-*n*-hexyloxyazoxybenzene and 4-methoxybenzylidene-4′-*n*-propylaniline with CCl_4[141] indicate the definite advantages of the mean field theory over the lattice models. The mean field theory was successfully used to study mixtures of CLC,[142] and in addition the effectiveness of this approach in studying the pitch of the helical twisting in solutions of CLC was established.

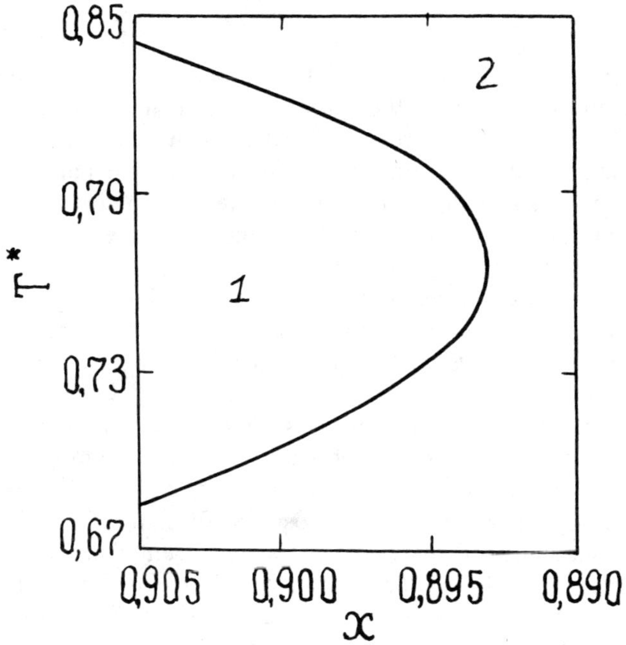

Figure 14.41. Part of phase diagram demonstrating appearance of secondary region of separation of phases at low reduced temperatures:[140] 1—NLC, 2—NLC + isotropic liquid.

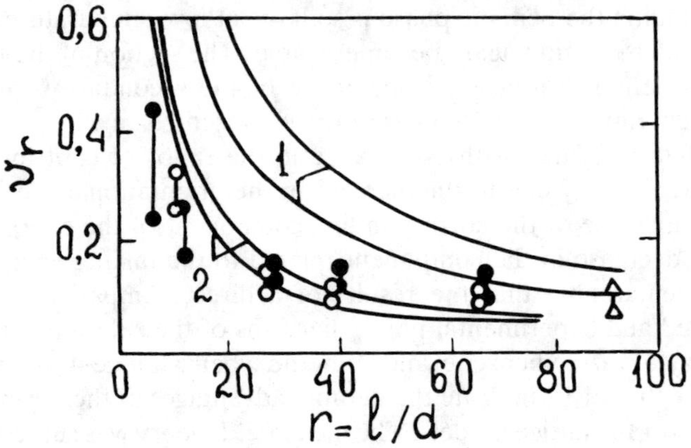

Figure 14.42. Volume contents of equilibrium phases in systems of rod-like particles (v_r—volume fraction of r-mer, r—the quotient obtained when molecular length l is divided by molecular width d);[31] 1—results of calculations by Flory's method;[146] 2—results of calculations by Onsager;[141] opened circles, triangles and black circles correspond to experimental data[147] for poly-γ-benzylglutamate in dioxane, dimethylformamide and dichloroethane respectively at 25°C.

The second approach is based on the idea that steric factors and the entropy character of the phase transition into the anisotropic phase play the dominant role. The system consisting of rod-shaped particles in an inert structureless solvent, in which the particles could be oriented in space in an arbitrary manner, was investigated in the fundamental paper by Onsager.[143] Starting from an expansion of the free energy of such a system with respect to the concentration for fixed orientations, Onsager showed that below a definite critical value of the density, the system remains isotropic and for concentrations exceeding a critical value the system separates and an anisotropic phase appears.

This method was used to analyze the properties of solutions containing asymmetrical particles of arbitrary shape[144] and in studying systems of particles with a discrete set of orientations.[145] Theoretical[143,146] and experimental[147] compositions of equilibrium phases in systems containing rod-shaped particles are presented in Fig. 14.42.[147] Although, at first glance, the computed and experimental data agree qualitatively, the expansions used in the theory in Ref. 143 are strictly valid only for dilute solutions. For this reason, the interpretation of the experimental data for a wide range of concentrations from the point of view of such theoretical assumptions encounters definite difficulties.

The extension of Onsager theory to include the solution containing rod-like and disc-like particles was carried out by A. Stroobants and H.N.W. Lekkerker.[148] This method showed that 3 liquid crystalline nematic phases of symmetrical type could exist in such a system: two of them are uniaxial phases and one is a biaxial phase, which agrees with the results of paper 149. It was noted that the biaxial phase can be observed in a rather narrow range of concentrations, because of which it is difficult to find this phase experimentally. In addition, attention is paid to the tendency of formation of aggregates of rod-like and disc-like particles.

The third direction is based on the use of lattice models. Such models have characteristic drawbacks: as a rule, only a discrete set of orientations of particles can be examined and, in addition, these models usually ascribe a higher degree of ordering than the true ordering to the systems being analyzed. On the other hand, using lattice models, it is possible to take into account not only the anisotropy of particle shape but also the van der Waal's attractive forces, which can play a definite role in the stabilization of an anisotropic phase. Lattice models are effectively used in calculating thermodynamic mixing functions.

The number of possible configurations of r-mers with limited flexibility in a solvent was calculated by Flory.[150] Making the assumption that the partition functions of the internal states of a r-mer and of the solvent do not depend on the configuration, Flory determined the compositions of the ordered and disordered phases, starting from the equality of the chemical potentials of the components in these phases. One of the results of this work is the conclusion that the probability for the appearance of the anisotropic phase is higher the less flexible the molecules are. In other words, the transition into the liquid crystalline phase is due primarily to the anisotropy of the shape and the limited flexibility of molecules. In addition, an attempt was made to examine the continuous distribution of orientations of r-mers around some direction chosen in space.[151]

A search for equilibrium distributions of particles over orientations and a calculation of the number of configurations of the cubic lattice containing rigid r-mers were performed in Refs. 152 and 153. The equilibrium distributions were found by comparing the values of the free energy of the system with a distribution of orientations corresponding to the isotropic case with the values of this same energy for the case of complete orientational ordering. We note that the free energy of mixing ΔF^M can be expressed in terms of the orientational order parameter $\langle P_2 \rangle$.

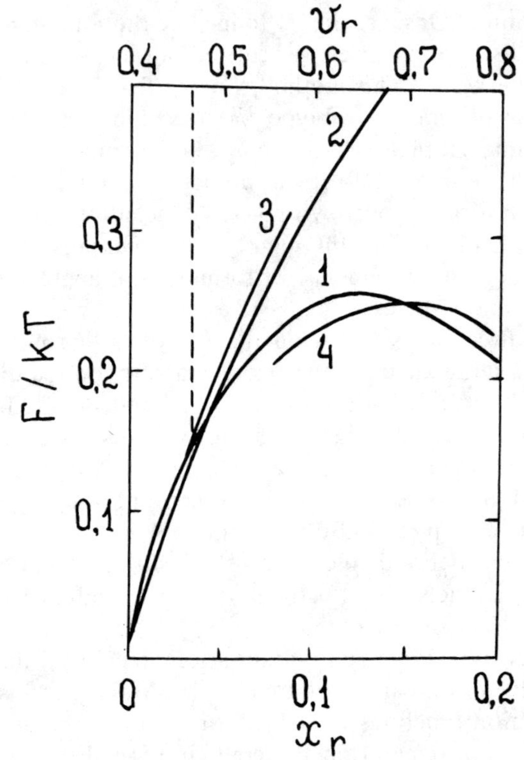

Figure 14.43. Reduced free energy of mixing in lattice system of rigid *r*-mers ($r = 10$, $z = 6$):[72,154] 1—isotropic distribution, 2 and 3—completely and partially ordered phases respectively, 4—at $N_1 = N_2 = N/2$.

Some computational results[154] are presented in Fig. 14.43.[72] It is evident that if all particles are oriented perpendicular to a fixed direction, then the state of the system is always metastable; a partially ordered anisotropic system has a lower free energy than a completely ordered system, and it is more stable than the latter. The concentration dependence of the reduced pressure is presented in Fig. 14.44.[72] As far as the compositions of equilibrium phases are concerned, they depend considerably on the coordination number of the lattice (i.e., on the number of possible orientations of rod-shaped particles in space). The results of calculations performed by different investigators[146,150,152,154,155] disagree considerably with one another and

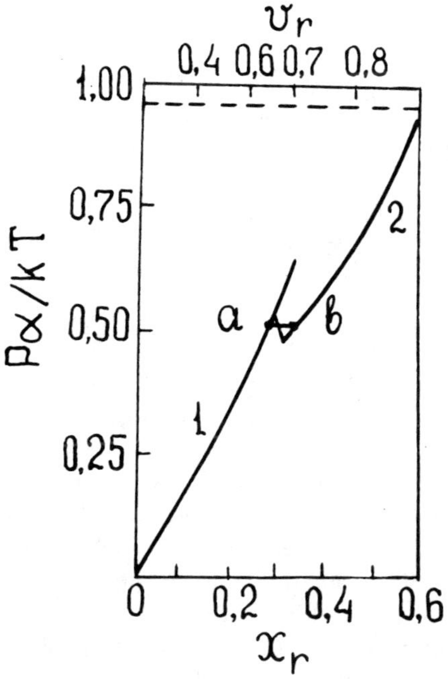

Figure 14.44. Reduced pressure $P\alpha/kT$ versus concentration in the system of rigid r-mers ($r = 5$, $z = 6$):[72] 1—curve of isotropic phase, 2—curve of anisotropic phase, a-b—extent of the heterogeneous region.

with Onsager's results.[143]* If orientational ordering is taken into account accurately within the framework of the lattice model, then the difference between the densities of the coexisting phases decreases and the agreement with the experimental data improves.[72] The common and most important result of these works, based on lattice models, is the conclusion that a first-order phase transition exists in systems of asymmetric rigid particles.

The method for calculating the number of configurations[152] is useful for studying solutions of two types of r-mers with different length.[153] The activity coefficients of the components of the system were calculated under the condition that the solvent does not form a stable anisotropic phase and, in addition, the concentration dependences of

*These results are compared in the review in Ref. 72.

the activity coefficient of the solvent in the anisotropic phase of the solution agreed qualitatively with the data.[124]

Within the framework of the quasichemical model for systems of rigid r-mers, it is also possible to include the anisotropic dispersion intermolecular interaction.[156] This was realized by introducing a parameter characterizing the mutual exchange energy in order to analyze the thermodynamic properties.[72] It was possible to analyze in detail the effect of the length of rod-shaped particles, the magnitude of the mutual exchange energy (characterizing the interaction of r-mers with one another and with monomers), as well as the density of the system on the ordering of the anisotropic phase and the activity coefficients of the components. A comparison of the computed and measured dependences shows that an appropriate choice of parameters, characterizing the energetics of molecular interactions, gives a qualitative agreement between the experimental and computed data. Further development of this model approach[157] permitted calculating the compositions of coexisting isotropic and anisotropic phases and obtaining satisfactory qualitative agreement between the computed and measured quantities for solutions of C_6H_6, n-C_6H_{14}, and CCl_4 in methoxybenzylidene-propylaniline. Inclusion of the dispersion energy of mutual attraction of particles greatly affects the position of the stability boundaries and the extent of the heterogeneous region.[158]

Flory's model was also used to analyze theoretically the behavior of mixtures consisting of rod-shaped and plate-like molecules.[159] It was possible to show that the increase in the concentration of plate-like molecules increases the temperature of the NLC–IL transition. The concentration dependence of the jump in the order parameter accompanying this transition and other features of similar model solutions was discussed. An interesting theoretical discussion of possible phase diagrams in such mixtures is presented in Ref. 160. According to the results of this analysis, the phase diagram of a mixture of elongated and flat molecules contains a region of an optically biaxial phase, arising as a result of the interaction of such particles. In addition, the transition from the "positive" NLC (which is understood to be the anisotropic system of elongated molecules, situated parallel to one another) or from the "negative" NLC (systems of flat molecules, whose short axes are oriented parallel to one another) to the biaxial phase is a second-order transition.

A simple lattice-gas model was proposed for investigating the miscibility of nematogens in binary systems.[161] It is assumed that the thermodynamic equilibrium in such systems is determined by the ratio of phase separation effects and nematic ordering. An equation of state is obtained for such a system in the self-consistent field approximation.

The interest for the application of the lattice type theories to study properties of liquid crystals solutions did not become less in recent years.[115,149,162] The general case of a multicomponent mixture was studied by Japanese scientists on the basis of a simple cubic lattice model.[162] It becomes possible to draw a complete phase diagram for the system of molecules having rigid cores and flexible end "tails," to find the value of transition entropy and the orientational order parameter for rigid cores and "tails." The generalized lattice model of nematic solutions was used to define binary and triple mixtures of oligomers of *p*-oxybenzoate.[115] The lattice model was also effective for a binary mixture of molecules of oblong and oblate forms.[149] This model made it possible to show the existence of 4 different types of phase diagrams with stability ranges of 2 uniaxial phases (with different signs of anisotropy) as well as biaxial phase and isotropic ones.

It is interesting to check the applicability of the thermodynamic relations, which have well proved themselves in the case of solutions of nonmesogenic substances, to mixtures of liquid crystals. Such relations were used, for example, to study phase equilibria in solutions containing 4-alkyl and 4-alkoxy-4′-cyanobiphenyl and compounds of a number of *n,n′*-dialkoxyazoxybenzenes.[163] The applicability of the Chatelier–Shroeder–van Laar equations to the determination of the eutectic temperatures in mixtures of nematic liquid crystals based on experimental data was analyzed for the example of nine mixtures (binary, ternary, and quaternary), whose composition included eight nematogenic substances.[164] It was found that the conditions of applicability of this equation are: a) absence of formation of solid solutions; b) ideality of the solution in the nematic phase; independence of heats of the transition solid crystal–nematic on the temperature; and, d) absence of a smectic phase. If even one of these conditions is not satisfied, then the computed and experimental data will disagree.[164] On the basis of an investigation of the phase diagrams of the mixtures of *p*-cyanophenyl-*n*-alkylbenzoates and *p*-alkoxycarbonyloxy-phenyl-*n*-butyloxybenzoates, it was shown that the Shroeder–van Laar equations can be used to calculate the compositions, melting temperatures of eutectic systems, and the heats of fusion.[165]

Extrapolation of the temperatures curves of transitions in mixtures containing Schiff bases and nematic *p*-propoxybenzene-*n*-arisidine and *p*-propoxybenzene-*n*-phenetidine (differing by the terminal polar groups) permitted determining the values of the temperatures of transitions into the mesophase for ten Schiff bases.[166] The agreement between the limiting values of the transition temperatures in mixtures with either nematic liquid crystals as well as the correspondence with other experimental data is viewed as proof of the validity of this method of extrapolation.

The analogy between the behavior of liquid helium, solutions ^{3}He–^{4}He, and some magnetic systems on the one hand and the possibility of the appearance of tricritical points (points of intersection of lines of phase transitions) in binary systems of liquid crystals[167,168] on the other, has been discussed specially in the literature. Thus, using the analogy to liquid helium, a phenomenological study of NLC with an impurity of "neutral" molecules was performed and the lines of phase transitions, characterizing the separation of phases and the NLC–IL transition, were found.[167] Basing on the analogy with the behavior of ^{3}He–^{4}He solutions, Alben[168] predicted that a tricritical point will exist under certain concentrations in mixtures of liquid crystals, in which the nematic–smectic first and second order transitions occur.

The nature of the phase transition in binary smectic solutions between A and C phases was investigated theoretically on the basis of the expansion of the free energy with respect to the C-smectic order parameter (tilt angle of the director relative to the smectic axis).[169] Here, the dependence of the expansion coefficients on the composition of the solution is included. Conditions, involving the derivatives with respect to the concentration, whose satisfaction determines the existence of the first-order phase transition, are introduced.

The statistical modeling approach developed by Japanese scientists[162] also proved to be effective for developing a molecular statistical theory of smectic A liquid crystalline mixtures.[170] Phase diagrams were drawn for some binary mixtures and temperature relationships of order parameters were calculated for phase transitions of smectics–nematics and nematics–isotropic liquid.

A theory of cholesteric mixtures has been constructed based on the molecular-statistical representations with the use of the general form of the pair interaction potential.[142] In writing down the total free energy of the system, the mean molecular field approximation is used, and terms up to fourth order, inclusively, are retained in the expansion of the order parameter and interaction in terms of the Legendre polynomials. It was established that at a temperature which is fixed for each solution, the pitch of the helix of the mixture is determined by the ratio of the squared-rational functions of the concentration. The results of the theoretical calculations agree with the experimental data for mixtures CBAC–CC, CBAC–CEEC, and EBBA–CEEC.

So far, only very few theoretical studies of the transport characteristics in liquid crystalline systems have been performed. A simple model of diffusion of an impurity in the smectic A phase is based on the representation of the impurity molecule as a hard sphere, whose dimensions are comparable to the dimensions of the hard core of the SLC molecules.[171] The freely bending side chains of the smectic molecules fill the gap between the smectic layers. Diffusion in the direction of dominant orientation of molecules, characterized by the coefficient $D_{\parallel}$, corresponds to the acoustic mode, while diffusion in the transverse direction ($D_{\perp}$) corresponds to diffusion in a liquid. The results of calculations performed with the help of this model for diffusion of tetramethylsilane in 4-*n*-butoxybenzylidene-4′-*n*-octylaniline at 53°C, led to the ratio $D_{\parallel}/D_{\perp} = 8.6$, which agrees well with the experimental data.

The solution of equation for rotational and translational diffusion of a particle dissolved in the smectic A phase was found by using the pseudopotential of McMillan taking into account orientational, translational, and mixed order parameters.[172] The influence of interaction between orientational and translational powers of freedom on diffusion was thoroughly discussed. The mixture of rigid spherocylinders modeling a nematic system was studied in paper 173. Integral Onsager equation for the orientational function of particle distribution was solved by iteration method. The resulting solution enabled not only the range of existence for the nematic phase to be found but the modulus of elasticity and viscous properties as well.

References for Part V

CHAPTER XIII

1. J. L. Fergason, *Sci. Amer.*, **211**, 76 (1964).
2. J. L. Fergason, N. N. Goldberg and R. J. Nadalin, *Molec. Cryst.*, **1**, 309 (1966).
3. H. Balssler and M. M. Labes, *Mol. Cryst. Liq. Cryst.*, **6**, 419 (1970).
4. H. Finkelmann and H. Stegemeyer, *Z. Naturforsch.*, **28A**, 1046 (1973).
5. N. Oron, K. Ko, L. J. Yu and M. M. Labes. *Molec. Cryst. and Order. Fluids*, N.Y.-London, **2**, 403 (1974).
6. Masubuchi Sadao, Akahane Tadashi, Nakao Kasuhara and Tako Toshiharu, *Mol. Cryst. Liq. Cryst.*, **38**, 623 (1977).
7. H. Kozawaguchi and M. Wada, *Mol. Cryst. Liq. Cryst.*, **44**, 97 (1978).
8. G. S. Chilaya, S. N. Aronishidze, K. D. Vinokur, M. D. Mhatvrishvili and M. I. Brodzeli, *Kristallografiya*, **22**, 1280 (1977).
9. G. Heppke and F. Oestreicher, *Z. Naturforsch.*, **A32**, 899 (1977).
10. B. Gajewska and B. Radomska, *Zecz. nauk Plódz.*, **271**, 13 (1977).
11. T. P. Kudryashova, V. V. Polihov and V. V. Tuturina, *V sb. Sintez vysokomolekulyarnykh produktov na osnove sapropelitov i kremniiorganicheskhikh soedininii* (Synthesis of high-molecular products on the base of sapropelytes and organosilicium compounds). Part 2. Irkutsk (1976), p. 16.
12. Shu-Hsua Chen and L. S. Chou, *Mol. Cryst. Liq. Cryst.*, **40**, 223 (1977).
13. P. Keating, *Mol. Cryst. Liq. Cryst.*, **8**, 315 (1969).
14. G. M. Zharkova and V. M. Khachaturyan, *Aerodinamicheskie issledovaniya*, **5**, 36 (1975).
15. K. A. Suresh, *Mol. Cryst. Liq. Cryst.*, **40**, 847 (1977).
16. H. Hanson, A. J. Dekker and F. Van der Woude, *Mol. Cryst. Liq. Cryst.*, **42**, 15 (1977).
17. Marsh Dana and Pochan Johu, *J. Chem. Phys.*, **58**, 2835 (1973).
18. P. Pollmann and H. Stegemeyer, *Ber. Bunsenges Phys. Chem.*, **78**, 843 (1974).
19. I. V. Vladimirskii, N. L. Kremarenko, N. M. Shkabara, *Kristallografiya*, **31**, 532 (1986).
20. Sato Sussumu and Wada Masanobu, *Jap. J. Appl. Phys.*, **11**, 1566 (1972).
21. W. Greubel, U. Wolff and H. Krüger, *Mol. Cryst. Liq. Cryst.*, **24**, 103 (1973).
22. V. N. Chirkov, D. F. Aliev and A. Kh. Zeinally, *Zh. Eksp. Teor. Fiz.*, **73**, 761 (1977).
23. I. I. Gorina and M. Yu. Rubtsova, *V sb. Zhidkie kristally* (*Liquid Crystals*), Ivanovo, 59 (1977).
24. C. Motoc, I. Cuculescu and M. Honciuc, *Mol. Cryst. Liq. Cryst.*, **45**, 215 (1978).
25. A. A. Abbas-Zade, N. S. Hanukaeva, *Kristallografiya*, **31**, 820 (1986).
26. W. H. Toliver, C. G. Roack, R. W. Roundy and P. F. Hoffmann, *Aerospace Med.*, **40**, 35 (1969).
27. J. Voss and E. Sackmann, *Z. Naturforsch*, **28**, 1496 (1973).
28. C. W. Graw, *Liquid Crystals and Plastic Crystals*, Ellis Harwood Ltd. Chichester, England, 327 (1974).
29. J. Adams and W. J. Haas, *Electrochem. Soc.*, **118**, 2026 (1971).

30. T. P. Kudryashova and L. N. Lisetskii, *Ukrainskii Fizicheskii Zhurnal*, **23**, 246 (1979).
31. I. I. Gorina and I. G. Chistyakov, *Uchenye zapiski Ivanovskogo universiteta* (Transactions of Ivanovo University), **128**, 90 (1974).
32. E. K. Galanov, R. I. Melnik and G. K. Kostyuk, *Kristallografiya*, **22**, 880 (1977).
33. Z. Yu. Gotra, Ya. I. Dutchak, Z. M. Mikityuk and M. I. Fedyshin, *Kristallografiya*, **26**, 630 (1981).
34. Z. Yu. Gotra, Z. M. Mikituk, M. G. Pavlyak, V. T. Fechan and V. V. Parkhomenko, *Ukrainskii Fizicheskii Zhurnal*, **30**, 1818 (1985).
35. E. H. Korte and P. Chinduang, *Liq. Cryst.*, One-and Two Dimens. Order. Proc. Conf. Garmisch-Partenkirchen. Berlin e.a. (1980), p. 212.
36. F. D. Saeva, *Mol. Cryst. Liq. Cryst.*, **23**, 171 (1973).
37. P. R. Gerber, *Phys. Lett.*, **A78**, 285 (1980).
38. K. Radley and A. Saupe, *Mol. Phys.*, **35**, 1405 (1978).
39. C. L. Khetrapal, *Bull. Magn. Reson.*, **8**, 158 (1986).
40. G. S. Bates, P. A. Beckmann, E. E. Burnell, G. L. Hoatson and P. Palffy-Muhoray, *Mol. Phys.*, **57**, 351 (1986).
41. G. S. Bates, E. E. Burnell, G. L. Hoatson, P. Palffy-Muhoray and A. Weaver, *Chem. Phys. Lett.*, **134**, 161 (1987).
42. S. Diele, L. Nasts and H. Sackmann, *Krist. und Techn.*, **12**, 1063 (1977).
43. W. Z. Urbach and J. C. Billard, *C. r. Acad. Sci.*, **274**, B1287 (1972).
44. C. S. Yannoni, *J. Amer. Chem. Soc.*, **91**, 4611 (1969).
45. K. Bergmann and H. Z. Stegemeyer, *Z. Naturforsch.*, **34A**, 251 (1979).
46. J. E. Adams and W. E. L. Haas, *Liq. Cryst.*, 3rd Int. Conf. Berlin, London e.a. Part. 2 (1972), p. 843.
47. J. E. Adams, W. E. L. Haas, *Mol. Cryst. Liq. Cryst.*, **16**, 31 (1972).
48. C. S. Bak and M. M. Labes, *J. Chem. Phys.*, **63**, 805 (1975).
49. C. S. Bak and M. M. Labes, *J. Chem. Phys.*, **62**, 3066 (1975).
50. H. Finkelmann and H. Stegemeyer, *Ber. Bunsenges Phys. Chem.*, **78**, 869 (1974).
51. J. E. Adams, G. Dir and W. E. L. Haas, In *Liq. Cryst. and Ordered Fluids*, Ed. by J. F. Johnson and R. S. Porter, Plenum, N.Y., Vol. 2 (1974), p. 421.
52. R. G. Priest and T. C. Lubensky, *Phys. Rev.*, **A6**, 452 (1972).
53. A. Wulf, *J. Chem. Phys.*, **60**, 3994 (1974).
54. H. Bauligaud, *J. Phys. (Fr.)*, **34**, 1011 (1973).
55. I. V. Ioffe and B. I. Lembrikov, *Pis'ma Zh. Eksp. Teor. Fiz.*, **3**, 404 (1977).
56. W. J. A. Goossens, *Mol. Cryst. Liq. Cryst.*, **12**, 237 (1970).
57. H. Stegemeyer and H. Finkelmann, *Chem. Phys. Lett.*, **23**, 227 (1973).
58. H. Stegemeyer and H. Finkelmann, *Naturwissenschaften*, **62**, 436 (1975).
59. A. G. Khachaturyan, *J. Phys. Chem. Solids*, **36**, 1055 (1975).
60. H. Hanson, A. J. Dekker and F. Van der Woude, *J. Chem. Phys.*, **62**, 1941 (1975).
61. V. V. Volovik, L. N. Lisetskii and V. G. Tishchenko, *V. Sb. nauchnykh trudov instituta monokristallov* (In: Collection of Scientific Works of Monocrystal Institute), 6 (1980), p. 97.
62. L. N. Lisetskii, B. L. Timan, V. G. Tishchenko and O. D. Kolotii. *Ukrainskii Fizicheskii Zhurnal*, **23**, 94 (1978).
63. Y. R. Lin-Liu, Shi Yu Ming, Woo Chia-Wei and H. T. Tan, *Phys. Rev.*, **14A**, 445 (1976).
64. M. Tsukamoto, T. Ohtsuka, K. Morinoto and V. Murikami, *Jap. J. Appl. Phys.*, **14**, 1307 (1975).

65. L. N. Lisetskii, A. V. Tolmachev and V. G. Tishchenko, *Pis'ma Zh. Eksp. Teor. Fiz.*, **27**, 205 (1978).
66. L. N. Lisetskii, G. S. Chilaya, K. D. Vinokur, V. G. Tishchenko and V. D. Panikarskaya, *Ukrainskii Fizicheskii Zhurnal*, **24**, 1582 (1979).
67. H. Kozawaguchi and M. Wade, *Jap. J. Appl. Phys.*, **14**, 651 (1975).
68. V. B. Mogalinskii and N. A. Glushkov, *Kristallografiya*, **25**, 1104 (1980).

CHAPTER XIV

69. H. Sackman, *Acht Arbeiten iiber Flüsige Kristalle. Wiss. Beitr.*, M. Luther Univ. Halle-Wittenberg, **4**, 85 (1976).
70. H. Arnold and H. Sackmann, *Z. Phys. Chem.* (DDR), Bd. **213**, 143 (1960).
71. E. C. Hsu and J. F. Johnson, *Mol. Cryst. Liq. Cryst.*, **20**, 177 (1973).
72. E. P. Sokolova and A. G. Morachevskii, *V sb. Khimiya i termodinamika rastvorov* (Chemistry and Thermodynamics of Solutions), Leningrad State University, **4**, 117 (1977).
73. J. W. Park, C. S. Bak and M. M. Labes, *J. Amer. Chem. Soc.*, **97**, 4398 (1975).
74. W. Woycicki and J. Stecki, *Bull. Acad. Polon. Sci. Ser. Chim.*, **22**, 241 (1974).
75. E. V. Generalona, Z. P. Dobronevskaya, T. N. Ignatovich, E. L. Kitaeva, B. I. Ostrovskii, A. S. Sonin, N. B. Titova and N. A. Tukmanova, *V sb. Zhidkie kristally* (Liquid Crystals), Ivanovo (1976), p. 126.
76. I. G. Chistyakov, L. A. Gusakova and G. G. Maidachenko, *Izvestiya vysshikh uchebnykh zavedenii. Khimiya i khim. technologiya* (News of High School. Chemistry and Chemical Technology), **14**, 1433 (1971).
77. G. M. Zharkova and V. M. Khachaturyan, *Izvestiya Sibirskogo Otdeleniya AN SSSR. Seria tekhnicheskikh nauk* (News of Siberia Division of Academy of Sciences of U.S.S.R. Section of Technical Sciences), **3**, 102 (1976).
78. B. E. North and D. M. Small, *J. Phys. Chem.*, **81**, 723 (1977).
79. L. T. Creagh, *Proc. IEEE*, **61**, 814 (1973).
80. R. Steinsträsser, *Z. Naturforsch.*, **B27**, 774 (1972).
81. V. A. Molochko, G. M . Kurdyumov and B. M. Bolotin, *V sb. Zhidkie Kristally* (Liquid Crystals), Ivanovo (1976), p. 111.
82. R. Steinsträsser, *L. Pohl. Z. Naturforsch.*, **268**, 87 (1971).
83. R. A. Beznheim, *J. Phys. Chem.*, **76**, 925 (1972).
84. A. V. Ivashchenko, V. V. Titov and E. I. Kovshev, *V sb. Zhidkie kristally* (Liquid Crystals), Ivanovo, 116 (1976).
85. D. L. Johnson, D. W. Allender, R. Dehoff, C. Maze, E. Oppenhim and P. Reynolds, *VI-th Int. Liq. Cryst. Conference*, Kent, 118 (1976).
86. G. Sigaud, F. Hardouin and M. F. Achard, *Solid State Communs.*, **23**, 35 (1977).
87. M. F. Achard, G. Sigaud and F. Hardouin, *Liq. Cryst. One- and Two-Dimens. Order. Proc. Conf.* Garmisch.-Partenkirchen, Berlin e.a. 149 (1980).
88. G. Grinstein and H. Pelcovits, *Phys. Rev.*, **A26**, 2196 (1982).
89. J. Szabon and S. Diele, *Cryst. Res. and Technol.*, **17**, 1315 (1982).
90. W. H. Müller and H. Stegemeyer, *Ber. Bunsenges Phys. Chem.*, **77**, 20 (1973).
91. J. P. Schoroeder and D. C. Schoroeder, *J. Org. Chem.*, **33**, 591 (1968).
92. B. K. Vainshtein, I. G. Chistyakov, G. G. Maidachenko, L. A. Gusakova, V. D. Belilovskii, V. M. Chaikovskii, L. K. Vistin' and S. P. Chumakova, *Doklady AN SSSR*, **220**, 1349 (1975).

93. B. K. Vainshtein, I. G. Chistyakov, B. N. Makarova, G. G. Maidachenko, L. A. Gusakova, V. M. Chaikovskii, L. A. Mineev, V. D. Belilovskii, L. K. Vistin' and S. P. Chumakova, *Kristallografiya*, **22**, 592 (1977).
94. I. G. Chistyakov, B. N. Makarova, L. K. Vistin' and S. P. Chumakova, *Doklady AN SSSR*, **229**, 1350 (1976).
95. G. Heppke and E. J. Richter, *Z. Naturforsch*, **A33**, 185 (1978).
96. B. Engelen, G. Heppke, R. Hopt and F. Schneider, *Ann. de Phys.*, **3**, 403 (1978).
97. L. Kotler and A. Kotler, *Thermo-Mikro-Metoden. Zur Kennzeichnung Organischer Staff und Stoffgemigche. Verlag Chem. Weinhein*, 1954.
98. A. Kolbe and G. Pelzl, *Adv. in Molec. Relaxation and Interaction Processes*, **24**, 251 (1982).
99. S. Chan, *Mol. Cryst. Liq. Cryst.*, **42**, 1 (1977).
100. J. W. Park and M. M. Labes, *J. Appl. Phys.*, **48**, 22 (1977).
101. G. Sigaud, F. Hardaun, M. F. Achard and H. Gasparoux, *Conference on Physics and Applications of Smectic and Lyotropic Liquid Crystals*, Madonna di Campiglio (1978), p. 205.
102. G. W. Gray and D. G. McDonnell, *Mol. Cryst. Liq. Cryst.*, **37**, 189 (1976).
103. G. W. Gray and D. G. McDonnell, *Mol. Cryst. Liq. Cryst.*, **48**, 37 (1978).
104. J. Billard and B. K. Sadashiva, *Pramana*, **13**, 309 (1979).
105. J. Billard, J. C. Dubois, N. H. Tinh and A. Zann, *Nouv. J. Chem.*, **2**, 535 (1978).
106. R. E. Goozner and M. M. Labes, *Mol. Cryst. Liq. Cryst. Letters*, **56**, 75 (1979).
107. C. Vauchier, A. Zann, P. Barny, J. C. Dubois and J. Billard, *Mol. Cryst. Liq. Cryst.*, **66**, 103 (1981).
108. J. D. Bunning, J. E. Lydon, C. Eaborn, P. M. Jackson, J. W. Goodby and G. W. Gray, *J. Chem. Soc. Farad. Trans.*, **78**, 713 (1982).
109. G. M. Schneider, *Thermochim. acta*, **88**, 1 (1985). [Calorimetry and Therm. Anal. Collec. Selec. Pap. 20 Anniversary Conf., Osaka, 13–16 Nov., 1984, 17–34].
110. Gerald Van Hecke, *J. Phys. Chem.*, **89**, 2058 (1985).
111. F. Hardouin, *Physica*, **A140**, 359 (1986).
112. K. Czupryński, R. Dabrowski, J. Baran, A. Zywocinski and J. Przedmojski, *J. Phys. (Fr.)*, **47**, 1577 (1986).
113. V. D. Panikarskaya, L. N. Lisetski, and V. G. Tishchenko, *Zh. Fiz. Khim.*, **60**, 2978 (1986).
114. J. Szabon, W. Weissflog, G. Pelzl, S. Diele and D. Demus, *Cryst. Res. and Technol.*, **21**, 1097 (1986).
115. M. Ballauff and P. J. Flory, *Ber. Bunsenges. Phys. Chem.*, **88**, 530 (1984).
116. P. Palffy-Muhoray et al., *Liquid Cryst. and Ordered Liquids*, **4**, 615 (1984).
117. P. Palffy-Muhoray, A. J. Berlinsky, J. R. De Bruyn and D. A. Dunmur, *Phys. Lett.*, **A104**, 159 (1984).
118. William R. Krigbaum, *Faraday Discuss. Chem. Soc.*, **79**, 133 (1985).
119. S. N. Mochalin, *Uchenye zapiski Ivanovskogo universiteta*, **123**, 86 (1974).
120. I. P. Zhuk and V. A. Karolik, *Izvestiya AN SSSR, Seriya fiziko-energeticheskikh nauk.*, **2**, 76 (1978).
121. K. W. Sadowska, A. Zywocinska and J. Stecki, *J. Phys. (Fr.)*, **43**, 1673 (1982).
122. W. H. Müller and H. Stegemeyer, *Chem. Phys. Lett.*, **27**, 130 (1974).
123. D. E. Martire, *J. Phys. Chem.*, **72**, 3489 (1968).
124. H. T. Peterson and D. E. Martire, *Mol. Cryst. Liq. Cryst.*, **25**, 89 (1974).
125. J. M. Sehnur and D. E. Martire, *Mol. Cryst. Liq. Cryst.*, **26**, 213 (1974).
126. T. G. Churyusova, E. P. Sokolova, M. M. Balakina and V. P. Belousov, *Zh. Fiz. Khim.*, **52**, 1757 (1978).

127. K. J. Lushington, G. B. Kasting and C. W. Garland, *Phys. Rev.*, **22B**, 2569 (1980).
128. C. Kartna, J. R. Otia and A. R. Padmini, *Indian J. Pure and Appl. Phys.*, **12**, 648 (1974).
129. I. Muscutariu, S. Bhattacharya and J. S. Ketterson, *Phys. Rev. Lett.*, **35**, 1584 (1975).
130. M. Groucher, B. Kronberg, D. Gilson and D. Patterson, *J. Chem. Soc. Chem. Communs.*, **16**, 686 (1965).
131. P. Ekwall, *Adv. Liq. Cryst.*, New York. Vol. 1. (1975), p. 1.
132. S. K. Kor and A. K. Srivastava, *J. Phys. Soc. Jap.*, **54**, 613 (1985).
133. R. Mahmood, M. Lewis, R. Biggers, V. Surendranath, D. Johnson and M. E. Neubert, *Phys. Rev.*, **33A**, 519 (1986).
134. M. Ballauff, D. Wu, P. J. Flory and E. M. Barall, *Ber. Bunsenges. Phys. Chem.*, **88**, 524 (1984).
135. V. W. Meier and A. Saupe, *Z. Naturforsch.*, **14A**, 882 (1959).
136. A. Saupe, *Ber. Buns. Phys. Chem.*, **78**, 848 (1974).
137. R. L. Humphries, P. G. James and G. R. Luckhurst, *Faraday Symp. on Liq. Crystals, London*, 12 (1971).
138. R. L. Humphries, P. G. James and G. R. Luckhurst, *Symp. Faraday Soc.*, **5**, 107 (1971).
139. R. L. Humphries and G. R. Luckhurst, *Chem. Phys. Lett.*, **23**, 567 (1973).
140. R. L. Humphries and G. R. Luckhurst, *Proc. Roy. Soc. London*, **A352**, 41 (1976).
141. G. W. Gray et al. In: *Liquid Crystals and Ordered Fluids*, (Eds. J. F. Johnson and R. S. Porter), Plenum Press, New York, Vol. 2 (1974), p. 617.
142. Y. R. Lin-Liu, Shin Yu Ming and Woo Chia-Wei, *Phys. Rev.*, **15A**, 2550 (1977).
143. L. Onsager, *Ann. N.Y. Acad. Sci.*, **51**, 627 (1949).
144. A. Isihara, *J. Chem. Phys.*, **19**, 1142 (1951).
145. S. L. Brenner, D. A. McQuarrie and D. Olivares, *J. Chem. Phys.*, **59**, 2596 (1973).
146. P. J. Flory, *J. Chem. Phys.*, **10**, 51 (1942).
147. J. P. Straley, *Mol. Cryst. Liq. Cryst.*, **22**, 333 (1973).
148. A. Stroobants and H. N. W. Lekkerker, *J. Phys. Chem.*, **88**, 3669 (1984).
149. Robert G. Caflisch, Zhong-Ying Chen, Nihat A. Berker and John M. Deutch, *Phys. Rev.*, **30A**, 2562 (1984).
150. P. J. Flory, *Proc. Roy. Soc. London*, **A234**, 60 (1956).
151. P. J. Flory, *Proc. Roy. Soc. London*, **A234**, 73 (1956).
152. E. A. Di Marzio, *J. Chem. Phys.*, **35**, 658 (1961).
153. H. T. Peterson, D. E. Martire and M. A. Cotter, *J. Chem. Phys.*, **61**, 3547 (1974).
154. A. Wulf and A. de Rocco, *J. Chem. Phys.*, **55**, 12 (1971).
155. R. Zwanzig, *J. Chem. Phys.*, **39**, 1714 (1963).
156. M. A. Cotter and D. E. Martire, *Mol. Cryst. Liq. Cryst.*, **7**, 295 (1969).
157. E. P. Sokolova, *V sb. Khimiya i termodinamika rastvorov* (Chemistry and Thermodynamics of Solutions), Leningrad State University, 4, (1977), p. 140.
158. F. Dowell, *J. Chem. Phys.*, **69**, 4012 (1978).
159. R. Alben, *J. Chem. Phys.*, **59**, 4299 (1973).
160. R. Alben, *Liq. Cryst. and Order. Fluids*, New York, London, Vol. 2, 81 (1974).
161. J. Sivardière, *J. Phys. (Fr.)*, **41**, 1081 (1980).
162. Masahiro Nakagawa and Tadashi Akahane, *J. Phys. Soc. Jap.*, **54**, 4133 (1985).
163. R. Cox and J. F. Johnson, *IBM J. Res. and Develop.*, **22**, 51 (1978).
164. G. W. Smith, *Mol. Cryst. Liq. Cryst. Lett.*, **49**, 27 (1978).

165. A. V. Ivashchenko, V. V. Titov and E. I. Kovshev, *Mol. Cryst. Liq. Cryst.*, **33**, 195 (1976).
166. J. M. Lohar and G. H. Patel, *Curr. Sci. (India)*, **44**, 887 (1975).
167. M. Papoular and J. P. Laheurte, *Solid Communs.*, **12**, 71 (1973).
168. R. Alben, *Solid State Communs.*, **13**, 1783 (1973).
169. F. C. Frank, *J. Phys. Lett. (Fr.)*, **38**, 207 (1977).
170. Makagawa Masahiro and Akahane Tadashi, *J. Phys. Soc. Jap.*, **54**, 69 (1985).
171. Chu Kwo-Sun, N. K. Ailawadi and D. S. Moroi, *Mol. Cryst. Liq. Cryst.*, **38**, 403 (1977).
172. Giorgio Moro and Pier Luigi Nordio, *J. Phys. Chem.*, **89**, 997 (1985).
173. Lee Sin-Doo and Robert B. Meyer, *J. Chem. Phys.*, **84**, 3443 (1986).

Conclusions

According to de Vries, computational-theoretical investigations of the structure and dynamic behavior of liquid crystals (and therefore thermophysical and physicochemical properties as well) can be separated into three levels: at the lowest level, the molecules are viewed as rigid rods; at the second level, the molecules are represented as flat plates, rods with tilted dipoles, etc.; the real structure of molecules is examined at the third level. Each level has its own range of interesting problems related to the physics of the liquid-crystalline state, and only the problems belonging to the first level can be considered adequately studied. Most modern theoretical ideas concerning the physics of liquid crystals correspond to the second level (even for the structurally simplest and most studied nematic liquid crystals), or occupy an intermediate position between the second and third levels. This indicates that the theory of and methods for computing the properties of liquid crystals have not yet been completed and will continue to be developed and improved.

The thermophysical properties of SLC and CLC have interesting qualitative features that distinguish them from isotropic liquids as well as from nematic liquid crystals. Although at present the range of synthesized and promising (for practical applications) SLC and CLC has greatly increased, there are still few experimental data on their thermophysical characteristics. Even available information, which has greatly expanded in recent years, requires careful analysis and critical selection. For this reason, the experimental study of a number of properties of both pure substances and liquid-crystalline mixtures is an important problem in the thermophysics and physical chemistry of SLC and CLC (for example, there still are practically no experimental data on the surface tension of SLC and CLC).

The theoretical methods for studying the structure and thermophysical properties of SLC and CLC are unfortunately not well developed. This is especially true for the transport coefficients. For this reason, the expansion of computational–theoretical investigations of SLC and CLC is of great scientific and practical value.

Chiral nematics as well as liquid crystals consisting of disc-like molecules and their solutions are interesting materials of great practical importance whose study is only just beginning.

Computer modeling methods, whose role will apparently increase in the future, occupy a prominent place among the various directions taken by the theory of the liquid-crystalline state. Even this powerful approach, however, does not yet permit taking into account all the characteristics of the molecular structure of liquid crystals and the diverse types of intermolecular interactions. The search for appropriate model potentials to describe correctly the interaction of molecules and their groups on the basis of information obtained from quantum mechanics and special physical experiments therefore remains an important problem.

Of the problems of immediate concern related with the thermophysics of the liquid-crystalline state, we can single out the following: development of methods for calculating the properties of liquid crystals, especially SLC, CLC, and systems of disc-like molecules, as well as experimental investigation of their thermophysical properties; study of the properties of solutions of liquid crystals and liquid-crystal films; and, formulation of paths for investigation of the structure and thermophysical properties of liquid crystals in electric and magnetic fields. Computer modeling will very likely play a large role in the solution of the problems of calculating thermophysical characteristics, predicting the structure of liquid crystals, and searching for liquid-crystalline systems with the desired properties.

List of Abbreviated Chemical Names of Liquid-Crystalline Substances*

AAOP—2-[4-*n*-alkylphenyl]-5-[4-*n*-alkoxyphenyl]-pyrimidines
ABAA—*n*-[4-*n*-alkoxybenzylidene]-4-*n*-alkylanilines
ADBC—*n*-amyl-4-[4-*n*-dodecyloxybenzylidene-amino]-cinnamate
AFYA—4-alkoxy-*N*-(fluoren-2-ylmethylene)-anilines
AMC-11—azoxy-4,4′-di-undecyl-α-methylcinnamate
ANMA—4-alkoxy-*N*-(2-naphthylmethylene)-anilines
AOFYMA—4-alkoxy-*N*-(9-oxofluoren)-2-ylmethylene-anilines
7AB—4,4′-di(*n*-heptyl)-azoxybenzene

BAB—4,4′-di(*n*-butyl)-azoxybenzene
nBABA—*p*-*n*-alkoxybenzylidene-*p*-aminobenzoic acids (n = 6–9)
BBAA—butoxybenzylidene-*p*,*n*-aniline acetophenone
BBBA—*N*-(*p*-*n*-butoxybenzylidene)-*p*-*n*-butylaniline
BBOA—butyloxybenzylidene-octylaniline
BBPAA—butoxybenzylidene phenylazoaniline
BHA** (or BHnA**)—benzene-hexa-*n*-alkanoates
BMAOB—4-butyl-4′-methoxyazoxybenzene
BMAB—*n*-butyl-4-methoxybenzene
BN—alkoxybenzylidene-amyno-benzonitrile
BPBAC—*n*-butyl-4-(4′-phenylbenzylidene-amino)-cinnamate
BPC—*n*-butyric acid ester of (hydroxypropyl)cellulose

CA—cholesteryl acetate
CB—cholesteryl benzoate
CBAC—4-cyanobenzal-amino-cinnamate
CBn (or nCB)—4-alkyl-4′-cyanobiphenyls
CBNA—*n*-*p*-cyanobenzylidene-*p*-nonylaniline
CBOOA—4-cyanobenzylidene-4′-*n*-octyloxyaniline
CBr—cholesteryl brassidate
CC—cholesteryl chloride
CCH7—1-heptyl-4-(4′-cyanocyclohexyl)cyclohexane
CD—cholesteryl decanoate

*Only the most common abbreviations and chemical names of liquid crystals are given.

**The disc-like molecules.

CE—cholesteryl erucate
CEEC—cholesteryl-2-(2-ethoxy-ethoxy)-ethyl-carbonate
CE1—cholesteryl elaidate
CE8—4-(2′-methylbutyl)phenyl-4′-*n*-octylbiphenyl-4-carboxylate
CF—cholesteryl formiate
CH—cholesteryl heptylate
CHOC (or COC)—cholesteryl oleylcarbonate
CL—cholesteryl lynoate
CLa—cholesteryl laurate
CM—cholesteryl myristate
CN—cholesteryl nonanoate
CO—cholesteryl oleate
CP—cholesteryl propionate
CPB—*n,n′*-cyanopentylbiphenyl
CV—cholesteryl valerate
CnBAB—4,4′-*bis*-[*n*-alkylamino]-biphenyls
CnHATX**—members of the hexa-alkanoyloxy truxene homologous series
5CB—4-*n*-pentyl-4′-cyanobiphenyl
6CB—4-*n*-hexyl-4′-cyanobiphenyl
7CB—4-*n*-heptyl-4′-cyanobiphenyl
8CB—4-*n*-octyl-4′-cyanobiphenyl
9CB—4-*n*-nonyl-4′-cyanobiphenyl

DBAE—*N*-*p*-dymethyl aminobenzene-1-aminonaphthalene in asobenzene ethyl
DBC—*n*-*h*-decyloxybenzoyl-*n*-cresole
D-BPBAC—partly deuterated *n*-butyl-4-(4′-phenylbenzylidene amino)-cinnamate
DBPBB—phenyl-benzoyloxybenzoate
DB7NO_2—heptylphenyl nitrobenzoloxybenzoate
DEAB—diethyl azoxybenzene
DIBAB—*p,p′*-dibutylazoxybenzene
DIISB**—di-isobutylsilanediole
D(or L)MBMBA—D-(or L-)-*n*-methoxybenzylidene-*n′*-(2-methylbutyl)-aniline
DMOAP—dimethyl octadecyl aminopropyl trimethoxysilyl chloride
DOBAMBC—*p*-(*n*-decyloxybenzylidene)-*p*-amino-(2-methylbutyl)-cinnamate
DOBA-1-MPC—*p*-(*n*-decyloxybenzylidene)-*p′*-amino-(1-methylpropyl)-cinnamate
DOBHOP—*p*-*n*-decyloxybenzoic acid-*p*-*n*-hexyloxyphenyl ester

DPAB—4,4′-dipentylazoxybenzene
DPPC—dipalmitoyl phosphatidyleholine

EAAC—ethyl-*p*-anisalaminocinnamate
EABAC—ethyl 4-(4′-acetoxybenzylidene)aminocinnamate
EBBA—*N*-(*p*-ethoxybenzylidene)-*p*-*n*-butylaniline
EBCA—*N*-(*p*-ethoxybenzylidene)-*p*-*n*-cyanoaniline
EBMBA—*N*-(*n*-ethoxybenzylidene)-*n*-2-methylbutylaniline
EEBAC—ethyl-*p*-[(*p*-ethoxybenzylidene)amino]cinnamate
EMBAC—ethyl-*p*-[(*p*-methoxybenzylidene)amino]cinnamate
EO6T—ethoxy-hexyle-tolane
EO7T—ethoxy-heptyle-tolane
EO8T—ethoxy-octyle-tolane
EPAB—4-ethyl-4′-pentylazoxybenzene

HAB—4-4′-heptyloxyazoxybenzene
HB—hexamethylbenzene
HBAB—*p*-hexyloxybenzylidene-*p*-aminobenzonitrile
HBT—(*p,n*-hexyloxybenzylidene)-*p*-toluidine
HpBBA—*N*-(*p*-*n*-heptyloxybenzylidene)-*p*-*n*-butylaniline
HBCA—4-heptyloxybenzylidene-4′-cyanoaniline
HCCF—4-fluorophenyl-*trans*-4′-heptylcyclohexan carboxylate
HEPTAB—4,4′-di-heptylazoxybenzene
HEPTOAB—4,4′-di-*n*-heptyloxyazoxybenzene
HHTP—hexa(hexyloxy)triphenylene
HOAB—4,4′-di-*n*-heptyloxy-azoxybenzene
HOBA—*n*-heptyloxybenzoic acid
HOBACPC—*p*-hexyloxybenzylidene-*p*′-amino-2-chloropropyl-cinnamate
HOBPA (or HpBPA)—*N*-(*p*-*n*-heptyloxybenzylidene)-*p*-*n*-pentylaniline
HOPDOB—*p*-*n*-hexyloxyphenyl *p*′-*n*-decyloxybenzoate
HOPHOB—*p*-*n*-hexyloxyphenyl *p*′-*n*-hexyloxybenzoate
HPC—hydroxypropylcellulose
HPDB—4-*n*-hexyloxy-phenyl-4-*n*-decyloxybenzoate
HPHB—4-*n*-heptyloxy-phenyl 4-*n*-hexyloxybenzoate
HPNHB—4-hexyloxyphenylen-4′-(3′-nitro)-hexyloxybenzoate
HTAB—hexadecyltrimetylbromide of ammonium

IBPBAC—isobutyl 4-(4′-phenyl-benzylidene-*n*-amino)-cinnamate
IBPC—isobutyric acid ester of (hydroxypropil) cellulose

K15—4-cyano-4′-*n*-pentyldiphenyl

MBA—*N*-(*p*-methylbenzylidene)-*p*-butylaniline
MBBA—*N*-(*p*-methoxybenzylidene)-*p*-*n*-butylaniline
MBCA—*N*-(*p*-methoxybenzylidene)-*p*-*n*-cyanoaniline
MBMBA—*N*-(*p*-methoxybenzylidene)-*p*-(2-methyl-butyl)aniline
MBMBAC—β(methyl-butyl)-*p*[(p-methoxybenzylidene)amino]-cinnamate
MOAT—*p*-methoxy-amyle-tolane
MPMB—*p*-methoxy-phenyl-*p*-methocryloyloxy-benzoate
MPMEB—*p*-methoxy-phenyl-*p*-methocryloyloxy-ethoxy-benzoate
MPT—*p*-methoxy-4-pentyl-4′-tolane

NBBA—*p*-nitrobenzylidene *p*′-*n*-butylaniline
NOBA—*p*-*n*-nonyloxybenzoic acid
NPOB—4-nitrophenyl 4′-octyloxybenzoate
NPOOB—4-nitrophenyloxy-4-*n*-octyloxybenzoate

mOAB—di-alkoxyazoxybenzenes ($m = 1–7$)
7OBACB—4-*n*-heptyloxybenzylideneamino-4′-cyanobiphenyle
OBBC—*N*-[4-(4-octyloxy-benzoyloxy)benzylidene]-4-cyanoaniline
46OBC—*n*-butyl-4′-*n*-hexyloxybiphenyl-4-carboxylate
nOBCAB—4-alkoxybenzoyloxy-4′-cyanoazobenzenes
65OBC—*n*-hexyl-4′-*n*-pentyloxybiphenyl-4-carboxylate
8OBOBCA—*N*-[(4-*n*-octyloxybenzoiloxy)benzylidene]4′-cyanoaniline
6OBPD—*bis*-(4-*p*-hexyloxybenzal)-1,4-phenylenediamine
9OBPD—*bis*-(4-*p*-nonyloxybenzal)-1,4-phenylenediamine
10OBPD—*bis*-(4-*p*-decyloxybenzal)-1,4-phenylenediamine
OBT—(*p,n*-octyloxy-benzylidene)-*p*′-toluidine
5OCB—4-*n*-pentyloxy-4′-cyanobiphenyl
6OCB—4-*n*-hexyloxy-4′-cyanobiphenyl
7OCB(or 7CBP)—4-*n*-heptyloxy-4′-cyanobiphenyl
8OCB—4-*n*-octyloxy-4′-cyanobiphenyl
9OCB—4-*n*-nonyloxy-4′-*n*-cyanobiphenyl
12OCB—4-*n*-dodecyloxy-4′-cyanobiphenyl
OOAB—4,4′-di-*n*-octadecyloxy-azoxybenzene
OOBA—*p*-*n*-octyloxybenzoic acid
OOBBA—*p*-(*n*-octoxy) benzylidene-*p*-(*n*-butyl)aniline
OOHCS—4-octyloxy-4′-heptyl-α-cyanostilbene
8OPCDC—di-(4′-*n*-octyloxy-phenyl)-*trans*-cyclohexane-1,4′-dicarboxylate
OPCTS—octaphenylcyclotetrasiloxane
OxCB—*p,n*-octoxy-*p*′-cyanobiphenyl

PAA—*p*-azoxyanisole
PAH—4,4′-di-heptyloxy-azoxybenzene
PAP—*p*-azoxyphenetole
PBEA—*N*-(*p*-*n*-pentyloxybenzylidene)-*p*-*n*-ethylaniline
PBHA—*N*-(4-*n*-pentyloxybenzylidene)-4-*n*-hexylaniline
PBLG—poly-(γ-benzyl L-glutamate)
PBPAA—*N*-(4-*n*-pentyloxybenzylidene)-4-*n*-phenylazoaniline
PCB—4-*n*-pentyl-4′-cyanobiphenyl
PCH7—1-heptyl-4-(4′-cyanophenyl)cyclohexane
4PCH—4-*trans*-butylcyclohexyl-4-benzonitrile
PDOBAC–*n*-pentyl-4(4′-*n*-decyloxybenzylidene-amino)cinnamate
PEBAB—*N*-(*n*-ethoxybenzylidene)-*n*-aminobenzonitrile
PEDG—poly-(γ-ethyl-D-glutamate)
PHT—propoxy-4-heptyl-4′-tolane
PMBAB—*N*-(*n*-methoxybenzylidene)-*n*-aminobenzonitrile
POBOA—*N*-(*p*,*n*-pentyloxybenzylidene)-*p*-*n*-octylaniline
POBPAA—pentyloxybenzylidene-phenylazoaniline
POAOB—4,4′-di-*n*-alkoxyazoxybenzene
PPBA—poly-(*p*-biphenyl acrylate)
PPC—propionic acid ester of (hydroxypropyl) cellulose
P/PC—phthalic acid ester of (hydroxypropyl) cellulose
PP5CC—4-propionylphenyl-*trans*(4-*n*-pentyl)cyclohexane carboxylate

RHO**—hexa-*n*-octanoate of rufigallol

7S5—heptyloxypentylphenylthiobenzoate

TBAA—terephthalylidene-*bis*-(4-*n*-alkyl)-aniline
TBACA—*p*-terephthal-*bis*-amino-methyl-butyl-cinnamate
TBBA—terephthal-*bis*-(4-*n*-butyl)-aniline
TBDA—terephthal-bis-4-*n*-decylalanine
TBPA—terephthalylidene-*bis*-(4-*n*-pentyl)-aniline
TBTHOBA—terephthal-*bis*-4-(3′,4′,5′-triheptyloxybenzoyloxy)-aniline
THA**—triphenylene 2,3,6,7,10,11 hexa-*n*-alkanoates
THE**—triphenylene 2,3,6,7,10,11 hexa-*n*-esters

7.05—4-*n*-pentylphenyl-4′-*n*-heptyloxybenzoate
40.8—4-*n*-butoxybenzylidene-4′-*n*-octylaniline
70.7—4-*n*-heptyloxybenzylidene-4′-*n*-heptylaniline

SUBJECT INDEX

Absorption of light 252
Absorption of sound 95, 207, 208, 217–220, 225, 379, 381, 382
Activity coefficients 375, 376, 380, 381, 391, 392
Adiabatic compressibility 93–99, 202, 203, 207
Amphiphilic systems 3
Analogy to helium 164, 187, 188, 384, 394
Anisotropy of properties 6, 58, 63, 70, 81, 83, 84, 107, 116, 118, 133, 137
Apparent viscosity 110, 113, 114
Application of liquid crystals 153, 239, 242, 245, 250, 265, 266, 322, 329
Arrhenius' law 228

Basis specimen 172
Bathnager-Gross-Krook model 62
Bethe's approximation 36, 41
Binary distribution function 145, 146
"Blocked" texture of cholesterics 189, 190
Blue phases 129, 157, 158, 196, 202, 229, 262, 337, 338
Bogolyubov-Tyablikov approximation 36
Born's theory 25–27
Bose condensation 187
Bottcher's theory 154
Bragg parameters 267, 270, 271
Bragg reflection 152, 251
Bragg scattering 143, 161, 267, 335
Brillouin zones 381

Carbonized mesophases 265, 281
Chandrasekhar-Madhusudana theory 32, 33, 49, 50
Chatelier-Shroeder-van Laar equations 393
Chemical potential 368, 389
Chiral molecules 8, 242, 248
Chiral nematics 143, 239–251, 329, 403
Chiral smectics 171, 190, 248–250
Cholesteric (chiral) additives 242, 337
Cholesteric liquid crystals 7–10, 13, 129, 131, 152–158, 169–171, 181–187, 189, 190, 191, 195–197, 199–201, 204–227, 241, 403
Circular dichroism 337
Classical theory of nucleation 258
Classification of liquid crystals from disc-like molecules 293–316
Classification of smectics 136–141, 193
Clausius-Clapeyron equation 87
Clusters (domains) 255, 293, 370
"Columns" of disc-like molecules 270, 272, 279–281, 295–297, 312, 314, 320, 324
Compressibility 75, 93–99, 202, 203, 207, 215, 216, 323
Computer experiments 41–61, 95, 97, 98, 104–106, 171–187, 258, 291–317
Configurational statistical sum 169
Continuum model 25
Correlation times 266, 307, 308
Cotton transmission 336
Critical index 229
Croxton-Chandrasekhar theory of surface tension 211
Cybotactic clusters 255

Debye theory 26
Defects in liquid crystals 140, 278
Density and volume 70–82, 85–92, 196–199, 203–209, 317–324, 371, 372, 374, 375, 378, 379, 385, 386, 391
Diamagnetic susceptibility 81, 83, 84
Dielectric anisotropy 70
Diffusion coefficient 55, 395
Differential scanning calorimetry 260, 317–319, 345, 347, 353, 384
Differential thermal analysis 345
Dilatometric method 375
Diogo and Martins theory 112
Dipole moment 5, 26, 132
Director 7, 8, 10, 13, 21, 26, 112, 116, 136, 142, 171, 182, 183, 187, 189, 314, 336, 337
Director field and its defects 140
Disclinations in smectics 140
Dislocations in smectics 140
Dispersion interaction forces 27
Dispirations in smectics 140
Distribution function of molecular mass centers 9, 147, 298, 301
Double layer structures of smectics 361, 362
Double refraction measurements 157

Elastic energy 26
Elastic moduli (coefficients) 190, 242, 273, 286, 341, 355, 357, 395
Electrical conductivity 217
Electric birefringence 37
Electron paramagnetic resonans 20
Electro-optical effect 251, 334
Ellipsoidal binary distribution function 147
Energy of activation 110, 221, 224, 225, 228
Energy of interface 258
Enthalpy, change of enthalpy 40, 85, 87, 157, 158, 200, 201, 212, 213, 260, 261, 317, 318, 320, 321, 349, 366, 367, 371, 373, 376, 378, 382, 383, 385
Entrainment effect of rotational viscosity 108
Entropy, change of entropy 29, 35, 82, 85, 87, 88–91, 166, 188, 200, 212, 260, 261, 377, 379, 383, 388, 393
Equation of state 72, 75, 81, 393
Ericksen-Leslie-Parodi theory 63
Expansion coefficient 95

Films of liquid crystals 55, 58, 60, 70, 181, 185, 259, 403
Flory and Ronca lattice model 35, 36
Flory's model 34–36, 370, 392
Fluctuation clusters 285
Fluctuations of order parameter 283
Fowler-Kirkwood-Waff approximation 211
Frank energy 283
Franklin's method 61, 102
Frank-Oseen approximation 341
Frank's moduli 283, 286
Freedericksz's effect 58, 60, 84, 286
Free energy 30, 34, 37, 66, 94, 285, 341, 384, 387, 389, 390
Frequency of nucleation 254
Frozen glass-type states 253, 254, 258

Gas-liquid chromatography 375
Gauss noise 60
de Gennes method 283
Gibbs canonical ensemble 42
Gibbs free energy 29, 77, 268, 384, 385
Ginzburg-Landau parameter 284
Ginzburg-Landau theory of nucleation 60
Goossens theory 153, 169, 170, 181, 293, 341

Hamiltonian 47, 339
Heat capacity 35, 57, 85, 92–95, 97, 98, 163, 166, 201, 202, 213, 214, 342, 377, 382
Heat thermal conductivity 118, 189, 217, 229, 353
Heat of evaporation 60, 100
Heat of the phase transitions 60, 85, 93, 100, 133, 134–136, 199, 200, 211, 244, 248, 249, 317, 319, 320, 393

Heat pulse method 219
Helfrich's theory 217
Helix of cholesterics 142, 143, 154–156, 182, 189, 258, 259, 337, 341, 387
Helmholtz free energy 29, 384, 385, 386
Hubbard-Stratanovich transform 189

Immiscibility of phases 245
Incoherent quasielastic neutron scattering 218
Induced phases 329, 356–359
Infrared absorption 20
Infrared spectroscopy 68, 118, 261, 332, 362
Integral-equation method 65
Integral Onsager equation 395
Interlayer distance of smectics 360, 361, 364
Internal energy, 57, 85, 87–89, 95, 98, 307
"Isokinetic" relation 262
Isothermal compressibility 96, 215, 216, 323

Keating's theory 153, 334
Kerr's effect 37
Kerr's constant 168
Kinetic theory of crystalline phase formation 254
Kinetics of phase transition 60
Kirkwood theory 61, 102
Kobayashi-McMillan theory 159–164, 170, 185
Kubo's linear response theory 54, 62

Langevin's theory 26
Landau-de Gennes model 36–41
Landau-Peierls theorem 287
Landau's rules 188
Landau's theory of phase transition 187, 188, 289
Lattice cubic model 393
Lattice-gas model 392
Lattice generalized model 392
Lattice model 36, 56, 58, 169, 369, 387, 389, 391, 393
Lattice parameters 267, 270, 271
Layers of smectics 136–141, 145, 149, 169
Lennard-Jones potential 33, 42, 56, 172
Leslie coefficients of viscosity 107, 111, 112, 115–118, 189
Liquid crystals consisting of disc-like molecules 12, 265–324
Lyotropic liquid crystals 12

Magnetic birefringence 37
magnetic susceptibility 278, 322
Maier-Lubensky model 33, 188
Maier Saupe theory 20, 27–33, 36, 40, 48, 50, 51, 55–57, 112, 157, 163, 170
Markov chain 44
Maxwell equilibrium distribution 44, 293
McMillan pseudopotential 395
McMillan theory 164, 169, 289
Mean molecular field theory 27, 33, 49, 56, 72, 84, 343, 384, 387, 394
Mean-square displacements 22, 44, 55
Meyer-McMillan theory 139, 149, 165, 166, 167, 186
Metastable states of liquid crystals 253–259, 260–262, 390
Method of calorimetry 272, 364, 377, 384
Methods of investigation of phase diagrams 193, 196
Methods of molecular movement investigation 229
Methods of self-diffusion measurements 102
Method of stress relaxation 228
Methods of surface tension measurements 82, 210, 211
Methods of viscosity measurements 110, 115, 116, 189, 228, 384
Microstructure of liquid crystals 335, 336
Miesowicz's coefficients of viscosity 107, 116
Mixed order parameters 148, 149, 160, 161, 175, 288, 395
Mixtures of liquid crystals 117, 156, 245, 246, 248, 251, 258, 260, 322, 329, 331–395

Molecular dynamic method 19, 21, 22, 44–59, 65, 95, 97, 99, 100, 102, 104–106, 118, 153, 171–185, 189, 219, 258, 259, 266, 276, 279, 291, 403
Molecular clusters (domains) 29, 31, 370
Molecular conformation 69, 376
Molecular field approximation 27, 154, 163
Molecular flexibility 118, 132, 343, 389, 393
Molecular layers of smectics 136–141, 145, 149, 169, 177
Molecular models 168, 173, 296, 298, 312
Molecular movement 229
Molecular reorientation 151
Molecular rotations 167, 170, 171, 180, 376
Molecular-statistical theories 25, 154, 394
Molecular theories of transport (kinetic) properties 61–63, 100–106, 178–180, 189, 190, 303–310, 395
Monte-Carlo method 42–43, 56, 57, 58

Nematic liquid crystals 1–23, 25–128, 136, 145, 178–181, 189, 191–199, 200–204, 211, 220–223, 241, 245, 403
Neutron diffraction 271, 318
Neutron scattering 20, 101
Newton's low 189, 222
Nonequilibrium molecular dynamics method 58, 116
Nonequilibrium states 255
Nonmesogenic additives 335, 394
Nuclear magnetic resonance method 20, 50, 62, 102, 157, 221, 229, 272, 280, 323, 337, 364

Onsager's linear theory 62
Onsager's method 31, 34–36, 388
Optical activity 158, 338
Optically active admixtures 343
Optical microscopy method 272, 278, 322, 324, 332, 345, 347, 355
Optical rotation dispersion 196
Orientation ordering 7–20, 70, 115, 136, 157, 168, 181, 265, 271, 276, 277, 291, 296, 298, 337, 366, 381, 387, 389, 391
Orientation order tensor 15, 37
Orientational distribution functions 14–20, 51–54, 147, 175, 182, 185, 186, 296, 297, 302, 395
Orientational order parameters 14–20, 28, 30–41, 48, 49, 53, 56–59, 62, 66, 73, 82, 104, 118, 149, 151, 157, 160, 161, 166,, 175, 176, 182, 184, 188, 288, 296, 298, 307, 310, 342, 343, 370, 382, 389, 394, 395
Orientational phase transition 163

Parson's theory 211
Packing fraction 82, 104, 105
Parameters of interaction potential 46, 48, 105, 118, 177, 181, 184, 392
Percolation effect 189
Periodical modulation of lattice 289
Perturbation theory 170, 188
Phase diagrams of pure nematics 66–70
Phase diagrams of pure smectics, nematics and cholesterics 166, 191–203
Phase diagrams of discotics 288, 317–321
Phase diagrams of mixtures of liquid crystals 260, 261, 332, 345–379, 384, 386, 387, 392, 393, 394, 395
Phase separation effect 393
Phase transition theory 188
Phenomenological theory 25, 93, 341
Phase transitions 7, 26, 29, 30, 33, 35, 36, 40, 41, 57, 60, 69, 71, 81, 85, 87, 90, 95, 100, 129, 149, 154, 165, 166, 176, 185–188, 191–203, 225–228, 242, 244, 247–249, 253, 255, 278, 281, 282, 311, 317, 320–324, 335, 338, 342, 345–377, 382, 386, 391, 394
Pitch of helix 9, 143, 152–158, 171, 183, 184, 190, 196, 244, 245, 248, 250, 259, 275, 313, 329, 331–334, 338–342, 381, 384, 394

Poiseuille's low 189, 222
Polarizability 133, 195
Polarization 151, 202
Polarizational microscopy 196, 278, 353, 384
"Positive" liquid crystals 392
Potential of molecular interaction 1, 21, 22, 28–32, 44–47, 56, 57, 63, 97, 100, 118, 155, 159, 163, 179, 172–174, 181, 182, 189, 195, 292, 294, 299, 342, 376, 382, 384, 392, 394, 403
Preparation of supercooled specimens 262
Pretransitional phenomena 26, 80, 154, 163, 167, 188, 207, 382
Priest and Lubensky model 339

Quasielastic neutron scattering 167
Quasinematic layers in cholesterics 8, 156, 170, 171, 182
Quasichemical model 392

Raman scattering 15, 19, 20, 23, 51, 261
Rayleigh scattering 20, 23, 102
Reentrant behavior (reentrant phases) 59, 60, 69, 188, 194, 201, 322, 368
Rotation diffusion coefficients 307, 370
Rotation diffusion of molecules 61
Rotational relaxation times 118
Rotational viscosity 108–110, 116

Scaled particle method 65
Scanning electron microscopy 259
Schiff bases 393, 394
Selectively reflected light 152, 155, 158, 196, 241, 245–251, 259, 331–334, 338, 365
Self-consistent field approximation 36, 166, 169, 186
Self-diffusion coefficients 20, 61, 62, 100–106, 178–180, 189, 213, 217–223, 266, 303, 313, 323
Self-diffusion tensor 61, 179
Shock freezing 258, 259
Skoulious-de Gennes model 289
Smectic liquid crystals 7, 9, 11, 13, 68, 114, 115, 129–143, 145–151, 159–169, 171–181, 185–190, 215, 217–229, 312, 314, 403
Sound velocity 85, 93, 95, 97–99, 207, 208, 216–220, 379, 382
Spectrum of elementary excitations 36
Spin label method 150
Spin-lattice relaxation 190, 229
Spiral column-like structures 279
Stability of liquid crystals 66, 168, 170, 188, 196, 239, 241, 242, 384, 391
Stroobants and Lekkerker method 89
Structure of induced mesophases 358
Structure of liquid crystals 3–12, 136–158, 266, 267, 270–277, 280, 303, 335, 336, 403, 404
Structure of mesogenic molecules 1, 3–5, 131–135, 143, 173, 242, 243, 277–281
Supercooled states of liquid crystals 239, 241–262, 349, 351, 353
Surface adsorption 70
Surface tension 33, 60, 81, 82–84, 208–211, 213, 220, 221

Tait's equation of states 72, 75
Temperature coefficients of pressure 97–99
Theory of cholesterics destruction 261, 262
Theory of nuclear spin relaxation 189, 200, 201
Thermal analysis 278
Thermal coefficients of isobaric expansion 374, 380
Thermal conductivity 118, 189, 217, 229, 353
Thermal diffusion coefficients 179, 217, 219, 223
Thermal effect of phase transitions 60, 85, 93, 100, 133–136, 199, 200, 211, 212, 244, 248, 249, 317, 319, 320, 393
Thermal expansion 85, 323
Thermodynamical relations 393
Thermodynamic interpretation of metastable states 255–257

Thermodynamic potential 256–258
Thermodynamic properties 35, 44, 54, 56, 59, 80, 85, 87, 95, 98, 100, 163, 167, 168, 172, 212, 242, 245, 317, 323, 324, 368, 372, 381, 384, 392, 403, 404
Thermooptical characteristics 152, 155, 158, 196, 241, 245–251
Time correlation functions
of angular velocities of molecules 22, 23, 55, 179, 180, 305, 308, 309
of linear (translational) velocities of molecules 20–22, 54, 55, 177, 178, 300–306
Thickness of smectic layers 149–151, 170, 193, 360, 361
Tilt angle of smectic molecules 150, 151, 167, 171, 187, 202, 394
Tilted columnar structure 275, 315
Time correlation reorientation functions 15, 179, 307, 310
Time of dielectric relaxation 189, 200, 201
Translational order parameters 148, 149, 160, 161, 175, 176, 180, 288, 395
Translation ordering 7, 9, 136, 137, 145–147, 168, 271, 291
Transport (kinetic) properties 61–63, 100–118, 178–180, 189, 190, 213, 217–219, 303–310, 324, 395
Tricritical points 191
Tsvetkov's model 25–27
Twist capability 338
Twist discotic liquid crystals 275, 276, 313, 314
Twist force 244, 250, 251, 339, 340
Twist viscosity 112
Two-dimensional crystal 315
Two-dimensional liquid 315
Types of polymorphism 347
Types of smectics 136–141, 193

Van der-Waals interaction 60, 389
Van der-Waals model 59
Viscosity coefficients 20, 58, 62, 63, 106–118, 189, 211, 213, 217, 222, 224, 225–229, 253–255, 324, 353, 384, 395
Vlasov-type kinetic equation 61

Weiss approximation 27
Wulf model 153, 164, 165

X-ray diffraction 106, 149, 197, 260, 267, 270, 273, 275, 278, 281, 292, 335, 345

AUTHOR INDEX

Alben, R. 394
Andrushkevich V.V. 52
Arnold H. 85, 93, 199, 347
Arrhenius S. 228
Avogadro A. 29, 42, 239
Avrotinskaya S. P-2 (Preface, page 2 of Original)

Bagmet A. D. 95, 172
Bailey L.W. 210
Barall E.M. 85
Baskakov V.Ya. 199
Beeman D. 48
Belousov V.P. P-1
Belyakov V.A. P-1
Berne B.J. 44, 172
Bethe H.A. 36
Billard J. 320
Blake W. 3, 159
Blinov L.M. I-1 (Introduction page 1 of Original)
Bogatskii A.V. P-1
Bogolyubov N.N. 36
Boltzmann L. 28, 42
Born M. 25
Böttcher B. 154, 170
Bragg G. and L. 143, 152, 161, 251, 335
Brown G.H. P-1, P-2
Brillouin M. 381
Bunning J.D. 278
Byankin V.M. 199
Byron G.G. 253

Carroll L. 265
Chaikovskii V.M. 52
Chandrasekhar S. P-1, I-1, 32, 33, 49, 50, 66, 85, 92, 95, 211, 265, 273, 275, 279, 281, 317, 324
Cherkashina R.M. 136
Chistyakov I.G. P-1, I-1, 7, 52
Chochtayaev 33
Chow L.C. 85
Chu K.S. 61, 62
Churchill D. 210
Clapeyron B. 87
Clark N.A. 257
Clausius R. 87
Cotton E. 336
Croxton C.A. 211
Curie P. 37

Dasannacharya B.A. 217
Debye P.J.W. 26
Demus D. P-1
Destrade C. 275
Diogo A.C. 112
Dmitriev S.G. 171
Doucet J. 260
Dubini B. 260
Dzyaloshinskii E.I. 171

Eiseley L. I-1, 13
Elmer A. 317
Eremenko A.I. P-2
Ericksen J.L. 61, 63, 106

Fergason J.L. 331
Few G.A. 43
Filev V.M. 154

Fisher I. Z. P-1
Flory P.J. 35, 36, 370, 389, 392
Foex G. 37
Fourier J.B.J. 284, 288
Fowler R. 211
Francis R. 129
Frank F.C. 275
Franklin W. 61
Freedericksz V. 58, 60, 286
Fredrickson A.G. 101
Frost R. 239, 241, 291, 345

Gajewska B. 333
Gal'tzev V.V. 195
Gasparous A. 281
Gauss C.F. 60
Gennes P.G. de I-1, I-3, 7, 32, 36–41, 84, 154, 164, 169, 187, 188, 190, 283, 289
Gibbs J.W. 29, 42, 77, 268, 384, 385
Ginzberg A.V. 194
Ginzburg V.L. 60, 284
Goethe J.W. von 65, 331
Gohwiller C. 107
Goodby J.W. 136
Goossens W.J.A. I-3, 153, 169, 170, 171, 181, 293, 341
Grandjean F. 27
Gray G.W. P-1 136, 245
Guillon D. 149, 261

Helfrich W. 189, 217
Helmholtz G. 29, 384, 385
Hirakawa K. 93
Hubbard J. 169
Humphries R.L. 33

Kai S. 93
Kalinina A.M. 258
Kapustin A.P. P-1, I-1
Kats E.I. P-1, 171, 281
Keating P.N. 153, 334, 338
Kelker H. 321
Kerr J. 37, 168
Khachaturyan A.G. 171
Khutornowa N.I. P-2
Kirkwood J.G. 61, 102, 147, 211
Kirov N. 261
Kobayashi K. I-3, 159, 163, 164, 170, 185
Kofman V.I. P-2
Krishnaswamy S. 211
Kronecker L. 16
Krüger G.J. 217
Kubo R. 54, 62
Kushick J. 44
Kuss E. 114

Landau L.D. I-3, 36–41, 60, 84, 89, 93, 164, 169, 187, 188, 282, 284, 287, 288
Langevin P. 26
Larkin A.I. 7
Lasher G. 42
Lee F.T. 188
Legendre A.M. 14, 170, 288, 394
Lekkerker H.N.W. 389
Lennard-Jones J.E. 33, 42, 56, 172, 257
Leslie F.M. 61, 63, 106, 189
Levelut A.M. 272
Lifshitz E.M. 39
Lisetskii L.N. 157
Lubensky T.C. 33, 188, 339
Luckhurst G.R. 31, 56, 57

Madhusudana N.V. 32, 33, 49, 50, 85, 92, 95
Maidachenko G.G. 136
Maier W.G. I-2, 20, 27–36, 40, 50, 51, 55–57, 112, 170, 188
Markov A.A. 42
Martins A.F. 112
Martire D.E. 85
Maxwell J.K. 44, 293
Mc Donnell D.G. 136, 245
McMillan W.L. I-3, 139, 159, 160, 164, 165, 170, 186, 188, 200
Meyer R.B. I-3, 139, 164, 165, 186
Miesowicz M. 107, 116
Mikhailov G.M. 107
Moroi D. 62
Murakami J. 60

Nedostup N.A. 195
Neumann A.W. 210
Newton I. 189

Onsager L. 31, 34–36, 165, 388, 389, 390, 395
Ostrovskii B.I. 85, 93
Oswald P. 278
Otani S. 281
Otnes K. 101

Park J.W. 347
Parodi O. 63, 107
Pauling L. 83
Pechukas P. 172
Peierls R.E. 287
Perkin W.H. 317
Pikin, S.A. P-1, 7, 188
Poe E.A. 1
Poiseuille J.L.M. 189, 222
Priest R.G. 32, 284, 339

Rayleigh J. 20, 102
Raman Ch. 15, 19, 20, 51
Ranganath G.S. 140
Rigby M. 43
Ronca G. 35
Romano S. 56, 57
Ryumtzen E.I. P-1

Sakevich N.M. 85
Sackmann H. 347
Saupe A.Z. I-2, 20, 27–36, 40, 50, 51, 55–57, 112, 170, 188
Schiff H.J. 393, 394
Selle P.J. 210
Semenchenko V.K. 33, 199
Shakespeare W. 25, 267, 329
Shashidhar R. 211
Shubnikov A.V. 7
Simpson P. 57
Sosnovskii A. 81
Stecki J. 70
Stephen M.J. I-1
Stishov S.M. 54
Straley J.P. I-1
Stroobants A. 389
Sukhodol'skaya N.E. P-2

Tabidze A.A. 224
Tait P. 72
Tan H.T. 188
Tikhomirova N.A. 66, 194
Tishchenko V.G. 136, 157
Tolmachev A.V. 157
Topler J. 101
Tsuji K. 259
Tsvetkov V.N. 15, 25, 81, 107, 108
Tyablikov S.V. 36
Tyutchev F.I. 145, 1991

Vainshtein B.K. P-1, I-1, 7
Vasserman V.A. P-2
Vigman P.V. 154
Vieillard-Baron J. 43
Vistin' L.K. 66
Vlasov A.A. 61
Voss J. 259
Vries A.de 4303

Waals J.D. van der 5, 59, 60, 171, 389
Weiss P. 27
Wilhelmy L.F. 221
Wirzing U.G. 321
Wojtowicz P.J. 32, 147
Wulf A. 153, 164, 171

Yun C.K. 101

Zawisza A.C. 70